ORNAMENTAL CROP PRODUCTION TECHNOLOGY

About the Editor

Prof. S. K. Bhattacharjee is an eminent scientist in the field of horticulture, retired as Project Co-ordinator, All India Co-ordinated Research Project on Floriculture, Indian Council of Agricultural Research, Indian Agricultural Research Institute, New Delhi, on January 31, 2003. He obtained his M.Sc. (Ag.) in Horticulture with first class and first rank from Kalyani University and Ph.D. in Horticulture from Calcutta University. He is also a diploma holder in Business Management and Industrial Administration, United Nations Studies and Landscape Architecture.

He has 38 years research, teaching, management, developmental and administrative experience in private organisations, Indian Botanic Gardens, Botanical Survey of India, State Agricultural Universities and ICAR Institutes. He was Head of the Division of Medicinal and Aromatic Crops, Indian Institute of Horticulture Research, Bangalore. He has also served as Professor of Horticulture and Professor-in-Charge, North Bengal Campus, Bidhan Chandra Krishi Viswavidyalaya, Cooch-Behar, West Bengal, which has later converted into a full-fledged Agricultural University presently named as Uttar Banga Krishi Viswavidyalaya

While serving in private organisations, he has laid out many landscape gardens (industrial gardens, parks, bungalow gardens and institutional gardens) in different agro-climatic conditions of the country, collected a wide range of indigenous and exotic plants from India and abroad and propagated them on a large scale for use in gardens. He has conducted extensive research work on standardisation of agro-techniques in ornamental crops, plant propagation and post-harvest technology of cut flowers. He is the pioneer in the research work on post-harvest technology and agro-technology of flower crops in India. He has served as a national leader of ornamental crop research and development in India, prepared research programmes on various ornamental crops for different agro-climatic conditions of the country and monitored the research work of different agricultural universities and the national institutes at 20 different centres.

Prof. Bhattacharjee has visited several foreign countries like England, Germany, France, Italy, Belgium, The Netherlands, Thailand, Malaysia and Singapore. He has published about 400 research papers and popular scientific articles in journals of national and international repute, and contributed chapters in several books. He is the author of 10 popular, informative and voluminous priced books titled *Advanced Commercial Floriculture; Orchids of India; Garden Plants; Advances in Horticulture—Ornamental Plants; Cacti—Their Botany, Culture and Uses; Handbook of Medicinal Plants; Handbook of Aromatic Plants; Landscape Gardening and Design with Plants; Medicinal Herbs and Flowers* and *Post-Harvest Technology of Flowers and Ornamental Plants.* He has also written 15 bulletins on ornamental crops and prepared many reports.

Prof. Bhattacharjee was the Vice-President of the Indian Society of Ornamental Horticulture and served as a member of the selection committee of CSIR, ASRB and agricultural universities. He was one of the editors of *Annals of Agricultural Research.* He is a member of several scientific societies of India and abroad, referee of scientific projects of different organisations, referee of foreign and Indian journals, Ph.D. examiner of agricultural universities and traditional universities. Prof. Bhattacharjee has a long teaching experience at post-graduate level in the universities and has guided 6 M.Sc. and 8 Ph.D. students in floriculture as chairman.

ORNAMENTAL CROP PRODUCTION TECHNOLOGY

Prof. Supriya Kumar Bhattacharjee
M.Sc. (Ag,), Ph.D.
Formerly
Scientist S-3 and Head
Division of Medicinal and Aromatic Crops
Indian Institute of Horticultural Research, Bangalore
Professor of Horticulture and Professor-in-Charge
Bidhan Chandra Krishi Viswavidyalaya
N. B. Campus, Coochbehar, W. Bengal
Principal Scientist, ICAR, New Delhi
Principal Scientist, IARI, New Delhi
Project Coordinator
All India Coordinated Research Project on Floriculture
Indian Council of Agricultural Research, New Delhi

Pointer Publishers
Jaipur 302 003 (Raj) India

Distributed by
Prem C. Bakliwal for
Aavishkar Publishers, Distributors
807, Vyas Building, Chaura Rasta
Jaipur 302 003 (Raj.) India
Phone : 0141-2578159
e-mail : aavishkarbooks@hotmail.com

ISBN 978-81-7132-473-6

First Published in 2006
Reprinted in 2022 by
Mrs. Shashi Jain for
Pointer Publishers
Vyas Building, Chaura Rasta
Jaipur 302 003 (Raj.) India
Phone : 0141-2578159
e-mail : pointerpub@hotmail.com
Visit us at : www.pointerpublishers.com

Printed at
Harish Art Printers
Jaipur (Raj.)

PREFACE

ORNAMENTAL horticulture is essentially a science-based technology, hence the recent scientific breakthrough is inevitably the most interesting and stimulating. It is an industry which relies and thrives on diversity and advancement of science. Many sectors in floriculture require an innovation and change to create a market demand for products. Scientific advances are often implemented first in ornamental crops, whether it is biotechnology, tissue culture, biofertiliser, molecular biology, plant breeding, biological control, plant physiology or biochemistry. However, there are inherent dangers in trying to implement laboratory research immediately into industry, and putting business at risk in testing new technology is irresponsible. Frequent and rapid changes are occurring in this field and there is an increasing demand of various nature from society. There will be need to provide information in an encouraging and exciting manner so as to stimulate the development of an enquiring mind.

In recent years, there has been rapid growth in the world demand for floricultural products which provide unlimited export opportunities. To enhance export and achieve leadership, intensive research work on various disciplines of floriculture is to be attempted systematically. In order to ensure supply of high quality uniform material in large quantity, development of low cost, environmental friendly and economically viable production technology is essential. There is a very large untapped potential that can be unlocked with the right triggers.

Although India is blessed with favourable climate and soils, skilled manpower and the largest pool of technically qualified people, India's share in the world flower market is 0.7 per cent only. Many export-oriented projects on floriculture were approved in the last 15 years, but most of them did not contribute successfully in spite of various concessions given by the Government of India. The causes of failures of the floriculture industry were reviewed often, and suggestions emerged in favour of further support by the Government in many ways. It has, however, been observed that the concessions and support from Government for establishing industry were not given to the traditional flower growers in India who are engaged in flower cultivation and business for generations, nor were they encouraged in any way to take up such ventures. Those who entered in this business and obtained support were non-technical people, very conservative in their outlook and ultimately displayed an indifferent attitude in their dealings with the public. They were not sympathetic to other persons' ideas and desires. Initially, for several years, they had never allowed people interested in floriculture to visit their farms and avoided free discussion. Honest and sincere appreciation of the work already done in India was lacking. In such a suffocating environment, the floriculture industry can not prosper. It must be realised that people are more valuable than capital and equipment. An industry can not be built by taking away man's initiative and independence. Progressive realisation of the worthy goal was missing all over. There is no doubt that candour is the better option. There must be definite programmes and commitments. Persons with high self-esteem should lead this industry, then only this enterprise will be successful on a long-term basis. In such important and satisfying ventures, there is need to motivate the workers by making the work more interesting, encouraging good setting, providing opportunities for growth and training, giving respect and recognition and at the same time by throwing challenges. The enthusiasm must be contagious. The entrepreneur should have a finely calibrated sense of their minimum requirements. Moreover, they must love what they are doing, otherwise they probably would not succeed at all.

Education, research and training in ornamental horticulture has developed and expanded steadily around the world over the years to serve the needs of industry and the best interests of the students. The social, industrial and technological changes in more recent years, including heightened environmental awareness and health consciousness, have however determined the pace of development and change across the whole spectrum of the floriculture industry, allowing new sectors to emerge. Education, research and training in ornamental horticulture must keep pace with these developments. The education and training in this sector in India must be internationally competitive and should be responsive to the challenges and opportunities flowing from changes in markets, materials, technologies and competitors. Education must facilitate students to effectively participate in the latest developments in floricultural research and other branches of science relevant to it. Graduate and post-graduate courses should be formulated in such a manner that students get adequate knowledge in floriculture science, technology and business. Students, in turn, are required to gain a high level of ability in the acquisition, interpretation and application of knowledge. To meet the demand of industry, a high level of proficiency has to be imported through education and training on production technology, post-harvest management and handling, plant protection, tissue culture, ecology, economics, computer application, business and management studies. Real analysis of problems can be enhanced by the use of case studies where the students become decision-makers. The academic programme must provide courses on autecology and flori-economics. Computer application can be used to develop students' professional skill and also impart greater effectiveness, cost-efficacy and flexibility. The courses must facilitate the process of making value judgements on relative merits of floricultural practices for a specific need. Training and education must also recognise local issues, local teaching material and resource personnel. Apart from academic and research institutes there is need to establish a National Training Board on Ornamental Horticulture whose purpose would be to raise national competency standards endorsed by the industry.

Flower growers, gardeners, landscape designers, educators and communicators—all whose economic well-being depends on people's involvement and satisfaction with plants—will have a significant role in implementing the research focussed on ornamental horticulture and human well being. All over the world, there is destruction of natural and human environment and we have not yet awakened to the seriousness of global warming. Faced with political decisions about urban migration, food production, agrarian reform, employment generation, national security, economic structural adjustment and other issues that demand their attention, many governments have opted to ignore deforestation. Deforestation of forests by the agents of deforestation like infrastructure developers, land settlement planners, commercial planters, firewood collectors, commercial farmers, cattle ranchers, loggers, livestock herders, slash and burn farmers, mining and petroleum industrialists, are causing extinction of thousands of species. Millions of people today are worse off than they were when our modern development process was started. Tribals have been alienated from their land and forests, nomadic communities have had their pastures taken away by mega-projects, marginal peasants and the landless have become economically more insecure. Lost of biodiversity has increasingly marginalised and impoverished many tribals and other forest dependent communities, which directly depend on various components of biodiversity for food, drink, medicine, culture and spirituality. The localised and national impact of environmental degradation is now becoming increasingly clear. Loss of topsoil, salinity, pollution, shortage of water, are making the food security situation bleak. Although environmental degradation ultimately affects every one, the immediate and most severe impact is felt by the poor. Millions of workers in India are exposed to dust, heat, noise, dirt, gases and dangerous machinery. Deforestation increases the amount of CO_2 and other trace gases in the atmosphere. The loss of forests has a profound effect on the global carbon cycle and there is worldwide release of 1.6 billion metric tons of carbon in the atmosphere per year. In comparison, fossil fuel burning releases about 6 billion metric tons of carbon per year. Release of CO_2 in the atmosphere enhances the greenhouse effect, and could contribute to the increase in global temperature. Humans have done so much of damage to the atmosphere that even if they stop burning all fossil fuel immediately, they risk leaving an impoverished Earth for their descendants. A Swiss team examined the heat wave that killed 20,000 people in Western Europe in 2003 and concluded that it was a consequence of global warming. European scientists reported that the entire Greenland ice cap would begin to melt irreversibly if average temperatures rise to 2.7° C . This melting would take 1,000 years to complete but raise the sea level by seven metres. Recent research suggests that about half of the precipitation that falls in a tropical rain forest is a result of its moist, green canopy. Evaporation and evapotranspiration processes from the trees and plants return large quantities of water to the local atmosphere, promoting the formation of clouds and precipitation. Less evaporation means that more of the sun's energy is able to warm the surface and, consequently, the air above, leading to rise in temperatures.

In India, there is a major attack on forests, water bodies, land, mineral-rich areas and other natural resources for the purpose of economic growth. During colonial times, forests were considered to be commercial and industrial raw materials, and the land was viewed as a revenue generator. Even after independence, between 1947 and 1980, we lost 4.2 million hectares of forests to non-forest activities, while during 1981 to 2003, about 847,000 hectares of forestland was diverted to non-forest use. The Indian subcontinent has been identified as one of the top 12 hotspots of biodiversity in the world. India harbours 8 per cent of the total number of species in the world, estimated to be 1.6 million, with possibly twice or thrice the number still to be discovered. Conservative estimates suggest that at least 10 per cent of India's recorded flora are on the threatened list. Our environmental needs help and we must safeguard water, land, forests and biodiversity.

Ornamental horticulture has the potential to render substantial benefits in enriching our environment. In the graduate and post-graduate courses of landscaping under ornamental horticulture, studies must be initiated on the impact of adding plants to a stressful setting, economic values of urban parks, greenbelts and residential property values, economic values of city parks and community gardens, contribution of trees to the appraised value of unimproved residential areas, psychological values of trees, the response of trees and nature in the inner city, inspiration of people from display gardens, economic values of urban plants, people plant interaction and horticultural therapy, people plant relationship, and role of ornamental horticulture in human well-being and social development.

This publication is planned to reveal advances on multifarious aspects of ornamental horticulture so that innovations made so far can be used judiciously for this sector. Floriculture, however, requires the further attention of researchers, industrialists and entrepreneurs. It was my sincere endeavour to present a complete and comprehensive scenario on ornamental horticulture through this book. In spite of my best efforts, I find there are some omissions; a few aspects, however, I have tried to touch very briefly, through this preface. Suggestions and criticisms on this book will be most welcome, so that the next revised issue will be a better one. I take this opportunity to express my heartfelt thanks to all the authors of the various chapters of this book. I hope this book will be useful to scientists, teachers, researchers, students, commercial flower growers, floriculturists, perfumers, traders, landscape designers, industrialists, entrepreneurs, technocrats and planners.

S. K. Bhattacharjee

CONTENTS

Colour Plates

CONTRIBUTORS

Bhattacharjee, S. K.
Division of Floriculture and Landscaping, Indian Agricultural Research Institute, Pusa, New Delhi, 110 012.

Chandra, Pitam
Division of Agricultural Engineering, Indian Agricultural Research Institute, Pusa, New Delhi 100 012.

Das, S. P.
National Research Centre for Orchids (I.C.A.R.), Pakyong, East Sikkim 737 106 (Sikkim).

Datta, S. K.
Floriculture Section, National Botanical Research Institute, Lucknow 226 001 (Uttar Pradesh).

Dhaduk, B. K.
Department of Horticulture, ASPEE College of Horticulture and Forestry, G.A.U., Navsari 396 450 (Gujarat).

Dhatt, K. K.
Department of Floriculture and Landscaping, Punjab Agricultural University, Ludhiana 141 004 (Punjab).

Gupta, M. J.
Division of Agricultural Engineering, Indian Agricultural Research Institute, Pusa, New Delhi 100 012.

Jadhav, A. S.
College of Agriculture, M.P.K.V., Pune 411 005 (Maharashtra).

Janakiram, T.
Division of Ornamental Plants, Indian Institute of Horticultural Research, P.O. Hessaraghatta, Bangalore 560 089 (Karnataka).

Karuppuchamy, P.
Department of Agricultural Entomology, Tamil Nadu Agricultural University, Coimbatore 641 003 (Tamil Nadu).

Kumar, R.
Department of Floriculture and Landscaping, Punjab Agricultural University, Ludhiana 141 004 (Punjab).

Mandal, A. K. A.
Floriculture Section, National Botanical Research Institute, Lucknow 226 001 (Uttar Pradesh).

Misra, P.
Floriculture Section, National Botanical Research Institute, Lucknow 226 001 (Uttar Pradesh).

Nagesh, M.
Project Directorate of Biological Control (I.C.A.R.), Hebbal, Bangalore 560 024 (Karnataka).

Palaniswamy, S.
Department of Agricultural Entomology, Tamil Nadu Agricultural University, Coimbatore 641 003 (Tamil Nadu).

Patil, M. T.
Department of Horticulture, College of Agriculture, M.P.K.V., Pune 411 005 (Maharashtra).

Patil, P. V.
College of Agriculture, M.P.K.V., Pune 411 005 (Maharashtra).

Raghava, S. P. S.
Division of Floriculture and Landscaping, Indian Agricultural Research Institute, Pusa, New Delhi 110 012.

Reddy, O. R.
Plant Quarantine Division, Directorate of Plant Protection, Quarantine and Storage, Faridabad 121 001 (Haryana).

Sharma, Pratibha
Division of Plant Pathology, Indian Agricultural Research Institute, Pusa, New Delhi 110 012.

Singh, M. C.
Division of Floriculture and Landscaping, Indian Agricultural Research Institute, Pusa, New Delhi 110 012.

Sudha, M.
Division of Agricultural Economics, Indian Institute of Horticulture Research, P.O. Hessaraghatta, Bangalore 560 089 (Karnataka).

Wakchaure, N. Y.
Department of Horticulture, ASPEE College of Horticulture and Forestry, G.A.U., Navsari 396 450 (Gujarat).

1

Seed Production in Ornamental Crops

R. Kumar and K. K. Dhatt

INTRODUCTION

FLOWER seed production seems to have great scope for expansion in developing countries like India, especially under northern Indian climatic zones having favourable growing conditions, skilled and cheap labour, marginal resources from small landholdings and need for crop diversification. Recently, the concept of contractual farming was set up, which is quite profitable for both foreign companies and farmers due to low volume commodity and reduced air freight charges as compared to perishable cut flowers.

Presently, annual flowers such as petunia, *Coreopsis*, *Helichrysum*, phlox, nasturtium, marigold, *Gaillardia*, *Salvia*, ice plant, *Verbena*, *Nemesia*, pansy, poppy, larkspur and chrysanthemum, etc. have been widely accepted for seed production. The total value of seed export from India has been reported to be around Rs. 3 crore in 1993-1994 and 80-90 per cent of the production comes from three main centres, namely, Ludhiana, Bangalore and Kalimpong (Kaul and Dadlani, 1995). At present, it is estimated to be around Rs. 5-6 crore from PunjabState alone.

Information about some of the flowering annuals for seed production from the experiments conducted at Punjab Agricultural University, Ludhiana, with relevant review is given below:

SEED PRODUCTION OF DIFFERENT CROPS

Alyssum maritimum: This is commonly known as sweet alyssum and grown as a transplant from end-October to the first week of November for good seed yield. The plants are planted 30 × 20 cm apart in flat beds in sandy soils and on slightly raised beds in heavy soils; plant growth and seed yields are comparatively better on raised beds due to adequate supply of water and lower mortality rate. The plants start producing flowers in every phase, but the peak flowering period falls during end-February to end-March and branches mature for harvesting after mid-April. Simple crushing with hands and winnowing help to separate the seed which are passed through a muslin cloth or a fine plastic sieve to separate the dust particles. The average clean seed yield ranges between 50-60 kg/acre.

***Coreopsis lanceolata*:** This is a very popular flower among farmers for seed production due to its good seed set potential and easy cleaning procedure. Dwarf varieties are suitable for pot culture, bedding purpose or transplanted in flat beds at 60 × 45 cm distance by mid-November for seed crop. The flowering commences in April and continues till May and seed pods get mature for picking by the end of April, of which continued plants are harvested till early June. The yield of clean seed varies between 200-250 kg/acre. The studies revealed that planting done on 10 November and 25 November gave seed yield of 49.25 g/m², which was significantly higher over the 13.06 g/m² from planting done on 10 December. Similarly, nitrogen application in *Coreopsis* resulted in increase in spread, branch number and seed yield. The maximum seed yield (37.41 g/m²) has been recorded when nitrogen was applied at the rate of 30 g/m² as compared to 25 g/m² under control (Kumar and Kaur, 2001).

***Coreopsis tinctoria*:** The cultural operations of this flower are almost the same as *Coreopsis lanceolata,* except the transplanting can be delayed by one week without lowering the seed yield. The flowering period prevails from April to mid-June; harvesting starts by end-April. Cleaning of seeds of *Coreopsis tinctoria* requires a little extra labour as compared to *Coreopsis lanceolata* seed. Pal (1988) reported that the highest number of flowers was obtained with 40g N/m² and 15 g P_2O_5 *Coreopsis* sp. The yield of clean seed varies between 200-275 kg/acre. The studies on seed germination revealed that the maximum germination percentage with the most rapid and uniform germination occurred at constant storage temperatures of 15° C and 20-35 per cent RH (Carpenter and Ostmark, 1992).

***Gaillardia pulchella*:** For seed production purposes, it is transplanted by mid-December at a spacing of 60 × 45 cm. The flowering duration is more than three months from May-July, even longer than this, but the seed quality deteriorates due to the humid weather and the seed pods, which are prickly, turn whitish grey on maturity and dry up, are picked individually by wearing hand gloves, during June-July. Seed can be conveniently collected at 15-day intervals without losing yield as there is no seed shattering problem. The yield per acre ranges between 200-250 kg.

***Cosmos* sp.:** *Cosmos sulphureus* and *Cosmos bipinnatus* are widely grown for seed production under sub-tropical north Indian climatic conditions as rainy season crops. Since the duration of the *Cosmos bipinnatus* crop is longer, it has not been adapted by growers for seed production as convenient. Samant reported that the seed sowing on 15 July resulted in best growth, flower production and seed yield in *Cosmos*. Experiments were conducted at PAU to standardise the planting time of *Cosmos sulphureus* for obtaining highest seed yield. Four plantings were done starting from 15 May at one-month intervals. The best time for transplanting has been observed to be end-August and flowering is at peak in October. Flowering and seed picking is done concurrently after mid-October till end-November. August planting resulted in the highest seed yield of 48.46 g/m². The yield of clean seed varies between 120-150 kg/acre, if the seed picking is done regularly after the third or fourth day, otherwise the seed yield is reduced due to shattering. Seed stored in different packages established that aluminium foil was best in maintaining seed germination (69.67 per cent) after six months of storage period as compared to 57 per cent germination under storage in butter paper bags (Dubey *et al*, 2002). Apart from this, high seed yield per hectare (9.1 q) has been reported with 20 g N/m², 10 g K/m² in *Cosmos* (Jana and Pal, 1991). Also, application of higher concentration of Paclobutrazol increased the number of branches and number of flowers/plant as reported by Mohd. Ahmad *et al* (1988).

Zinnia elegans: Zinnia transplanted on 15 March, 15 April, 15 May and 15 June established that early plantings resulted in tallest plants (74.37 cm), widest spread (49.55 cm) and more branches (126.44/plant). The highest seed yield (29.83/m^2) was recorded from March planting as compared with 22 g/m^2 and 1.70 g/m^2 from April and May plantings, respectively. Earlier planting (15 April) of zinnia doubled the seed yield and increased the ultimate size of the plants, besides extending the period of vegetative growth, in comparison to the late planting as reported by EI-Gamassy *et al* (1965). Further, Jhon *et al* (1991) reported that maximum number of flowers per plant (43.80 and 4453 for first and second year, respectively) were obtained at a spacing of 30 × 30 cm, while more number of branches and flowers per plant at wider spacing was reported by Rajanna and Khalak (1992). Application of NPK has also been observed to increase the seed yield by 24.62 per cent (Oberthora, 1981). High relative humidity in July-August at the ripening time of late planted crop resulted in the rotting of inflorescence and ultimately reduction in seed yield. The seeds of zinnia packed in muslin cloth bags kept in desiccators and those packed in aluminium foil resulted in increased germination percentage after six months of storage, as reported by Poonam *et al*, 2002. Miyazima (1996) reported that there was about 100 per cent germination after 120 hours of water inhibition of pericarp-removed zinnia.

Impatiens balsamina: The optimum time for transplanting is end-July and mid-August and planting density and spacing is 45 × 30 cm. Seager (1980) reported that there is a three-fold increase in flower production of balsam when planted in February rather than in July. Rani (1997) also reported that the maximum seed yield was obtained under July planting. Jayanthi *et al* (1987) reported noticeable height reduction, maximum number of shoots (7.5) and flowers (44.1) per plant at a spacing of 30 × 22 cm and plucked after 27 days of sowing. Investigations undertaken to optimise the nutrient requirement of balsam revealed that nitrogen and phosphorus at the rate of 20 g and 10g/m^2, respectively, resulted in highest seed yield of 45.11 g/m^2 (Kumar and Kaur, 1996). The flowering starts after 40-45 days of transplanting and lasts for one month. The pods containing seeds are harvested after 15-20 days of flowering when the pod colour turns pale green and seeds turn brown. The seeds shatter on pod splitting due to over maturity, therefore delay in picking reduces the seed yield. Bold size of seed results in easy cleaning by winnowing. The seed yield of a good crop varies between 100-125 kg/acre. Aery and Tiagi (1989) reported 100 per cent germination in 'Rose Flowered Double Mixed' *Impatiens balsamina* after 48 hours in control and with treatment of 1 mg/ml of IAA and IBA. Pramesh *et al* (1994) reported that KNO_3 and IAA (100 ppm) can be used to improve the germination of substandard balsam seeds.

Mesembryanthemum criniflorum: The seedlings are transplanted by end-October to mid-November at 30 × 25 cm distance on slightly raised beds. The flowering is at peak in March and seed pods get mature by the second week of April. During the period, the whole plants dry up and the pod turns very light brown in colour and seeds inside pods change to off-white colour one week prior to harvesting. Irrigation is withheld and the whole plant is harvested with a sickle. The yield of clean seed varies 90-12 kg/acre.

Helichrysum bracteatum: Delay in planting beyond mid-November results in decreased yield, hence for seed production, transplanting is started in the last week of October at a spacing of 60 × 45 cm. Cireasa and Cozucu (1985) reported that *Helichrysum bracteatum* raised from seed of the third and fourth order branch produced best flowers. Fully ripe pods are picked manually every day during end-March to mid-May to avoid the loss of yield, as the seed is light in weight and is generally blown off with the wind. The seed yield varies at 70-90 kg/acre.

***Papaver rhoeas*:** This is commonly known as Shirley Pappy and bears red or pink single, semi-double or double flowers on terminals in the months of March-April. For growing seed crop, the seeds are directly sown on raised beds at a distance of 45 × 30 cm. The seeds are light brown and very small and generally mixed with sand for sowing *in situ* to ensure even germination. The flowering is at peak after mid-March and the seed pods mature for picking after three weeks of flowering. The seed pods have apical pores which start opening as the pods mature and change their colour to pale brown. Therefore, picking before that stage ensures good seed yield and the pods are picked manually during end March to end-April at an interval of three days. The seed yield procured varies between 100-120 kg/acre, depending on management and weather.

***Eschscholzia californica*:** The seeds are directly sown *in situ* at 40 × 30 cm spacing by end-October for seed crop. The flowering commences in March and lasts up to mid-April. Seeds are harvested when the legumes turn pale yellow in colour and seeds are brownish black in colour. Pod legumes split after drying; the seed is then collected and winnowed to remove dust and inert matter. The yield varies between 60-80 kg/acre.

***Gazania splendens*:** Month old seedlings are transplanted on 70 cm wide beds in the centre at a distance of 45 cm, preferably during the first fortnight of November. The plants attain a height of 30-40 cm and a spread of 50-60 cm during the reproductive phase. The flowering peak is produced in April and simultaneously the seed collection starts in April and continues till the end of May. The flowers are borne on long stalks which tend to droop outwards and downwards as the seeds mature for harvesting. The shattered seeds are then collected and cleaned. The seed yield varies from 200-250 kg/acre.

***Phlox drummondii*:** Seedlings should be transplanted latest by mid-November when it is to be grown as a seed crop. It produces flowers from March till May and likewise the flowering branches are harvested twice or thrice depending on their maturity in April-May. Jauhari *et al* (1972) reported that plants of *Phlox*, *Calendula* and *Dianthus* resulted in better flowering in early sowings (21 August, 31 August and 10 September) as compared with late sowings (20 September, 30 September and 10 October). The experiment conducted to standardise planting time established that November planting resulted in vigorous growth in terms of plant height (53.65 cm), plant spread (55.25 cm) and number of branches (21.65/plant). The longest flowering duration (91.88 days) and seed yield (141.85 g/m^2) were also recorded in November planting as compared to December and January plantings where seed yield was 2-3 times lower than that of the November planting. (Kumar and Kaur, 2000).

***Petunia hybrida*:** The seedlings are transplanted in the first fortnight of November on raised beds at 60 × 45 cm leaving 50-60 cm fallow space in between beds. The seed pods are initially hand picked during the fortnight of the start of the harvesting season due to irregular flowering. But when the flowering reaches its peak by mid-March, the flowers mature uniformly, so the plants are shaken over metal trays to collect ripe seed. This practice results in quick collection of clean seed and also acts as a labour-saving device. The optimum stage for seed picking is when the pod turns brown and shows the minute apical splitting. A little delay in picking will result in great loss, as the shattered seed beneath the plant cannot be collected due to its very small size. The yield of clean seed varies between 60-80 kg/acre depending on management ability and seed set. The experiment on hybrid seed production of petunia variety where 'cascade' N, P and K application at 400 ppm, 300 ppm and 200 ppm, respectively, have been reputed seed yield (Elengovan and Sullanmath, 1979).

TABLE 1
Flower Seed Yield and Income per Acre

Name of Flower Species	Seed Yield (kg/acre)	Price (Rs./kg)	Gross Income (Rs.)	Net Income (Rs.)
Helichrysum bracteatum	70-80	250	19,000	14,000
Coreopsis lanceolata	200-220	175	36,000	28,000
Coreopsis tinctoria	200-225	175	37,000	31,000
Gazania splendens	220-225	200	42,000	33,000
Petunia hybrida	60-70	500	33,000	25,000
Verbena hybrida	100-120	300	33,000	25,000
Phlox drummondi	100-120	200	22,000	18,000
Viola coernuta	65-70	700	47,000	34,000
Viola tricolour cv. Swiss Giant	20-25	1,000	25,000	15,000
Gaillardia aristata	220-220	150	32,000	28,000
Mesembryanthemum criniflorum	90-110	200	20,000	15,000
Cosmos sulphureus	100-120	150	17,000	14,000
Papaver naudicule	60-70	250	17,000	12,000

Verbena hybrida: Experiments conducted to standardise nutritional requirement of *Verbena* revealed that nitrogen application up to 20 g/m^2 increased the plant height to 38.68 cm as compared with the height of plants (30.29 cm) receiving no nitrogen. The longest flowering duration of 73.02 days and seed yield of 33.58 g/m^2 were recorded when nitrogen was applied at the rate of 30.8/m^2. The harvesting of seed is done manually during April-May at one-week intervals. The collected seed is then lightly beaten up and cleaned. The seed yield varies between 80-90 kg/acre.

Iberis amara: This is commonly called Candytuft and the flower colour is white and mauve. Transplanting is done on slightly raised beds at 40 × 30 cm distance in early November. Plants come into bloom by end February with the peak in March and last till end-April. The flowering branches get mature uniformly by early May and are harvested manually when the seed colour turns brown. Kaur and Kumar (2001) have reported that nitrogen application at the rate of 30 g/m^2 resulted in tallest plants (38.89 cm), widest spread (48.27 cm), more number of branches (51.41/plant), longest flowering duration (58.68 days) and maximum seed yield (39.75 g/m^2). However, phosphorus application up to 10g/m^2 resulted in significantly better growth andseed yield.

Lupinus hartwegii: Lupin seeds are sown in the prepared beds directly due their bold size or growth and transplanted in early November at 60 × 40 cm spacing. The plants attain a height of up to 90 cm at maturity and produce 10-14 branches. The flowering commences in March and continues till mid-April. Seeds become ready for picking in April. The mature legumes turn brown and get dry on maturity. Therefore, these are picked before they open to burst seeds. These pods are kept under sun for 2-3 days. The seed is then collected and cleaned.

Gamolepis elegans: Transplanting is done in early November on raised beds at 30 × 30 cm spacing. The flowering is greatly influenced by temperature and the peak is attained only for 15-20 days in March. The seed is ready for harvesting by mid-April. At maturity, the seeds turn brown and plants are harvested in a single lot because of uniform seed ripening. The harvested stock is kept in the shade for 3-4 days for shattering of seeds, which are then winnowed under a fan. The average seed yield varies between 125-150 kg/acre.

***Viola tricolour* (Pansy):** From the two cultivars tried for their seed set potential under sub-tropical conditions of Punjab, 'Swiss Giant' gave the maximum seed yield at a spacing of 20 × 20 cm (Singh, 1979). An experiment conducted on cv. Swiss Giant revealed that the number of seeds per capsule (55.41) was the highest under 25 October planting at 20 × 20 cm spacing (Kumar, 1996). Kaur (1997) reported maximum seed yield in pansy at a spacing of 20 × 20 cm. Singh (1978) recorded the highest spread of 41.1 cm with 15 g/m^2 N application in variety 'Hiemalis Mix' of pansy. Jhon and Saini (1984) tested the various spacing and nutrition and found that spacing of 20 × 30 cm and fertiliser application with N P 20:5 g/m^2 produced better growth and flowering on pansy. Kaur and Kumar (1998) reported that there was maximum flower and capsule development with the application of 30 g N and 20 g P_2O_5 in pansy. *Viola tricolour* does not set seed on selfing. The hand pollination is done with a soft camel hair brush at the three-fourth open stage of flower. This stage has been found to result in maximum seed set as compared to fully opened flowers. The studies also indicate that stigma is receptive one day before and two days after anthesis which resulted in 62.46 seeds/pod in cv. Punjab Velvet (Singh, 1987). Flowers are produced in succession from end-February till end-April and, likewise, the seed pods mature at different intervals. The seeds turn pale yellow to light brown on maturity. These seed capsules are picked individually along with floral pedicel, which are then kept in the shade for one week.

REFERENCES

Aery, N. C. and Tiagi, Y. D. 'Studies on the physiology of seed germination in the cultivar 'Rose flowered double mixed' and wild variety of *Impatiens balsamina*'. *National Academy Science Letters* 12(5): 137-139, 1989.

Carpenter, W. J. and Osmark, E. R. 'Growth regulators and storage temperature govern germination of *Coreopsis* seed'. *HortScience* 27: 1190-1193, 1992.

Cireasa, E. and Cazucu, C. 'Contribution to the cultivar technology of some floral species with persistent inflorescence'. *Lucrari Seiin Gfico Instituted Agronomic Ion Ionescude La brad Horticulturae* 25: 95-96, 1985.

Dubey, R. K., Kumar, R. and Poonam. 'Effect of planting time and spacing on *Cosmos*'. *J. Orn. Hort.* 5(2): 46-47, 2002.

Dubey, R. K., Kumar, R. and Poonam. 'Effect of storage material on germination and viability of *Cosmos* seed'. *J. Orn. Hort.* 5(2): 51-52, 2002.

Elengovan, R. and Sulladmath, U. N. 'Effect of nitrogen, phosphorus and potassium on production of hybrid seeds in *Petunia hybrida*, Vilmil'. *S. Ind. Hort.* 34(5): 94-99, 1979.

El-Gamassy, A. M., Hussain, M. F. and El-Bakly, S. A. 'Effect of planting dates and fertilisation levels on seed production of some summer flowering annuals'. *Ann. Agric. Sci. Cairo* 10(2): 361-374, 1965.

Jana, B. K. 'Zinnia'. In: *Commercial Flowers*. (Eds.). Bose, T. K. and Yadav, L. P. Kolkata, India: Naya Prokash, pp. 837-841, 1998.

Jana, B. K. and Pal, A. 'Response of nitrogen and phosphorous on growth, flowering and seed yield of cosmos (*Cosmos sulphureus*) cv. Super Sunset'. *Ind. Agric.* 35: 113-118, 1991.

Jauhari, O. S., Dubey, K. C. and Singh, N. 'Response of phlox, *Calendula* and *Dianthus* to different dates of sowing'. *Ind. J. Hort. Res.* 16: 310-312, 1972.

Jayanthi, R., Shiva Roy, B., Ramachandrappa. G. K. and Chandrashakarappa, K. N. 'Pinching induces branches in balsam'. *Curr. Res.* 16: 82-83, 1987.

Jhon A. Q., Paul, T. M. and Tanka, M. J. 'Effect of nitrogen and plant spacing on growth and flower production of *Zinnia elegans* Jacq.' *Adv. Pl. Sci.* 4(1): 1-7, 1991.

Jhon, A. Q. and Saini, S. S. 'Pollination and seed setting studies in pansy'. *Res. and Dev. Rep.* 4(2):189-193, 1984.

Kaul, G. L. and Dadlani, N. K. *Prospects of Floriculture in India.* New Delhi, India: Government of India, Department of Agriculture and Co-operation. Ministry of Agriculture, 1995.

Kaur, K. and Kumar, R 'Effect of nitrogen and phosphate on seed yield and its parameter in Candytuft (*Iberis amara*)'. Proc. *4th Punjab Science Congress,* February 9-20, 2001. Abst. AAMF 45, 2001.

Kaur, R, and Kumar, R. 'Effect of nitrogen, phosphorous and plant spacing on seed yield in pansy (*Viola tricolour* L.) cv. Swiss Giant'. *J. Orn. Hort.* New Series 1(1): 21-25, 1998.

Kaur, R. Studies on nutrition, plant spacing and pollination on seed yield and its parameters in pansy *(Viola tricolor)* cv. Swiss Giant. M.Sc. Thesis. Punjab Agricultural University, Ludhiana, India, 1997.

Kaur, R. and Kumar, R. 'Effect of nitrogen, phosphorus and potassium on seed yield in pansy (*Viola tricolour* cv. Swiss Giant)'. *J. Orn. Hort.* 1(1): 21-25, 1998.

Kumar, Poonam R. and Dubey, R. K. 'Effect of storage materials on seed germination and seed viability in zinnia'. *J. Orn. Hort.* 5(2): 59, 2002.

Kumar, R. and Kaur, K. 'Effect of nitrogen and phosphorus on plant growth, pod number and seed yield in balsam (*Impatiens balsamina* L.)'. *J. Orn. Hort.* 4(1-2): 23-26, 1996.

Kumar, R. and Kaur, K. 'Effect of planting time and cultivars on growth, flowering and seed yield in phlox (*Phlox drummondii*)'. *Seed Res.* 28(1): 23-26, 2000.

Kumar, V. Studies on the effect of planting dates and spacing on the seed yield in pansy *(Viola tricolour)* cv. Swiss Giant. M.Sc. Thesis. Punjab Agricultural University, Ludhiana, India, 1996.

Miyajima, D. 'Germination of zinnia seeds'. *Seed Sci. Tech.* 24: 465-473, 1996.

Mohd Ahmad; Savker, G. and Muthoo, A. K. 'Effect of Paclobutrazol on growth and flowering of cosmos (*Cosmos bipinnatus* Cau.)'. *The Pb. Hort. J.* 30: 2000-2002, 1988.

Oberthora, K. 'Nutrient content of zinnia plants grown for seed'. *Acta Fytotechnica* 37: 71-80, 1981.

Pal, P. 'Response of *Coreopsis* (*Coreopsis picta*) to various levels of nitrogen and phosphorus'. *Prog. Hort.* 20: 313-317, 1988.

Poonam, Kumar, R. and Dubey, R. K. 'Effect of planting time and spacing on zinnia'. *J. Orn. Hort.* 5(2): 49-50, 2002.

Pramesh, R., Deshpande, V. K. and Krishna, A. 'Methods of improving substandard germination in balsam (*Impatiens balsamina*)'. In: *Floriculture Technology, Trades and Trends.* (Eds.). Prakesh, J. and Bhandary, K. R., pp. 223-224, 1994.

Rajanna, D. and Khalak, A. 'Effect of inter-row spacing on yield of zinnia'. *Curr. Res. Univ. of Agric. Sci.,* Bangalore 21: 154-155, 1992.

Rani, R. Effect of planting time and spacing on growth, flowering and seed yield in balsam *(Impatiens balsamina* L.) and celosia (*Celosia plumosa* L*).* M.Sc. Thesis. Punjab Agricultural University, Ludhiana, India, 1997.

Seager, J. C. R. 'Evaluation of New Guinea *Impatiens* cultivars'. *Irish J. Agric. Res.* 19: 11-18, 1980.

Singh, A. Floral biology and seed set studies in pansy *(Viola tricolour hortensis).* M.Sc. Thesis. Punjab Agricultural University, Ludhiana, India, 1987.

Singh, A. Studies on the seed production of pansy (*Viola tricolour* L.) var. 'Hiemalis Mix'. M.Sc. Thesis. Punjab Agricultural University, Ludhiana, India, 1978.

Singh, A. Studies on the seed production of pansy'. M.Sc. Thesis. Punjab Agricultural University, Ludhiana, India, 1979.

Swarup, Vishnu In: *Garden Flowers.* New Delhi, India: National Book Trust of India, 1967.

—oo(O)oo—

2

F_1 Hybrid Seed Production in Ornamental Crops

S. P. S. Raghava and M. C. Singh

INTRODUCTION

F_1 hybrids are the results of crossing of two homozygous but genetically dissimilar parental lines. Hybrid vigour is defined as the increase in size or vigour of a hybrid over its parents. The increased vigour may be over the average of its parents or over the better parent. The term 'heterosis' was proposed to refer to the developmental stimulation resulting by the union of different gametes. Hybrid vigour is entirely due to bringing together, a large number of favourable dominant genes contributing to vigour in the first generation hybrid.

F_1 hybrids are in immense usage throughout the world with the rapid increasing interest in flowers; however, the total number of the hybrids developed in flowers is still less than those developed in vegetables. The first flowers to be used for F_1 hybrid seed production were those in which the parental lines and cultivars were homozygous or could be easily selfed and also gave large numbers of seeds from a small number of hand pollinations. Shull (1911) gave his idea on hybrid vigour in plant breeding and introduced the name 'heterosis' for this conception which carried the criterion, that is, hybrid vigour in connection with the optimal uniformity, a heterozygosity giving rise to the strong segregation of characters in F_2 and necessity of being reproduced by constant new crossing of two parental lines bred to homozygosity. F_1 hybrids in ornamental crops were known even before Shull in 1911 when he propounded the classical theory of hybrid vigour in plant breeding for the first time. Way back in 1909, F_1 hybrid variety 'Prima Donna' in begonia (*Begonia semperflorens* Link et Otto) was released by Benary Seed Company in Germany (Reimann Philip, 1983). In spite of the early start in heterosis breeding in floriculture, there was no further use of this phenomenon to other flower species after *Begonia* during the period of the early Nineties, not even to the tuberous Begonias. It was during World War II, in 1942, that Japan produced the first commercial F_1 hybrids in *Petunia*. From 1950 onwards, F_1 hybrids were produced in flower crops like *Ageratum, Anemone,*

TABLE 1
Promising Hybrid Series in Some Annual Flowers

No.	Flower Name	Botanical Name	Hybrid Variety Series
1.	*Ageratum*	*Ageratum houstonianum*	Blue Horizon
2.	Aster	*Callistephus chinensis*	Serene
3.	Balsam	*Impatiens balsamina*	Carousal, Swirl, Borneo, Tango , Show Topper
4.	*Begonia*	*Begonia semperflorens*	Primadonna
5.	Cockscomb	*Celosia argentea*	Century, Sparkler, Toreador
6.	*Dahlia*	*Dahlia variabilis*	Waverly series
7.	*Iris*	*Iris americana*	Blue Diamond , Ideal, Telstar, White Wedgewood
8.	Larkspur	*Delphinium ambigua*	Blue Cloud, Giant Imperial, QIS
	Larkspur	*Delphinium consolida*	Sydney
9.	*Lavendula*	*Lavendula hybrida*	Silver Streak
10.	Marigold	*Tagetes* sp.	Gold Coin Series, Climax Series, Space Age Series, Galore Series, Inca Series, Red and Gold, Nugget, Red Seven Star
11.	*Petunia*	*Petunia hybrida*	Multiflora-Silver Medal, Comanche, Orange Bells, Coral Bells, Summer Sun, Horizon, Merlin, Polo, Polaris, Lollipop, Satellite Grandiflora-Ballerina, Daddy, Dream , Blue Danube, Sonata, Allegro, Nocturne, Cherry Tart, Apple Tart, Apricot Tart, Grape Tart, Floribunda
		Madness, Milliflora-Fantasy	
12.	Pincushion flower	*Scabiosa stellata*	Blue Moon
13.	*Phlox*	*Phlox drumondii*	Patriot, Chanal
14.	*Plumbago*	*Plumbago auriculata*	Plumbago , Escapade
15.	Poppy, California	*Eschscholzia californica*	Mission, Bells
	Poppy, Iceland	*Papaver nudicaule*	Champagne, Bubbles, Wonderland Mix
16.	*Portulaca*	*Portulaca grandiflora*	Calypso Mix, Tutti Frutti Mix, Margaritta
17.	Snapdragon	*Antirrhinum majus*	Rocket, Sonnet Mix, Butterfly Series

Continued...

...*Continued*

No.	Flower Name	Botanical Name	Hybrid Variety Series
18.	Statice	*Limonium hybrida*	Misty
		Limonium sinuata	Excellent, Fortress Mix, QIS Pastel Shades, Arrow, New Era, New Wonder
19.	Sunflower	*Helianthus annus*	Jerusalem Gold, Jerusalem Seri, Orit, Sunrich Orange
20.	*Zinnia*	*Zinnia elegans*	Sunrise, Gold Sun, Red Sun

Gerbera, Primula, Petunia, Tagetes, Cyclamen, pansies, *Begonia, Geranium, Portulaca, Dianthus,* balsam, stock, wall flower, ornamental sunflower (*Helianthus annus*), *Gazania,* hollyhock, *Calceolaria* and *Zinnia* by several seed companies in the U.S.A., China, Japan, the Netherlands, Denmark, Germany, the United Kingdom and Israel. Flower breeders became familiar with the F_1 hybrid production of those crops also which were formerly propagated by vegetative parts such as *Pelargonium* and *Gerbera.* Inbred lines and their F_1 combinations are derived from the obligate outbreeders only after a long period of investigation. Thus, their production requires a vast knowledge and manipulation of a wide range breeding techniques. Hybrids exhibit some very desirable traits that non-hybrids do not exhibit and it could take years for the 'ideal' combination of desirable traits to happen by chance (Table 1).

ADVANTAGES OF F_1 HYBRIDS

Unlike vegetables, the ornamental crops still lack quantitative standards, except for number of flowers harvested per season, but traits like colour, appearance, texture, uniformity and fragrance are the breeders' priority. However, most of the cut flowers such as roses, carnations, chrysanthemums, etc. are propagated by vegetative parts and, therefore, do not represent true F_1 hybrids. In bedding plants such as *Petunia, Pelargonium, Begonia semperflorens,* etc., which are also to some extent used as pot plants, the superiority of F_1 hybrids over conventional varieties is overwhelmingly different in producing an effect like an intensively coloured carpet, while the conventional varieties often produce a shimmer, which disappears with distance, particularly in some cut annuals like aster, Calendula, Solidago and chrysanthemum and potted plants such as *Pelargonium zonale, Anthurium scherzerianum,* etc. An important feature is the time needed for the appearance of the first flower which depends specifically on the F_1 hybrid in question The same possibility applies to the duration of the 'flush', that is, the blooming period which is very important in the bedding plants and in some F_1 hybrids of *Begonia semperflorens, Ageratum houstonianum,* etc., exceeds that of non-hybrids considerably. Generally, we may say that uniformity is the most essential component of F_1 hybrid performance. There are individual plants of outstanding performance in each conventional variety, but they occur as a minority. To raise their frequency to a majority and to achieve uniformity, more or less all the individuals should be of outstanding performance; this has always been the fascination for breeders, which only F_1 hybrids areable to exhibit.

In addition to the above, F_1 hybrids have other advantages also like dwarf and compact plants, abundant basal branching, profuse free flowering, earliness, enhanced blooming season, doubleness, large-sized flowers, novelty in flower colour and shapes, tolerance/resistance to abiotic and biotic stresses. In *Petunia, Dianthus* and balsam, the F_1 hybrids are dwarf with compact growth, producing profuse basal branches, maturing early and producing large-sized flowers. *Petunia* hybrids also have the faster growth and longer blooming season. Doubleness in *Petunia* can only be derived through hybridisation. It has also been possible to grow hundred per cent double flowers in stocks (Hansen Series) by discarding dark green foliage seedlings and selecting pale green-leafed seedlings growing under cool conditions. The flowers produced by triploid hybrids like Nugget in marigold from the cross of *Tagetes erecta* (diploid) × *Tagetes patula* (tetraploid), being sterile with no fertilisation seed sets remain fresh on the plant for a longer duration. The unusual shape of the flower has been derived in Butterfly hybrids of *Antirrhinum,* as compared to the standard varieties. In such hybrids, each floret is like a colourful butterfly exposing more colour.

In carnation florist double is an unfixed hybrid from the single crossed with super double, doubleness being incompletely dominant. For example, the perpetual flowering carnation seems to have been developed in Europe even before 1750 and is reported to be a hybrid between *Dianthus caryophyllus* (2n = 30) and *Dianthus chinensis* (2n = 30, 60).

Hybrid vigour can be preserved where the plants can be propagated by vegetative means. In F_1 hybrids 'Carefree' geranium are more vigorous and uniform when grown from cuttings than F_2 plants obtained from their seeds.

In addition to the above advantages, F_1 hybrids are advantageous in the following sense:

1. Greater vigour expressed as yield and targeted desirable traits.
2. Greater adaptability to varying environmental conditions because all the genes are present in heterozygous state.
3. The expression of the advantageous character as governed by dominant genes.
4. Exclusivity of control by the breeder, since he alone retains the parents which together produce the commercial product.
5. Stability of the variety over a period of years.
6. Proprietary control of the variety by the originator.

DISADVANTAGES OF F_1 HYBRIDS

1. Sometimes F_1 hybrids have less adaptability because of a narrow genetic base and uniformity. Hybrids may be too uniform in maturity for home gardens.
2. Cost often more than offset by increased returns.

PRINCIPLES OF F_1 HYBRID SEED PRODUCTION

Most of the flower seed is produced in greenhouses. Both emasculation and pollination are done by hand, which enables the breeder to have a wider choice of parents in flowers like *Antirrhinum, Gloxinia, Petunia, Impatiens,* ornamental chillies and *Geranium.* In *Tagetes* and *Zinnia,* genic male-sterility facilitates efficient and economic hybrid seed production. Collection of the pollen with a vacuum pump, dehiscence of anthers under UV lamp, storage of pollen at cool temperature and use of a pollen gun for pollination are some of the useful innovations facilitating cost effective hybrid seed production. In many cases like *Petunia, Antirrhinum* or *Nicotiana,* single pollination produces numerous small seeds in a capsule. About 200 flowers are needed to produce 30 g of *Petunia* seeds. At every stage of production, care is taken to avoid selfingand to ensure complete hybridity. Harvesting, processing and storage are done carefully to avoid mechanical mixture.

TABLE 2
Generalised Breeding Systems in Some Flower Annuals

No.	Outbreeders	Preferentially Outbreeders	Preferentially Inbreeders	Inbreeders
1.	*Alyssum*	*Ageratum*	*Antirrhinum*	*Impatiens*
2.	*Aquilegia*	*Anemone*	Aster	Sweet pea
3.	*Begonia*	Cornflower	*Dahlia*	
4.	*Cineraria*	*Cyclamen*	*Salpiglossis*	
5.	*Petunia*	*Delphinium*	*Salvia*	
6.	*Polyanthus*	Marigold	Wallflower	
7.	*Primula*	*Verbena*		

Source: Watts (1980).

Inbreeding

An F_1 is the result of a cross between two genetically distinct parent plants, irrespective of their state of homozygosity. Thus, in an outbred population, because self-pollination is unlikely to occur, all plants will be F_1 hybrids. In the modern connotation, however, an F_1 hybrid is the result of a cross between two homozygous (but genetically different) parents or lines, and all F_1 plants resemble one another exactly. The importance of obtaining homozygous lines lies in the fact that the full effect of the genes can be observed only in this condition. Without homozygosity, recessive genes will be masked by their dominant alleles. Continuous inbreeding, on the other hand, allows identification and selection of particular gene combinations. The rate at which the homozygosity is achieved depends upon the original heterozygosity in the selected plant. The inbred lines having the tendency of gene segregation affecting the phenotypes, should not be crossed to form the F_1 hybrids but should be used as separate inbreds in crosses with progenies from other less closely related plants. The effect of inbreeding in cross-pollinating species (Table 2.) is to increase homozygosity levels within lines, which leads to reduction in the vigour or depression of yield. Hence, in this system very closely-related lines become so weak that further seed production by inbreeding is impossible.

Inbreeding is virtually unnecessary before hybridisation in homogeneous cultivars except where the cultivar is thought to consist of a number of very similar pure lines which can only be satisfactorily 'sorted out' by saving progenies of single plants separately. Under these circumstances, one generation of inbreeding is adequate and no breeding depression would be found even if inbreeding was continued over a number of years.

Assessing Hybrid Combinations

The level of outcrossing determines the results of inbreeding, but it is a fairly safe rule that single plants may be selected almost at random from the population before raising the first inbred generation. Inbred lines from the cross-pollinated ancestors should be selected for specific qualities during the selfing programme. In many instances, the characters selected will probably be only marginally acceptable and some may, in fact, appear to be worse than those of the parents. Selection, nevertheless, should be relative to all other lines and only the least desirable lines should be discarded in the early stages. Single plant selections in this stage and subsequent generations must be made from the best lines only. Tentative assessment of an F_1 hybrid can be made by crossing pairs of the representative plants from different lines. Although uniformity is one of the features of an F_1 hybrid, the lack of it must be allowed in the early experimental crosses, for example, in *Cyclamen* and *Begonia*.

F_1 hybrids usually show increased vigour over the conventional cultivars. This is best expressed in F_1 derived from a cross between inbred lines of different cultivars. Inbred lines which combine best in the early stages should be inbred further until complete visual homozygosity is obtained, although there is little need to repeat the crossing programme until the S_5 or S_6 generations.

TECHNIQUES FOR PRODUCTION OF F_1 HYBRID SEEDS

Besides the conventional method of selecting inbred lines for desired characters, testing their combining ability and production of hybrid seeds, evaluation of F_1 hybrid performance, there are special general techniques used in the flower crops for the development of F_1 hybrids. Some of these important techniques are as under:

Use of Self-Incompatibility

The self-incompatibility system operating in *Nicotiana* and *Petunia* is gametophytic and can be utilised in practical application of guaranteed cross pollination under open field conditions. However, one additional advantage in these two flowering plants is that there are enough seeds per pollination, which can adequately compensate the high cost of F_1 hybrid seeds. In *Gerbera*, sporophytic system of self-incompatibility operates and dominance relationship between two self-incompatibility (SI) alleles influences the seed set. In such species/cultivars, the degree of self-incompatibility can be assessed by examining flower styles with the ultra-violet fluorescent microscope, a few hours after the pollination. Pollen tubes which have penetrated the stigma, and are thus able to grow down the style unimpeded, are able to be counted since they identify themselves by fluorescence. The lack of, or a small number of pollen tubes throughout their length, indicate a strong incompatibility reaction. If the incompatibility reaction is weak, counting will prove impossible since numerous pollen tubes will be visible as a tangled mass within the stylar tissue. With the advent of tissue culture technology, maintenance of the female parent carrying SI alleles can be managed under protected conditions. *Myosotis* (Forget-me-not) tends to be self-incompatible but since this characteristic is rarely absolute, for certainty of cross-fertilisation it is first necessary to emasculate. This should be done in the bud stage by the removal of the five anthers followed by cross-pollination. In Sweet Williams, the head of the flowers comprises individuals which mature at different times, usually starting with the centre flower of a cluster. Therefore, pollination, either by all but a single cluster, or by pollinating members of the head which have reached the right stage, followed by removal of all other buds will be effective.

Pollen Sterility

Wherever pollen sterility is governed by a single recessive gene, maintenance of the genetic stock is difficult as there will be continuous segregation of the fertile and sterile individuals in 1:1 proportion. This phenomenon is present in *Tagetes*, *Zinnia, Delphinium, Antirrhinum, Calceolaria, Salvia* and *Impatiens*. In *Ageratum*, however, both the male-sterility and self-incompatibility systems are prevalent and a choice can be made on the basis of economy in the seed production. In Petunia, cytoplasmic male-sterility has been observed but the use of this type of male sterility is not so practically common because of breakdown of male-sterility in the maternal parent or malformation of flowers in F_1 plants.. Male sterility in Sunflower has been used to produce ornamental varieties like, 'Sunrich Orange' (Japan) and 'Orit' (Israel), which have no pollen grains and allergic effects like in other male fertile varieties grown for their seed.

Double Flower Condition

Many flower species in the family Asteraceae assume a double form because of numerous petals but a central disc may be seen when petals are pulled aside and make a 'composite' head such as in *Ageratum*, aster, *Chrysanthemum*, cornflower, *Dahlia*, daisy, *Gaillardia*, marigold, *Pyrethrum*, *Rudbeckia* and sunflower. Sunflower illustrates most clearly the satisfactory example of pollination where each floret bears only a single ovary. In most cases, double flower results from a transformation of the anthers into petals. Therefore, the double flower character can be regarded as a form of male-sterility. Further, this double flower character should be inherited as a dominant character to be manifested in F_1 hybrids. However, in the case of *Gerbera*, where the pistils are hidden by petals, manual hand pollination is tedious for maintenance of the female parent. Reimann-Philip (1969) developed a breeding scheme using the double flower character in garden carnation (*Dianthus caryophyllus*) as a form of male sterility. There are also some floral abnormalities, like the 'Cinderella' character in *Begonia* and 'Femina' in *Tagetes erecta* and *Zinnia elegans* resulting in the male-sterility.

Triploidy

The advantage of growing triploid varieties is that of continuous blooming period as there is little or no seed set. Among the flower species, in *Tagetes*, commercial triploid hybrid (2n = 36) results from a cross between diploid female *Tagetes erecta* (2n = 24) and tetraploid male *Tagetes patula* (2n = 48). The triploid hybrid 'Nugget' (*Tagetes erecta* × *Tagetes patula*) has the unique ability of holding the flowers on the plants for a longer period. However, in *Begonia semperflorens*, the loss of uniformity in triploids has been reported by Reimann-Philip (1983).

Somatic Embryogenesis

Perhaps in the future, somatic embryogenesis using liquid bioreactors may be an economically viable option for hybrid seed production in flowers. The use of this technique is of preference over others, requiring less space and labour, least cost involved and faster propagation.

METHODS OF MAKING THE CROSS

There must be some way to control pollination so that only the desired cross is obtained and selfing of the seed parent is eliminated. The method must be economically feasible. Various possibilities are as follows:

1. Hand emasculation and hand pollination.
2. Hand emasculation and natural pollination.
3. Hand elimination of male plants.
4. Genic male sterility as in marigold, *Zinnia, Ageratum* and *Calceolaria.*
5. Cytoplasmic male sterility as in *Petunia.*
6. Self-incompatibility as in *Petunia*, pansy, stocks and kale.
7. Chemical emasculation—selective elimination of pollen production, that is, use of gameticides.
8. Use of marker genes to identify the selfs so that they can be eliminated as seedlings.

TESTING OF THE INBREDS/VARIETIES

If a character is controlled by a recessive allele, it will be exposed during inbreeding. If this is a desirable character, inbreeding is, therefore, a positive event. But a recessive character can be a negative event too and in outbreeding species, the recessive alleles are frequently negative, even lethal, and the plant dies frequently, sometimes early as the embryo within the seed due to the occurrence of the lethal combination.

Modifications

1. F_1 hybrids are sometimes used where uniformity can be sacrificed and vigour is still high. The cost of the resulting seed would be much lower.
2. Double crosses ($F_1 \times F_1$) and other combinations such as the 3-way (inbred $\times F_1$), line crosses (2 sub-lines crossed × inbred) or (2 sub lines crossed × 2 other sub lines crossed), are used to reduce seed cost.

Explanations for Heterosis

1. Interaction of alleles—heterozygosity, over-dominance—additive, or separate function of different alleles (a1a2 superior to a1a1 or a2a2).
2. Interaction of dominant genes—masking the deleterious recessives by accumulating favourable dominants (AAbbccDD) × (aaBBCCdd) = (AaBbCcDd).

The suggested methods for testing of inbreds or varieties are outlined below:

Single Cross Method

In this method, single crosses are made and the resulting F_1 hybrids are evaluated. Earlier, the breeders were using the varieties for crossing but later in 1909, Shull suggested the inbred lines as parents for production of resulting F_1 hybrids and recommended single cross method for production of superior hybrids. However, the F_1 hybrids, so produced become expensive due to the low yielding female parent and poor production of the pollen by the male parent for short duration. The resulting F_1 hybrids set poor shaped seeds with low germination rate.

Double Cross Method

The double cross method to overcome the problems associated with the single cross method was suggested to produce seeds much better in shape with higher germination percentage. The resulting seeds were also found very uniform in shape and produced vigorous seedlings as compared to those obtained from the single cross method. In this method, the hybrids are produced by crossing two resulting F_1 hybrids involving four inbred lines as given below.

$$
\begin{array}{ccc}
A \times B & & C \times D \\
AB & \times & CD \\
 & ABCD &
\end{array}
$$

To predict the performance of the double cross hybrids, the following four different methods are used:

1. If we consider the four inbreds and take the average yield of six single crosses, that is,

   A × B , A × C, A × D, B × C, B × D, C × D.

2. The average yield of four non-parental single crosses, that is,

A × B
A × C, A × D, B × C, B × D non-parental crosses
C × D

3. Average performance of four inbreds in a series of single crosses.

4. Average performance of four inbreds in top cross.

Three Way Cross

Three inbred lines are involved in this cross. F_1 hybrid produced by two lines is further crossed with a third line as follows:

A × B
AB × C
ABC

Convergent Method

This method involves combining the desirable character from a number of inbreds as soon as possible. The desirable genes from each inbred converge upon one another and it carries a risk that all the desired characters may not be present in the final segregating population. This system can be used in partially cross-pollinated species where the resulting population will be highly heterozygous. If 8 inbreds are used together, the system is as follows:

1st crosses	A × B, C × D, E × F, G × H
2nd crosses	F_1's (no need of selection) AB × CD, EF × GH
3rd crosses	(segregating F_1's) ABCD × EFGH

Top Cross Method

This method is suitable to test the general combining ability of inbreds and in identifying promising ones. The inbreds are crossed with the varieties as follows:

A × variety, B × variety, C × variety, D × variety, E × variety, F × variety, G × variety, etc.

Diallel Cross Method

This is pair cross method which allows the use of every plant in pair crosses with every other plant in selected groups (Griffing, 1956). Homozygous diploid lines are essential to obtain satisfactory results. It also requires sufficient number of individuals or progeny to make a large number of crosses.

On the basis of number of parents (n), the number of crosses are calculated as n(n-1)/2. If 8 parents are involved in the diallel cross, it will lead to a total number of 28 crosses (Table 3).

PROCESS OF HYBRID SEED PRODUCTION IN SOME IMPORTANT FLOWER CROPS

African Marigold

In African marigold, genic male sterility system is being used for F_1 hybrid seed production. There are two types of male sterility in marigold, that is, apetalous form (with no stamens) and true

TABLE 3
Crosses Resulting from 8 Parents in Diallel Cross

Parents		Male Parents							
		A	B	C	D	E	F	G	H
	A	—	A × B	A × C	A × D	A × E	A × F	A × G	A × H
	B		—	B × C	B × D	B × E	B × F	B × G	B × H
	C			—	C × D	C × E	C × F	C × G	C × H
Female	D				—	D × E	D × F	D × G	D × H
Parents	E					—	E × F	E × G	E × H
	F						—	F × G	F × H
	G							—	G × H
	H								—

double form (produces no anthers in the disc florets). Apetalous type of male sterility has greater reliability and is being used for the production of F_1 hybrid seeds on a commercial scale in the U.S.A., the U.K., Holland, France, etc. The apetalous type of male sterility is controlled by recessive alleles (Gupta *et al*, 1999). Thus, a male sterile plant has a genetic constitution of ms ms (homozygous recessive), while male fertile plants may be of Ms Ms or Ms ms genotypes. The heterozygote Ms ms will segregate to give 1 Ms Ms:2 Ms ms:1 ms ms, the usual ratio for a single gene, but both homozygous (Ms Ms and ms ms) will, of course, breed true (except that the latter will produce no pollen to allow it to breed further).

Therefore, for use as a parent in F_1 hybrid production, the male sterile line is perpetuated by back-crosses of double recessive (ms ms) male sterile by heterozygous male fertile (Ms ms) maintainer. In the subsequent generations, 50 per cent of the progeny will be male fertile Ms ms, while the other 50 per cent will be male sterile ms ms. Male sterility phenomenon can be incorporated into standard varieties through five or six generations of back-crossing.

For the actual production of F_1 hybrid seed in the field, the back-cross generation us inter-planted with the other parental line (which is a normal male fertile inbred) and heterozygous male fertiles occurring in back-cross generation should be removed as soon as identification is possible, thus leaving male sterile plants only. The male sterile plants can be identified in the early stage by the shape of flower buds as these plants have pitcher type of flower buds, whereas the male fertile plants have normal flower buds.

To summarise, the mechanics of using genetic male sterility for F_1 hybrid seed production under open field conditions require (1) a normal fully fertile inbred line, and (2) an inbred line which is maintained by crossing together known heterozygous (Ms ms) and male sterile (ms ms) plants. Seeds should always be harvested from ms ms plants. The cross of ms ms × Ms ms should be repeated in every generation, as it will segregate in 1 Fertile:1 Sterile ratio. A ratio of male sterile line to pollinating fertile line is dependent on the size of hybrid block, but ratio of 3 male sterile:1 male fertile has proved to be the optimum.

Petunia

Goldsmith (1968) reported that double flower forms and large-flowered or Grandiflora types of petunia are completely dependent upon hybrid seed production. The double character is dominant to the single form and is essentially female sterile producing an occasional functional pistil. The F_1 hybrid petunias are produced by using a single-flowered line as the seed parent and a homozygous double clone as the pollen parent. The double flowered line is maintained by cuttings. Likewise, the Grandiflora class can only be produced by the hybrid method. Grandiflora type is dominant over Multiflora type. In hybrid seed production, Multiflora single type is used as seed parent, while Grandiflora single is used as pollen parent. Hybrid seed production in petunia is entirely by hand emasculation and hand pollination.

Saito (1959) reported that tetraploid plants (double-flowered) of DDDD, DDDd, DDdd, Dddd types are all expected to be double flowering. So, all the double strains were obtained only by crossing between Dddd or dddd and homodouble DDDD parents.

The gametophytic self-incompatibility is available in petunia, but the commercial production of F_1 hybrid is carried out by hand pollination rather than by insects. The plants of one inbred line are pollinated by pollen collected from another line which carries a different S allele and vice versa.

Antirrhinum

Although antirrhinums are fully self-fertile and tend to be more receptive to pollen from other plants of the same crop rather than their own, care is to be taken to avoid disturbing the self anthers. Cross pollination may easily be made in the glasshouse without resorting to emasculation. However, in the open, the flowers are pollinated by bees. For complete certainty, buds should be emasculated about two days before they are expected to open. Removal of four anthers is very straightforward, but the single style so closely resembles an anther filament in the confines of a bud that some practice is required.

A fairly recent mutation in the flower form of antirrhinum has produced an open-throated penstemon-type flower and although the bud stage still remains very similar to the normal 'bunny face' type, the young open flower is very much easier to emasculate and pollinate. Crosses between penstemon-flower and 'bunny face' produces a bunny face F_1 with a 3:1 segregation of bunny face-penstemon in the F_2 generation, thus indicating that the inheritance of the conventional flower form 'bunny face' is controlled by a single dominant gene.

The F_1 hybrid seed is produced by hand emasculation and pollination. An apetalous pollen sterile form with a thicker and fasciated pistil is quite common in genetic stalks, but it has not proved useful.

Zinnia

The apetalous character has been the source of pollen sterility and the production methods for hybrid seed are similar to those described for marigold. Pollination has been a problem in this crop. Hand pollination is used at large, resulting in the development of superior varieties having stronger growth, large-sized flowers in more number and better doubleness. The hybrids are available in medium and tall habits and also in cactus flower form.

Begonia

The commonly cultivated species (*Begonia rex* and *Begonia semperflorens)* are monoecious, bearing male and female flowers separately on the same plant. Plants with different sexes have a pre-potency for cross-fertilisation, although self-fertilisation can be carried out easily. Emasculation is unnecessary but female flowers must be protected from contamination by pollen from the male flowers of the same plant if controlled pollination is to be made.

Hollyhock

The anthers surround the stylar tube. Anthers are numerous and styles are separate from one another, resembling a tuft in the centre of the flower. The easiest way of crossing is to ignore emasculation and to apply pollen to the stigma in the bud stage, about two days before the flower opening. In this way, the foreign pollen is able to effect fertilisation before self-pollens are shed.

China Aster

The florets in the capitulum become sexually mature from the outside towards the inside. The ray florets are female and are used for artificial pollination. The process of emasculation is done when the flowers are fully open. The ray florets are trimmed up to the level of the stigma with scissors or a sharp blade and disc florets are removed completely with the help of forceps. Pollination is done on the third day after the emasculation.

Pansy

Since self-incompatibility exists in pansy, the hybrids are produced without emasculation. Like petunia, the hybrid is obtained from both the parents used.

WORK DONE ON HETEROSIS BREEDING IN INDIA

The work on heterosis breeding in flower crops has been carried out at different research institutes in India, especially at the Indian Agricultural Research Institute, New Delhi; Indian Institute of Horticultural Research, Bangalore; National Botanical Research Institute, Lucknow and State agriculture universities. These institutes have made significant contributions in crops like marigold, antirrhinum, petunia, balsam, aster and hollyhock.

African Marigold

The significant achievements on heterosis breeding in marigold have been reported by several workers. Singh (1970) and Singh and Swarup (1971 and 1973a, 1973b) carried out work on heterosis in African marigold in a diallel set (excluding reciprocals) involving seven parents. Maximum heterosis was recorded in a number of flowers. In the crosses, Alaska × Hawaii, Alaska × Cupid Orange Mum and Katrain Local × Cupid Orange Mum, highest heterosis was observed for flower size, flower weight and flower number, respectively. In a study conducted by Reddy *et al* (1989) in African marigold using seven parental lines in a diallel set of crosses (including reciprocals), the crosses African Giant Double × Spun Gold, Giant Double African Orange × Coimbatore Local and Giant Double African Orange × African Double Yellow were recommended for exploitation of heterosis for commercial use due to high specific combining ability (sca) effects and per cent heterosis for a number of characters.

Out of 15 F_1 hybrids developed by Barad *et al* (1993) through diallel crosses involving six parents(excluding reciprocals), one hybrid raised from the cross of Giant Double African Orange × Cracker Jack was found to be promising for commercial purposes. Using apetalous male-sterile line, several F_1 hybrids have been developed by Raghava (1995). One hybrid, MS-8 × Pusa Narangi Gainda has been found to be very promising in respect of higher flower yield.

Antirrhinum

In snapdragon, appreciable heterosis for all the characters studied was reported by Malik (1979). Over-dominance in the inheritance of all the characters was observed. Involving the parents *Antirrhinum majus* and *Antirrhinum glutinosum*, Mahal (1972) developed F_1 dwarf hybrids.

Balsam

Involving five parents in a diallel set of crosses, Swarup *et al* (1975) studied heterosis in balsam for various characters. The performance of all the F_1 hybrids was observed to be better over the superior parent. Highest heterosis was recorded for number of flowers per plant (145 per cent) followed by branches per plant (62 per cent).

China Aster

In the studies on heterosis in China aster carried out by Raghava *et al* (1988) and Raghava and Negi (1992) involving 12 parents in a diallel set of crosses (without reciprocals), considerable heterosis was recorded for all the characters. Three crosses, Shell Pink × Azure Blue, AST-20 × Azure Blue and AST-20 × AST-16 were recommended for commercial exploitation of heterosis on the basis of economic traits like flower size, flowers per plant and stalk length.

Hollyhock

Five F_1 hybrid varieties, namely, Pusa Pink Beauty, Pusa Yellow Beauty, Pusa Pastel Pink Supreme and Pusa Apricot Supreme were developed by the Indian Agricultural Research Institute, New Delhi (Kaicker and Singh, 1985; Swarup and Raghava, 1973). Genetical studies were carried out by Swarup *et al* (1973).

Petunia

In single Multiflora and Grandiflora types of Petunia, Singh (1985) and Singh and Swarup (1993) studied heterosis in a diallel fashion of crosses using eight parental lines. Observation revealed considerable heterosis for all the characters studied.

Apart from the government institutes, the Indo-American Hybrid Seeds, Bangalore, a private sector company, has made significant contributions towards F_1 hybrid seed production in *Petunia, Antirrhinum,* marigold, flowering tobacco, etc. on a commercial scale. This company is exporting F_1 hybrid seeds to the U.S.A.

REFERENCES

Attavar, M. 'Hybrid seed production in flowers'. In. *Ornamental Horticulture in India.* (Eds.). Chadha, K. L. and Choudhary, B. New Delhi: I.C.A.R, 1986.

Barad, A. V., Sachan, S. C. P. and Kukadia, M. U. 'Heterosis in African marigold (*Tagetes erecta*)'. Golden Jubilee Symp. *Hort. Res. Changing Scenario,* Bangalore, Abstr. 359, 1993.

Goldsmith, G. A. 'Current developments in the breeding of F_1 hybrid annuals'. *HortScience* 3(4): 269-271, 1968.

Griffing, B. 'Concepts of general and specific combining ability in relation to diallel crossing systems'. *Aust. J. Biol. Sci.* 9: 463-493, 1956.

Gupta, Y. C., Raghava, S. P. S. and Misra, R. L. 'Inheritance of male sterile apetalous inflorescence in African marigold'. *J. Ornamental Hort.* New Series 2(2): 65-66, 1999.

Kaicker, U. S. and Singh, H. P. 'Tall hybrids and dwarf mutants of hollyhock'. *Indian Horticulture* 29: 2-4, 1985.

Mahal, C. Cytogenetics of *Antirrhinum* and *Lantana*. Ph.D. Thesis. Kanpur University, Kanpur, 1972.

Malik, R. S. Studies on heterosis in antirrhinum (*A. majus* L.). Ph.D. Thesis. Agra University, Agra, 1979.

Negi, S. S. and Raghava, S. P. S. 'Breeding of ornamentals'. In: *Plant Breeding Theory and Practice.* (Ed.). Chopra, V. L. New Delhi, India: Oxford and IBH Publishing Co. Pvt. Ltd., 2000.

Raghava, S. P. S. 'Genetic improvement of African marigolds'. Nat. Sem. *Ornamental Horticulture and Environment* held at Kolkata, 1995.

Raghava, S. P. S. 'Hybrid seed production in flowers'. *Indian Horticulture* 46(4): 18-19, 24-25, 2002.

Raghava, S. P. S. and Negi, S. S. 'Combining ability studies in China aster (*Callistephus chinensis* Nees)'. *Indian J. Hort.* 50: 180-186, 1992.

Raghava, S. P. S. and Singh, Kanwar, P. 'Heterosis breeding in important annual flowers'. In: *Advances in Floriculture.* (Eds.). Choudhary, M. L., Singh, Kanwar P. and Hussain, Sakkeer. New Delhi, India: I.A.R.I., 2001.

Raghava, S. P. S., Negi, S. S. and Ramachander, P. R. 'Heterobeltiosis in China aster (*Callistephus chinensis* Nees)'. *Indian J. Hort.* 45: 336-343, 1988.

Reddy, N. T., Muthuswamy, S., Irulappan, I. and Abdul Khader, J. B. M. 'Heterosis and combining ability for yield and yield components in African marigold (*Tagetes erecta* L.)'. *South Indian Hort.* 36: 51-61, 1989.

Reimann-Philip. 'Heterosis in ornamentals'. In: *Heterosis Monograph. Theoretical and Applied Genetics.* (Ed.). Frankel, R. Berlin: Springer Verlag, Germany, 1983.

Reimann-Philip. In: *Die Zuchtung der Bluemen Parey*, Berlin, Hamburg, 1969.

Saito, K. 'Studies on the utilisation and peculiar seed productibility of the autotetraploid plants. I. On a new strain of giant flowering petunia with special reference to the genetical and environmental factors for the fertility'. *J. Hort. Asson., Japan* 28: 45-51, 1959.

Shull, G. H. 'Experiments with maize'. *Bot. Gaz.* 52: 480-485, 1911.

Singh, B. Morphological and genetical studies in marigold (*Tagetes erecta, T. patula and T. tenuifolia).* Ph.D. Thesis. Agra University, Agra, 1970.

Singh, B. and Swarup, V. 'Heterosis and combining ability in African marigold'. *Indian J. Genet. Pl. Breed.* 31: 407-415, 1971.

Singh, B. and Swarup, V. 'Inheritance of flowering, flower weight and flower number in African marigold'. *Indian J. Genet. Pl. Breed.* 33: 172-175, 1973a.

Singh, B. and Swarup, V. 'Inheritance of qualitative characters in African marigold'. *Indian J. Genet. Pl. Breed* 33: 373-377, 1973b.

Singh, H. P. Varietal assessment for improvement in petunia. Ph.D. Thesis. Agra University, Agra, 1985.

Singh, H. P. and Swarup, V. 'Heterosis in petunia (*Petunia hybrida* Hort.)'. Golden Jubilee Symp. *Hort. Changing Scenario,* Bangalore, Abstr. 3: 57, 1993.

Sundararaj, D. D. and Thulasidas, G. *Introduction to Cytogenetics and Plant Breeding*. Chennai, India: Popular Book Depot. 1972.

Swarup, V. 'Heterosis breeding in flowers. In: *Hybrid Research and Development*. (Eds.). Rai, M. and Mauria, S., 1996.

Swarup, V. *Ornamental Horticulture*. Macmillan India Limited, 1997.

Swarup, V. and Raghava, S. P. S. 'F_1 hybrids in hollyhock'. *Indian Horticulture* 18: 21-22, 1973.

Swarup, V., Raghava, S. P. S. and Balakrishnan, K. A. 'Genetical studies in hollyhock'. *Indian J. Genet. Pl. Breed.* 33: 446-451, 1973.

Swarup, V., Raghava, S. P. S. and Balakrishnan, K. A. 'Heterosis in balsam'. *Indian J. Genet. Pl. Breed.* 35: 69-75, 1975.

Watts, L. In: *Flower and Vegetable Breeding*. London, U. K.: Grower Books, 1980.

—oo(O)oo—

3

MICROPROPAGATION OF ORNAMENTAL CROPS

A. K. A. MANDAL, P. MISRA AND S. K. DATTA

INTRODUCTION

MUCH like the fashion industry, the floriculture industry must respond to frequently changing consumer demands. Changes in preferences, fashion and aesthetic values create an ever-increasing demand for flowers, which have unusual or novel characteristics. Therefore, development of newer and newer cultivars and their fast marketing are major challenges in the floriculture industry. Since most of the ornamental crops are highly heterozygous, their seed progeny is not true-to-type. Conventionally, in these plants the vegetative reproduction method is followed to produce genetically identical plants on a commercial scale. But the main constraint of this method is that only a limited number of plants can be raised in a year from a single mother plant and this often fails to meet the market demand. Therefore, most of the modern floriculturists follow the micropropagation technique for rapid multiplication of ornamental plants. Among the many facets of biotechnology, micropropagation is the most successfully commercialised. The most significant advantage of micropropagation is that a large number of plants can be produced in a year from a single mother plant, in a relatively short time and space. Moreover, disease- and virus-free plants can be produced by this method. A large number of ornamental crops like orchids, rose, chrysanthemum, gerbera, carnation, etc. have been commercialised through micropropagation. This review aims to discuss the updates on various aspects of micropropagation research in commercially important ornamental plants. Ornamental crops are normally propagated by cuttings or bulbs/corms. In this review, crops selected (that are propagated through cutting, bulb or corm) of commercial importance are included.

ROSE

Rose is one of the most important cut flowers. It belongs to the family Rosaceae. More than two hundred published papers on different aspects of tissue culture in different rose species are available, the earliest being the induction of callus and root in explanted buds (Nobecourt and Kofler, 1945). However, the development of shoot primordia from stem callus was first reported by Hill (1967).

Regeneration protocol has been developed from different types of explants. Shoot regeneration has been reported from stem node derived callus of *Rosa persica* × *Rosa xanthiana* hybrid (Lloyd *et al*, 1988) and *Rosa damascena* (Ishioka and Tanimoto, 1990), immature embryos of *Rosa hybrida* hybrids (Burger *et al*, 1990). Somatic embryogenesis has been reported from anther filament derived callus (Noriega and Sondahl, 1991) and stem derived callus (Rout *et al*, 1991). In addition to callus mediated regeneration, direct shoot regeneration from leaf midrib and petioles of *Rosa persica* × *Rosa xanthiana, Rosa laevigata, Rosa wichuriana* has been reported (Lloyd *et al*, 1988).

A lot of work has been done for large scale multiplication of different types of roses starting from either shoot tip, axillary buds or nodes with different degrees of success. Micropropagation from shoot tip explant has been reported by Avramis *et al* (1982a, 1982b), Bressan *et al* (1982), Delbard (1982), Hasegawa (1979), Horn *et al* (1988), Kosh-Khui and Sink (1982a, 1982b), Martin *et al* (1981) and Skirvin and Chu (1979a); from axillary buds, micropropagation has been reported by Avramis *et al* (1982a, 1982b), Bini *et al* (1983), Curir *et al* (1986), Davies (1980), Delbard (1982), Kosh-Khui and Sink (1982a, 1982b), Martin *et al* (1981), Sauer *et al* (1985), Walter *et al* (1979) and Valles and Boxus (1987a, 1987b) and from nodal explant it has been reported by Avramis *et al* (1987a, 1987b), Chakrabarty *et al* (2000), Davies (1980), Horn *et al* (1988), Kosh-Khui and Sink (1982a, 1982b) and Pittet and Moncousin (1982). In micropropagation work, growth regulator combinations of NAA and BA are mostly used. Depending upon the type of explant, one to five primary shoots developed prior to transfer to shoot multiplication medium (Hasegawa, 1979; 1980; Davies; 1980; Khosh-Khui and Sink, 1982a, 1982b; Horn, 1991). The bud size in the initial nodal explant had no effect on shoot numbers after three weeks; however, it was generally observed that larger buds regenerated more shoots than the smaller ones (Horn, 1991).

Different species and cultivars of rose have shown to have different nutritional and hormonal requirements and also have different regeneration potentiality (Davies, 1980; De Wit *et al*, 1990; Ishioka and Tanimoto, 1990; Horn, 1991; Chakrabarty *et al*, 2000).

Horn (1991) reported that a single node explant produced an average of 2.5 shoots within three weeks of culture initiation. When these shoots were removed from the initial explants and two successive subcultures were done on the same media composition, each shoot formed 3.9 side shoots, so that each source explant gave $2.5 \times 3.9 = 9.25$ shoots within about 20 weeks.

Rose micropropagation has faced serious setbacks from yellowing of leaflets and vitrification (Hasegawa, 1980; Horn, 1991). In our laboratory, this problem was averted by addition of $AgNO_3$ in the culture media which resulted in reduction of leaflet yellowing as well as enhanced growth of shoots (Chakrabarty *et al*, 2000).

Different authors have used different auxins for root induction. NAA in the range of 0.003 to 2 mg/l has induced roots (Hasegawa, 1980; Arvamis *et al*, 1982b; Khosh-Khui and Shink, 1982a, 1982b) while a supplement by IBA is used by others (Fiorino and Leva, 1981; Bini *et al*, 1983; Leffring *et al*, 1984; Curir *et al*, 1986; Valles and Boxus, 1987b; Alderson *et al*, 1988). A large number of workers have used IAA for root induction (Hasegawa, 1979; Khosh-Khui and Sink, 1982b; Collet, 1985; Sauer *et al*, 1985; Mederos and Enriquez, 1987). The auxin requirement for root induction in rose is cultivar specific. Besides auxin, successful rooting and field transfer of rooted shoots has been reported, to some extent, to be dependent on the level of cytokinin (Bressan *et al,* 1982; Podwyszynska and Hempel, 1988), vitamins (Skirvin and Chu, 1979b; Kosh-Khui and Sink, 1982a; Bini *et al,* 1983), phenylalanin and tyrosin (Avramis *et al,* 1982b) used in the multiplication stage. Also there is a report that superior results are obtained when no auxins or only 0.1 mg/l NAA are applied in the rooting medium (Horn, 1991).

In most of the cases, MS medium is used for culture initiation as well as multiplication; however, many have suggested modified MS as superior (Avramis *et al*, 1982b; Curir *et al,* 1986; Valles and Boxus, 1987b). Among different growth regulators used, NAA and BA are reported to be the best. A combination of cytokinin with low doses of auxin are effective (Skirvin and Chu, 1979a; Davies, 1980; Bressan *et al,* 1982; Kosh-Khusi and Sink, 1982a, 1982b). Walter *et al* (1979) reported that good results are obtained when kinetin (0.21-8.9 mg/l) is used, while 2iP is used by Bini *et al* (1983). Addition of GA_3 in the multiplication medium is reported to have a positive effect (Martin *et al,* 1981; Delbard, 1982; Davies, 1980; Valles and Boxus, 1987b) while no effect of GA_3 is observed by Pittet and Moncousin (1982).

For photoperiod requirement, 16-hour light is reported to be better than 8-hour light (Davies, 1980) while the effect of temperature has been studied by Alderson *et al* (1988), Bressan *et al* (1982), Davies (1980) and Kosh-Khui and Sink (1982b).

In our laboratory, micropropagation protocol for 10 rose cultivars (namely Mrinalini, Queen Elizabeth, Windy City, Chandrama, Raktima, Doris Tysterman, Sylvia, Dr. B. P. Pal and Contempo) has been standardised. We follow the following protocol for routine multiplication of rose. Nodal explants of 1.5-3 cm length are surface sterilised in 70 per cent ethanol for 30 seconds followed by in 0.1 per cent $HgCl_2$ for 2 minutes and rinsed thoroughly in sterile distilled water. For culture initiation, MS medium supplemented with BA (2 mg/l) and IAA (0.2 mg/l) are used. For shoot multiplication, the same media composition along with 10 mg/l $AgNo_3$ is used. Addition of $AgNO_3$ did not reduce the shoot multiplication rate but it enhanced the shoot growth and reduced leaflet yellowing. Roots are initiated in one-fourth strength MS media supplemented with IAA (0.1 mg/l) and IBA (0.1 mg/l). Then the rooted shoots are transferred to the soil for hardening.

GERBERA

Gerbera is a member of Compositae family, covers around 30 herbaceous and perennial species, but the commercial varieties are derived mainly from *Gerbera jamesonii.* The traditional vegetative propagation of *Gerbera* is too slow (7 to 6.5 cuttings per year) to be commercially viable. Micropropagation has come to the rescue and micropropagation methods have been developed for commercial propagation of this species (Murashige *et al*, 1974; Pierik *et al*, 1982).

Plant regeneration of *Gerbera* has been attempted from many different types of explants like shoot tip, leaf, petiole, disc florets and mature and immature flower heads. Most of the *Gerbera* cultures are started from meristems, because it is faster and plants are more likely to be true-to-type. Using shoot tip explants, which are pubescent and grown in close proximity to the soil surface, establishment of a contamination-free culture is a big problem. Contamination rate is usually high and only 3 per cent non-contaminated cultures can be obtained from shoot tip explants after following proper surface sterilisation procedure (Queralt *et al*, 1991).

A two-step disinfection protocol for direct establishment of shoot tips has been reported by Murashige *et al* (1974) to prevent very high contamination rate. Huang and Chu (1985) developed a single one-step protocol for disinfection while flaming the shoot tip briefly several times to obtain 30 per cent clean culture has been reported by Reynoird *et al* (1997).

In our laboratory, we used the following disinfection protocol, which gave 90 per cent contamination free culture. Immature flower heads (1.5 cm in diameter) are washed in 5 per cent Teepol (a liquid detergent) and running tap-water for 15 and 30 minutes, respectively. After removing the bracts, the flower head was pre-treated with 70 per cent ethanol for 1 minute followed by surface sterilisation with $HgCl_2$ (0.1 per cent w/v) for 2 minutes and finally thoroughly washed 3 times with sterile distilled water (Mandal and Datta, 2002a).

A method of culture establishment from bracts of immature capitula has been described by Maia *et al* (1983) on MS media containing 0.5 mg/l IAA. Establishment of culture from mature capitula has been followed by Pierik *et al* (1975) where segments of capitula were first placed under darkness at 25° C for four weeks and then transferred under continuous fluorescent light (2,100 Lux) at 23° C for another four weeks on basal medium with 10 mg/l BA. Except Peirik *et al* (1975), in almost all other cases, cultures were incubated at 16 hour light for culture initiation. Chu and Huang (1983) initiated bud regeneration from the upper part of the stem after 16 weeks on MS medium supplemented with 10 mg/l BA. Laliberte *et al* (1985) has established culture from capitula segmented into 20 pieces, which regenerated approximately 20 buds each after 12 weeks on 1 or 2 mg/l BA and 0.1 mg/l IAA.

We have established culture from segments of immature flower heads cultured on modified MS medium containing 0.5 mg/l IAA and 2 mg/l BA. Shoot regeneration was obtained from about 70 per cent cultures with 4-5 shoots per culture (Mandal and Datta, 2002a).

Irrespective of the method of culture establishment, mass propagation is always achieved via axillary bud development. Murashige *et al* (1974) used a modified MS medium with additional phosphate, adenine sulphate and high sucrose concentration. They found that high level of kinetin (10 mg/l) in the medium was necessary to increase the multiplication rate. Although no exogenous auxin was found to be essential, addition of 0.5 mg/l IAA seemed to improve the vigour of the cultures. Experiments related to illumination requirements for multiplication showed that a 12 to 16 hour photoperiod of 1,000 lux intensity was best (Murashige *et al,* 1974). Maia *et al* (1983) used a first medium differing from the established medium by lower growth regulator concentrations (0.2 mg/l IAA and 1 mg/l BA) and then the subsequent subcultures for mass propagation were carried out on a second medium with doubled growth regulator (IAA and BA) concentrations. Huang and Chu (1985) improved the multiplication rate using high BA concentrations (5 mg/l).

Although commercial production of *Gerbera* has never been achieved through adventitious bud regeneration, this has been reported from leaves (Hedtrich, 1979; Reynoird *et al,* 1993, 1994) and petioles (Jerzy and Lubomski, 1991; Ruffoni and Massabo, 1991; Elomaa *et al,* 1993; Orlikowska *et al*, 1999).

The efficiency of rooting was strongly affected by size of shoots (Pierik *et al*, 1975; Laliberte *et al,* 1985). Maia *et al* (1983) increased the duration of last subculture on multiplication medium and reduced the growth regulator concentration by half. Then the individual shoots were rooted on medium containing 2.5 mg/l IAA. Murashige *et al* (1974) used a more higher level of IAA for root induction and obtained 100 per cent rooting efficiency with 10 mg/l IAA which also had an effect on vigour of plantlets. They have also noted the strong influence of light intensity on rooting and vigour of the plantlets. Light intensities higher than 1,000 lux increased the percentage of rooting (Murashige *et al,* 1974) while Pierik *et al* (1975) has reported efficient rooting on 10 mg/l IAA containing modified MS medium under low light intensity (800 lux). *Ex vitro* root induction by dipping the shoots in 0.1 per cent IBA for 30 seconds has been described by Huang and Chu (1985). Strong effect of genotype on multiplication and rooting efficiency, indicating interactions between clone and medium has been reported (Renoird *et al*, 1997; Pierik *et al*, 1982; Meynet, 1983). In our laboratory, 100 per cent rooting of shoots was achieved by sub-culturing on modified MS medium without any growth regulators (Mandal and Datta, 2002a).

CHRYSANTHEMUM

Chrysanthemum (*Chrysanthemum morifolium* Ramat.) is valued for cut flowers and potted plants. It belongs to the family Compositae. Tissue culture of chrysanthemum has proven relatively flexible and plants have been regenerated from shoot apex stem nodes, ray floret-derived callus, leaves, immature flower heads and receptacles.

In vitro clonal multiplication from shoot tip and nodal explants has been reported by many authors (Ben-Jaacov *et al*, 1972; Earle and Langhans, 1974a, 1974b; Langhans, 1974; Roset and Bokelmann, 1975; Petru *et al*, 1981; Cossio and Bussi, 1982; Prasad *et al*, 1983; Bhattacharya *et al*, 1990; Mandal *et al*, 1997).

Earle and Langhans (1974a, 1974b) studied plant regeneration in detail to standardise plant regeneration from shoot tip explants. They used MS medium containing various levels of kinetin, NAA and GA_3. Multiple shoots were produced on kinetin (2 mg/l) and NAA (0.02 mg/l) while addition of GA_3 (10 mg/l) promoted elongation of leaves and shoots. Langhans (1974a, 1974b) reported differentiation of tissue into plants, shoot multiplication and establishment of regenerated plants in greenhouse. Propagation of a large number of chrysanthemum cultivars (65 large flowered cultivars) from shoot apices has been reported on modified MS medium supplemented with 2.5 mg/l IAA and 4 mg/l BA (Petru *et al*, 1981). They have also observed the effect of explant collection on *in vitro* shoot proliferation and found spring season as the best over other seasons. Among different cytokinins, BA in combination with IAA is reported to be superior over kinetin for *in vitro* proliferation from shoot tip explants (Prasad *et al*, 1983). Effect of different pH and agar concentrations has been studied by Oliviera *et al* (1999) who did not find any significant effect; however, 0.7 per cent agar and pH 5.8 has been recommended.

Micropropagation from nodal explants has been reported by many workers. Cossio and Bussi (1982) has reported culture initiation on MS medium containing 0.5 to 1.0 mg/l BA and 0.1 mg/l IBA and rapid growth and rooting of the minicuttings on IBA or NAA supplemented MS media. Bhattacharya *et al* (1994) have studied the effect of two gelling agents (herbal origin) and a few support matrixes (for example, filter paper, nylon cloth, polystyrene foam and glass wool) on growth and shoot production. Clonal propagation from nodal explants of two large flowered cultivars has been reported in MS media supplemented with kinetin (5 mg/l) and NAA (0.2 mg/l) by Mandal *et al* (1997). They have also observed variation in response among these two cultivars.

In addition to these, adventitious shoot regeneration from leaves (Sutter and Langhans, 1981; Kaul *et al*, 1990; Khehra *et al*, 1995; Mityushkina *et al*, 1995; Nagatomi *et al*, 1996; Jerzy *et al*, 1996; Lee *et al*, 1997; Oka *et al*, 1999), stem (Miyazaki and Tashiko, 1978; Kaul *et al*, 1990; Lu *et al*, 1990; Yang *et al*, 1995), flower bud (Tsukamoto and Fujime, 1971) and floret (Lizuka *et al*, 1973; Bush *et al*, 1976; Miyazaki and Tashiko, 1978; Khalid *et al*, 1989; Matsumato and Onozawa, 1989; Malure *et al*, 1991a, 1991b) and pedicel explant (Broertjes *et al*, 1976; Lazar *et al*, 1981; Khalid *et al*, 1989; Latado *et al*, 1996) have been reported in chrysanthemum. The shoot regeneration frequency of stem explant is reported to be better than leaf explant (Kaul *et al*, 1990). In most of the cases, shoots were differentiated through callus and loss of genetic homogeneity of such callus derived shoots has been reported (Malaure *et al*, 1991a).

In our laboratory, we have standardised an efficient protocol, mainly to use in *in vitro* chimera management, for direct shoot regeneration from ray and disc florets of several chrysanthemum cultivars, including mini, small and large flower types (Chakrabarty *et al,* 1999; Datta *et al,* 2001a, 2001b). In this system we obtained up to 79.2 per cent response with 16.6 number of shoots per responding explant on MS medium containing 0.2 mg/l NAA and 2 mg/l BA. Shoots were excised and easily rooted on MS medium without any growth regulators.

BOUGAINVILLEA

Bougainvillea, a member of the family Nyctaginaceae is one of the most showy ornamental plants in floriculture and is used for landscaping in tropical and sub-tropical regions. A perusal of literature shows that only a limited attempt has been made to micropropagate this plant (Chaturvedi *et al*, 1978; Sharma *et al,* 1981; Sharma and Chaturvedi, 1988). Shoot tip explants are largely used by these authors for establishment of aseptic cultures in two cultivars of bougainvillea, 'Magnifica' and 'Scarlet Queen Variegated' only. One-cm long shoot apices were multiplied on NAA/IAA and BA supplemented media. On an average, seven to 10 shoots were developed from each explant and approximately 100,000 plants could be regenerated from a single shoot tip within a year (Sharma and Chaturvedi, 1988; Sharma *et al*, 1981). In our laboratory (Mandal and Datta, 2002b), we have micropropagated one chlorophyll variegated mutant cultivar 'Los Banos Variegata' from stem node on MS medium supplemented with BA, NAA and GA_3 and obtained 100 per cent response with 8.3 number of shoots per shoot clump. We rooted the individual shoots on high NAA-supplemented medium and successfully transferred the micropropagated plantlets to the field. In case of cv. 'Magnifica', individual shoots were rooted by sub-culturing them either first in 0.75 mg/l NAA for 30 days and then transferring them on 0.5 mg/l 2,4,5-trichlorophenoxyacetic acid (2,4,5-T)-supplemented media (Chaturvedi *et al,* 1978) or by sub-culturing individual shoots on modified MS medium at pH 4.3 and supplemented with NAA or IBA (0.1 mg/l) and 2,4,5-T (0.1 mg/l (Sharma *et al*, 1981). In case of cv. 'Scarlet Queen Variegated', roots were induced by sub-culturing the individual shoots on high NAA containing medium for 15 days followed by their transfer on low NAA containing medium for 15 days, while pH of the medium was 4.5 (Sharma and Chaturvedi, 1988).

LILY

Out of different types of lilies, Oriental and Asiatic hybrid lilies and, to some extent, Easter and Tiger lilies are the popular ones. Propagation of lilies by conventional means is too little to meet the present demand for planting material. *In vitro* adventitious bud regeneration has been reported from bulb scales of *Lilium longiflorum* Thunb. (Gupta *et al*, 1978), while complete plantlets were produced from callus cultures of bulb scales (Priyadarshi and Sen, 1992) and they did not find any significant variation in chromosome number of 3-year-old Easter lily callus. Regeneration from other plant parts, like filaments with anther (Arzate-Fernandez *et al*, 1997), leaf explants, but to a less extent (Kato and Yasutake, 1977; Niimi, 1986), somatic embryogenesis in styles and pedicels (Tribulato *et al*, 1997) and from bulb scales in some lily hybrids (Haensch, 1996), stem nodes and pseudobulblet culture (Nhut, 1998) and *in vitro* bulbing (Marinengeli *et al*, 1998) have also been reported. A few other species of *Lilium* have also been micropropagated like *Lilium auratum* Lindl., *Lilium speciosum* Thunb. (Takayama and Misawa, 1983), and *Lilium rubellum* (Niimi, 1984). But only recently a cost effective, large scale multiplication scheme for Asiatic hybrids of lily has been reported for the first time by Varshney *et al* (2000) through liquid stationary cultures. However, acclimatisation of the Asiatic hybrid lily from temperate to subtropical climates has also been reported, after propagation through tissue culture (Misra and Datta, 2001). There are a few reports on clonal multiplication of Asiatic and Oriental lilies through bulb scales (Simmonds and Cumming, 1976a; 1976b; Dilta *et al*, 1999; Linn *et al*, 2002) and also through callus (Sindhu and Niimi, 1999). It has been reported by Misra and Datta (2001) that 0.5 mg/l BA + 1 mg/l IAA is sufficient to induce about 25 shoots from one bulb scale segment of Asiatic hybrid lilies within 3-4 months of establishment and then by a factor of 5-10 shoots every four weeks producing millions of shoots every year.

It has been estimated that from a single scale segment (explant), as many as 9.68 × 105 bulblets could be produced in one year costing as little as US$ 0.02 per bulblet (Varshney *et al*, 2000). As lilies lose their viability if preserved for long-term under cold storage (Bonnier *et al*, 1997), they could be stored *in vitro* for long-term by manipulating the concentration of nutrients, sugar and the temperature (Bonnier and van Tuyl, 1997; Shin *et al*, 2002). *Lilium* bulblets have been micropropagated on a pilot scale by Lim *et al* (1998). The fed-batch culture system using a bioreactor for the production of Oriental lilies has been developed by Seon *et al* (2000) and Lian *et al* (2002a, 2002b). This regenerative potentiality has been exploited in most of the bulbous crops for large scale micropropagation using liquid shake cultures in bioreactors (Paek, K.-Y. *et al*, 2001). Bioreactor techniques are useful both by reducing the production cost, by scaling up and automation. Virus-free bulblets of *Lilium longiflorum* have been micropropagated by tank culture using liquid culture method (Takahashi *et al*, 1992). A thin cell layer culture system has also been developed in *Lilium longiflorum* to establish successful organogenic and somatic embryogenic pathways (Nhut *et al*, 2001).

GLADIOLUS

Gladiolus is a monocot crop belonging to the family Iridaceae. Seeing its immense commercial importance, tissue culture proved an excellent tool for mass propagation of pathogen-free planting material. Tissue culture and plantlet regeneration of gladiolus had started three decades back (Ziv *et al*, 1970). Segments of different vegetative parts, like, corm slices, cormlets, shoot tips, young leaf bases as well as floral parts like inflorescence stalk have been used as explants for regeneration purposes. The regeneration may be through somatic embryogenesis or it may be through shoot bud formation (Stefaniak, 1994; Remotti, 1995). Initiation of multiple shoots was visible 7-8 days after inoculating the explants. The top slice of the cormlet having the apical meristem, had the highest regenerative potentiality. *In vitro* culture of gladiolus has already been reported in several instances in semisolid media (Bajaj *et al*, 1982; Ziv and Lilien-Kipnis, 1990; Bajaj *et al*, 1992; Jager *et al*, 1998; Nagaraju *et al*, 1998) and also in suspension cultures (Remotti, 1995; Kamo, 1995; Kamo *et al*, 1990). Plants have been regenerated from callus of gladiolus by several researchers (Simonsen and Hildebrandt, 1971; Kim *et al*, 1988; Kamo, 1994; Stefaniak, 1994; Remotti, 1995; Remotti and Loffler, 1995). The use of liquid cultures can increase the rate of bud proliferation and also the cost through mechanisation (Levin *et al*, 1988; Ziv, 1990a). Use of bioreactor was also in a true sense a successful proposition for commercial propagation of gladiolus (Ziv, 1990b, 1991; Ziv *et al*, 1998). But the proliferated shoots regenerated in liquid cultures were hyperhydrated (vitrified) and the running cost of the bioreactor was also very high. Further, the labour can be reduced by automation through computer vision analysis (Mcfarlane, 1993; Wang *et al*, 1999; Sundaram and Dutta Gupta, 2000). The shoots, after attaining the height of 6-8 cm, were transferred to rooting medium. Roots were visible after 15 days and initiated from the collar region of the shoots having cormels at their base (Nagaraju *et al*, 1998). There are reports available on successful acclimatisation and field evaluation of plants raised through *in vitro* produced cormlets (Ziv, 1979; Dantu and Bhojwani, 1995). Recently, the role of heat shock and sucrose has been demonstrated in shoot multiplication through somatic embryogenesis and rooting of microshoots in gladiolus (Kumar *et al*, 1999, 2002). The explants, isolated from the floral stem just below the expanding florets, enclosed by the bracts, were highly regenerative. When the ratio of Kn to NAA was high (10:5), 10 buds developed on a 3 to 4 mm inflorescence explant compared with three on apical bud explants (Ziv and Lilien-Kipnis, 2000).

In vitro cormlet production has also been found very practicable in gladiolus (Bhojwani, 1995). After multiplication when shoots were grown in quarter or half strength MS media having high concentration of sucrose along with m-inositol and ABA, the size of the corms started increasing and, within eight weeks of incubation, attained the size of 1.5 or 2.0 cm of diameter (Goo and Kim, 1994). The frequency of cormlet production could also be increased many-fold (Sen and Sen, 1995). This size was comparable to 1-year-old cormels regenerated through conventional methods. Thus, the use of tissue culture saved lots of time, both for increasing the rate of multiplication and for the size of the cormlet. Any hybrid developed by conventional breeding methods may take years to reach the market/consumer, while if multiplied though tissue culture, it can be made available in the market within two years and that too clonal plants.

TUBEROSE

Polianthes tuberosa L. is a slow to propagate herbaceous geophyte, where development of regeneration protocol is an initial step towards its commercial exploitation. There is very limited work done on micropropagation of tuberose. Meagre information is available only in abstract form on shoot regeneration from callus cultures of tuberose (Muralidhar and Mehta, 1982; Waithaka, 1986; Shen *et al*, 1991). A method for *de novo* induction of shoot buds and regeneration of complete plants from leaf callus cultures of tuberose has been developed by Sanyal *et al* (1998). Clonal propagation of *Polianthes tuberosa* var. 'Rajat Rekha' has been reported by Datta *et al* (2002), where shoot buds were regenerated in two steps directly from the leaf segments.

TULIPS

Tulipa gesneriana L. (Fam. Liliaceae) is a temperate crop and very sensitive to higher temperature. There are very limited reports on micropropagation of tulips (Famelaer *et al*, 1996; Kuijpers and Langens-Gerrites, 1997). It may be due to (1) less regenerative potentiality *in vitro*, (2) regeneration is mostly in floral stem explants with the result very limited experimental material, that too for a very short time (Taeb *et al*, 1987; Lenard, 1989; Alderson and Taeb, 1990), and (3) growing stem explants also show bud formation but no regeneration from bulb and simultaneously there are many problems related to bud and bulblet formation (LeNard and Chanleloube, 1992).

MARIGOLD

In normal African marigold (*Tagetes erecta* L., Asteraceae) there are a few reports on somatic embryogenesis from cotyledon (Bespalhok and Hattori, 1998) and on organogenesis from leaves (Kothari and Chandra, 1984a; 1986), florets (Kothari and Chandra, 1984b) and hypocotyl (Belarmino *et al*, 1992).

WHITE MARIGOLD

It is one of the important ornamental crops but due to non-availability of planting material, it has not been properly commercialised. Multiplication rate of White Marigold is not as fast as that of the yellow type because of low seed viability and poor germination (30 per cent). White Marigold is very delicate and requires extremely favourable climatic conditions for vegetative growth as well as for good blooms. It is difficult to maintain pure line seeds due to its cross-pollinated nature. Tissue culture is the only method to maintain genetically identical clones having snow white flower colour and rapid propagation as well as long-term preservation of germplasm (Misra and Datta, 1999).

F_1 Hybrids of Marigold

Hybrid varieties have been developed and commercially exploited in most of the cross-pollinated crops. Wherever it is commercially feasible, F_1 hybrids may be used directly as a variety. Cytoplasmic/genetic male sterile lines have been developed and used for synthesising F_1 hybrids of ornamental and horticultural crops. Breeders will have to maintain the parents and crossing has to be performed every year for raising F_1. Similarly, users will have to purchase fresh seeds every year from breeders as there is no seed setting in F_1 hybrids (Misra and Datta 2000). In *Broccoli* and *Brassica oleracea,* tissue culture studies have been found of immense importance in seed production of F_1 hybrids (Anderson and Carstens, 1977). In marigold two types of F_1 hybrids, namely, 'Pusa Narangi' and 'Pusa Basanti' have been used for tissue culture studies. In ray florets of 'Pusa Narangi' shoot buds differentiated directly in the presence of 1 mg/l of TDZ, while leaf segments of 'Pusa Basanti' differentiated shoot buds directly in presence of 0.2 mg/l of TIBA (Datta *et al*, 2002).

REFERENCES

Alderson, P. G. and Taeb, A. G. 'Effect of bulb storage on shoot regeneration from floral stems of tulip *in vitro*'. *J. Hortic. Sci.* 65: 65-70, 1990.

Alderson, P. G., McKinless, J. and Rice, R. D. 'Rooting of cultured rose shoots'. *Acta Hortic.* 226: 175-182, 1988.

Anderson, W. C. and Carstens, J. B. 'Tissue culture propagation of Broccoli, *Brassica oleracea* (Italica group) for use in F_1 hybrid seed production'. *J. Amer. Soc. Hortic. Sci.* 102: 69-73, 1997.

Arditti, J. and Ernst, R. In: *Micropropagation of Orchids.* New York, U.S.A.: John Wiley and Sons, 1993.

Arzate-Fernandez, A. M., Nakazaki, T. and Okumoto, Y. 'Efficient callus induction and plant regeneration from filaments with anther in lily (*Lilium longiflorum* Thunb.)'. *Plant Cell Rep.* 16: 836-840, 1997.

Avramis, T., Hugard, J. and Jonard, R. 'Exaltation des potentialites rhizogenes de pousses feuillees de rosier multipliees *in vitro*'. *CR Acad. Sci.,* Paris 294: 679-682, 1982a.

Avramis, T., Hugard, J. and Jonard, R. 'La multiplication *in vitro* du rosier porte-greffe *Rosa indica* major'. *CR Acad. Sci.,* Paris 294: 63-68, 1982b.

Bajaj, Y. P. S., Sidhu, M. M. S. and Gill, A. P. S. 'Micropropagation of *Gladiolus*'. In: *Biotechnology in Agriculture and Forestry,* vol. 20. (Ed.). Bajaj, Y. P. S. Berlin, Germany: Springer-Verlag, pp. 135-143, 1992.

Bajaj, Y. P. S., Sidhu, M. M. S. and Gill, A. P. S. 'Some factors affecting the propagation of *Gladiolus*'. *Sci. Hortic.* 18: 269-275, 1982.

Belarmino, M. M., Abe, T. and Sasahara, T. 'Callus induction and plant regeneration in African marigold (*Tagetes erecta* L)'. *Jpn. J. Breed.* 42: 835-841, 1992.

Ben-Jaccov, J. and Langhans, R. W. 'Rapid multiplication of chrysanthemum plants by stem tip proliferation'. *HortScience* 7: 289-290, 1972.

Bespalhok, J. C. and Hattori, F. K. 'Friable embryogenic callus and somatic embryo formation from cotyledon explants of African marigold (*Tagetes erecta* L.)'. *Plant Cell Rep.* 17: 870-875, 1998.

Bhattacharya, P., Dey, S. and Bhattacharya, B. C. 'Use of low-cost gelling agents and support matrices for industrial scale plant tissue culture'. *Plant Cell Tissue Org. Cult.* 37: 15-23, 1994.

Bhattacharya, P., Dey, S., Das, N. and Bhattacharya, B. C. 'Rapid mass propagation of *C. morifolium* by callus derived from stem and leaf explants'. *Plant Cell. Rep.* 9: 439-442, 1990.

Bhojwani, S. S. '*In vitro* corm formation and field evaluation of corm-derived plants of gladiolus'. *Sci. Hortic.* 61: 115-129, 1995.

Bini, G., Leva, A. R. C. and Nicese, F. P. 'Richere sulla micro-propagazione della rosa'. *Riv. Ortoflorofutt Ital.* 67: 1-13, 1983.

Bonnier, F. J. M. and van Tuyl, J. M. 'Long-term *in vitro* storage of lily: effects of temperature and concentration of nutrients and sucrose'. *Plant Cell Tiss. & Org. Cult.* 49: 81-87, 1997.

Bonnier, J. M. F., Hoeskstra, A. F., Vos, R. and Vantuyl, J. M. 'Viability loss and oxidative stress in lily bulbs during long term cold storage'. *Plant Sci.* 122: 133-140, 1997.

Bressan, P. H., Kim, Y. J., Hyndman, S. E., Hasegawa, P. M. and Bressan, R. A. 'Factors affecting *in vitro* propagation of rose'. *J. Amer. Soc. Hortic. Sci.* 107: 979-990, 1982.

Broertjes, C., Roest, S. and Bokelmann, G. S. 'Mutation breeding of *Chrysanthemum morifolium* Ram. using *in vivo* and *in vitro* adventitious bud techniques'. *Euphytica* 25: 11-19, 1976.

Burger, D. W., Lui, L., Zary, K. W. and Lee, C. I. 'Organogenesis and plant regeneration from immature embryos of *Rosa hybrida* L'. *Plant Cell Tissue Org. Cult.* 21: 147-152, 1990.

Bush, S. R., Earle, E. D. and Langhans, R. W. 'Plantlets from petal segments, petal epidermis and shoot tips of the periclinal chimera, *Chrysanthemum morifolium* 'Indianapolis'. *Amer. J. Bot.* 63: 729-737, 1976.

Chakrabarty, D., Mandal, A. K. A. and Datta, S. K. '*In vitro* propagation of rose cultivars'. *Ind. J. Plant Physiol.* 5: 189-192, 2000.

Chakrabarty, D., Mandal, A. K. A. and Datta, S. K. Management of chimera through direct shoot regeneration from florets of chrysanthemum (*Chrysanthemum morifolium* Ramat.)'. *J. Hort. Sci. Biotech.* 74: 293-296, 1999.

Chaturvedi, H. C., Prasad, R. N., Sharma, A. K. and Misra, P. 'Tissue culture propagation of ornamental plants'. In: *Floriculture: Technology, Trades and Trends.* (Eds.). Prakash, J. and Bhandari, K. R. New Delhi, India: Oxford and IBH Publishing Co. Pvt. Ltd., pp. 367-377, 1994.

Chaturvedi, H. C., Sharma, A. K. and Prasad, R. N. 'Shoot apex culture of *Bougainvillea glabra* 'Magnifica''. *HortSci.* 13: 36, 1978.

Chu, C. Y. and Huang, M. C. '*In vitro* formation of gerbera (*Gerbera hybrida* Hort.) plantlets through excised scape culture'. *J. Jpn. Soc. Hortic. Sci.* 52: 45-50, 1983.

Collet, G. F. 'Enracinement ameliore lors de la production *in vitro* de rosiers'. *Rev Suisse Vitic Arboric Hortic.* 17: 259-263, 1985.

Cossio, F. and Bussi, F. 'Micropropagation of chrysanthemum'. *Colture Protette* 11: 45-49, 1982.

Curir, P., Damiano, C. and Cosmi, T. '*In vitro* propagation of some rose cultivars. *Acta Hortic.* 189: 221-224, 1986.

Dantu, P. K. and Bhojwani, S. S. '*In vitro* corm formation and field evaluation of corm-derived plants of gladiolus'. *Sci. Hortic.* 61: 115-129, 1995.

Datta, S. K., Chakrabarty, D., Saxena, M., Manda, A. K. A. and Biswas, A. K. 'Direct shoot regeneration from florets of chrysanthemum cultivars'. *Ind. J. Genet. Plant Breed.* 6 :373-376, 2001b.

Datta, S. K., Mandal, A. K. A. and Saxena, M. 'Direct organogenesis from ray and disc florets of a newly evolved chlorophyll variegated chrysanthemum (*Chrysanthemum morifolium*)'. *Ind. J. Agric. Sci.* 71: 655-657, 2001a.

Datta, S. K., Misra, P., Mandal, A. K. A., Deepti and Chakrabarty, D. 'Direct shoot organogenesis from different explants of chrysanthemum, marigold and tuberose'. *Israel J. Plant Sci.* 2002.

Davies, D. R. 'Rapid propagation of rose *in vitro*'. *Sci. Hortic.* 13: 385-389, 1980.

Debergh, P. and Maene, L. 'Contribution of tissue culture techniques to horticultural research and production'. *Acta. Hortic.* 131: 23-37, 1983.

Delbard, H. 'Micropropagation of roses at Delbard nurseries'. *Plant Prop.* 28: 7-8, 1982.

DeWit, J. C., Esendam, H. F., Konkanen, J. J. and Tuominen, U. 'Somatic embryogenesis and regeneration of flowering plants in rose'. *Plant Cell Rep.* 9: 456-458, 1990.

Dilta, B. S., Sehgal, O. P. and Pathania, N. S. '*In vitro* propagation studies in hybrid lilies'. In: Natl. Sym. on *Emerging Scenario in Ornamental Horticulture in 2000 A.D. and Beyond*, 21-22 July, 1999, held at Division of Floriculture and Landscaping, IARI, New Delhi, India, 1999.

Earle, E. D. and Langhans, R. W. 'Propagation of chrysanthemum *in vitro*. 1. Multiple plantlet from shoot tips and the establishment of tissue culture'. *J. Am. Soc. Hort. Sci.* 99: 128-132, 1974a.

Earle, E. D. and Langhans, R. W. 'Propagation of chrysanthemum *in vitro*. 2. Production, growth and flowering of plantlets from tissue culture'. *J. Am. Soc. Hort. Sci.* 99: 352-358, 1974b.

Elomaa, P., Honkmen, J., Puska, R., Seppanen, P., Helariutta, Y., Mehto, M., Kotilainen, M., Nevalainen, L. and Terri, T. H. '*Agrobacterium*-mediated transfer of antisense chalcone synthase cDNA to *Gerbera hybrida* inhibits flower pigmentation'. *Biotechnology* 11: 508-511, 1993.

Famelaer, I., Ennik, E., Eikelboom, W., Van Tuyle, J. M. and Creemers-Molenaar, J. 'The initiation of callus and regeneration from callus culture of *Tulipa gesneriana*'. *Plant Cell Tiss. & Org. Cult.* 47: 51-58, 1996.

Fiorino, P. and Leva, A. R. 'La multiplicatione del melo con la tecnica della micropropagazione. I. Indagine sulla ottimizzazione dei substrati nutritivi durante la fase di moltiplicatione'. *Atti Cngr. I. fitoregolatori in agricoltura Firenze*, pp. 147-156, 1981.

Goo, D. H. and Kim, K. W. 'Influence of sucrose, ABA and day length on cormlet formation of gladiolus; *in vitro* histological observations'. *J. Kor. Soc. Hortic. Sci.* 35: 400-405, 1994.

Gupta, P. P., Sharma, A. K. and Chaturvedi, H. C. 'Multiplication of *Lilium longiflorum* Thunb. by aseptic culture of bulb scales and their segments'. *Indian J. Exp. Biol.* 16: 940-942, 1978.

Haensch, K. T. 'Plant regeneration through somatic embryogenesis in different genotypes of *Lilium* hybrids'. Gartenbauwissenscht 61: 214-218, 1996.

Hasegawa, P. M. 'Factors affecting shoot and root initiation from cultured rose shoot tips'. *J. Am. Soc. Hortic. Sci.* 105: 216-220, 1980.

Hasegawa, P. M. '*In vitro* propagation of rose'. *HortSci.* 14: 610-612, 1979.

Hedtrich, C. M. 'Regeneration of shoots from leaves and propagation of *Gerbera jamesonii*'. *Gartenbauwissenschaft* 44: 1-3, 1979.

Hill, G. P. 'Morphogenesis of shoot primordia in cultured stem tissue of a garden rose'. *Nature* 216: 596-597, 1967.

Horn, W. A. H. 'Micropropagation of rose (*Rosa* L.)'. In: *Biotechnology in Agriculture and Forestry.* (Ed.). Bajaj, Y. P. S. Berlin, Heidelberg, Germany: Springer, pp. 30-342, 1991.

Horn, W., Schlegel, G. and John, K. 'Micropropagation of roses'. *Acta. Hortic.* 226: 623-626, 1988.

Huang, M. C. and Chu, C. Y. 'A scheme for commercial multiplication of Gerbera (*Gerbera hybrida* Hort.) through shoot tip culture'. *J. Jpn. Soc. Hortic. Sci.* 54: 94-100, 1985.

Ishioka, N. and Tanimoto, S. 'Plant regeneration from Bulgarian rose callus'. *Plant Cell Tissue Organ Culture* 22: 197-199, 1990.

Jager, A. K., McAlister, B. G. and Van Staden, J. '*In vitro* culture of *Gladiolus carneus*'. *S. African J. Bot.* 64: 146-147, 1998.

Jerzy, M. and Lubomski, H. 'Adventitious shoot formation on *ex vitro* derived leaf explants of *Gerbera jamesonii*'. *Sci. Hortic.* (Amst). 47: 115-124, 1991.

Jerzy, M. and Zalewska, M. 'Polish cultivars of *Dendrathema grandiflora* Tzelev and *Gerbera jamesonii* Bolus bred *in vitro* by induced mutation'. *Mutation Breeding Newsletter* 42: 9, 1996.

Kamo, K. A. 'Cultivar comparison of plant regeneration from suspension cells, callus, and cormel slices of Gladiolus'. *In Vitro Cell and Dev. Biol.-Plant* 31: 113-115, 1995.

Kamo, K. A. 'Effect of phytohormones on plant regeneration from callus of *Gladiolus* cultivar 'Jenny Lee''. *In Vitro Cell and Dev. Biol.-Plant* 30: 26-31, 1994.

Kamo, K., Chen, J. and Lawson, R. 'The establishment of cell suspension cultures of Gladiolus that regenerates plants'. *In Vitro Cell and Dev. Biol.-Plant* 26: 425-430, 1990.

Kato, Y. and Yasutake, Y. 'Plantlet formation and differentiation of epidermal tissue in green callus cultures from excised leaves of *Lilium*'. *Phytomorphology.* 27: 390-396, 1977.

Kaul, V., Miller, M. R., Hutchinson, J. F. and Richards, D. 'Shoot regeneration from stem and leaf explants of *Dendranthema grandiflora* Tzvelev (syn. *Chrysanthemum morifolium* Ramat.)'. *Plant Cell Tissue Org. Cult* 21: 21-30, 1990.

Khalid, N., Davey, M. R. and Power, J. B. 'An assessment of somaclonal variation in *Chrysanthemum morifolium*: The generation of plants of potential commercial value'. *Scientia Hort.* 38: 287-294, 1989.

Khehra, M., Lowe, K. C., Davey, M. R. and Power, J. B. An improved micropropagation system for Chrysanthemum based on Pluronic F-68 supplemented media'. *Plant Cell Tissue Org. Cult.* 41: 87-90, 1995.

Khosh-Khui, M. and Sink, K. C. 'Micropropagation of new and old world rose species'. *J. Hortic. Sci.* 57: 315-319, 1982a.

Khosh-Khui, M. and Sink, K. C. 'Rooting enhancement of *Rosa hybrida* for tissue culture propagation'. *Sci. Hortic.* 17: 371-376, 1982b.

Kim, K. W. and De Hertogh, A. A. 'Tissue culture of ornamental flowering geophytes'. *Hortic. Rev.* 18: 87-169, 1997.

Kim, K. W., Choi, J. B., and Kwon, K. Y. 'Rapid multiplication of Gladiolus plants through callus culture'. (Korean) *J. Korean Soc. Hortic. Sci.* 29: 312-318, 1988.

Kothari, S. L. and Chandra, N. '*In vitro* propagation of African marigold'. *HortScience* 19: 703-705, 1984a.

Kothari, S. L. and Chandra, N. 'Plant regeneration from cultures disc florets of *Tagetes erecta* L'. *J. Plant Physiol.* 117: 105-108, 1984b.

Kothari, S. L. and Chandra, N. 'Plant regeneration in callus and suspension cultures of *Tagetes erecta* L. (African marigold)'. *J. Plant Physiol.* 122: 235-241, 1986.

Kuijpers, A. M. and Langens-Gerrits, M. 'Propagation of tulip *in vitro*'. *Acta Hortic.* 430: 321-324, 1997.

Kumar, A., Palni, U. T., Sood, A., Palni, L. M. S. and Gupta, A. K. 'The effect of heat-shock and sucrose on morphogenesis in cultures of *Gladiolus hybridus* Hort'. In: *Role of Plant Tissue Culture in Biodiversity Conservation and Economic Development.* (Eds.). Nandi, S. K., Palni, L. M. S. and Kumar, A. Nainital, India: Gyanodaya Prakashan, pp. 245-259, 2002.

Kumar, A., Sood, A., Palni, L. M. S. and Gupta, A. K. '*In vitro* propagation of *Gladiolus hybridus* Hort. Synergistic effect of heat shock and sucrose on morphogenesis'. *In Vitro Cell. and Dev. Biol.-Plant* 57: 105-112, 1999.

Laliberte, S., Chretien, M. and Vieth, J. '*In vitro* plantlet production from young capitulum explants of *Gerbera jamesonii*'. *HortScience* 20: 137-139, 1985.

Langhans, R. W. 'Chrysanthemum micropropagation'. *New York's Food and Life Sciences* 7: 3-7, 1974.

Latado, R. R., Tulmann, N. A. and Mendes, B. M. J. 'Breeding Chrysanthemum cv. Rupin Rosa by means of mutation induction *in vitro*'. *Pesquisa Agropecuria Brasileira* 31: 489-496, 1996.

Lazar, M., Cachita, C. D. and Bader, S. M. 'Micropropagation of Chrysanthemum. 1. *In vitro* multiplication of Chrysanthemum by floral peduncle explants'. *Productia Vegetala Horticultura* 30: 18-23, 1981.

Lee, T., Huang, M. E. E., Pua, E. C. and Pua, E. C. 'High frequency shoot regeneration from leaf disc explants of garland *Chrysanthemum coronarium in vitro*'. *Plant Science* 126: 219-226, 1997.

Leffring, L., Skovbogaard, M. and Esendam, H. 'Weefselkweek bij roos'. *Bloem Onderz Ned, Alasmeer,* pp. 364-365, 1984.

Lenard, M. '*In vitro* adventitious bud formation on floral stem explants of active growing tulips (*Tulipa generiana* L.)'. *CR Acad. Sci.*, Paris 308: 389-394, 1989.

Lenard, M. and Chanleloube, F. '*In vitro* culture of explants excised from growing stems of tulip (*Tulipa gesneriana* L.): problems related to bud and bulblet formation'. *Acta. Hortic.* 325: 435-440, 1992.

Levin, R., Gaba, V. and Tal, B. 'Automated plant tissue culture for mass propagation'. *Biotechnology* 6: 1035-1040, 1988.

Lian, M.-L., Chakrabarty, D. and Paek, K.-Y. 'Growth and uptake of sucrose and mineral ions by bulblets of *Lilium* Oriental Hybrid 'Casablanca' during bioreactor culture'. *J. Hortic. Sci. Biotech.* 77(2), 2002b.

Lian, M.-L., Chakrabarty, D. and Paek, K.-Y. 'Growth of *Lilium* Oriental Hybrid 'Casablanca' bulblet using bioreactor culture'. *Sci. Hortic.* 1822: 1-8, 2002a.

Lim, S., Seon,-J. H., Paek,-K. Y., Son, S. H., Han,-B. H. and Drew, R. A. 'Development of pilot scale process for mass production of *Lilium* bulblets *in vitro*'. *Acta. Hortic.* 461: 237-241, 1998.

Lizuka, M., Matsumoto, E., Doi, A., Madrigal, R. and Fukushima, A. 'Tubular floret culture of chrysanthemum and Cineraria *in vitro*'. *Jpn. J. Genet.* 48: 79-87, 1973.

Lloyd, D., Roberts, A. V. and Short, K. C. 'The induction *in vitro* of adventitious shoots in *Rosa*'. *Euphytica* 37: 31-36, 1988.

Lu, C. Y., Nugent, G. and Wardley, T. 'Efficient, direct plant regeneration from stem segments of chrysanthemum. (*C. morifolium* Ramat. cv. Royal Purple)'. *Plant Cell Rep.* 8: 733-736, 1990.

Maia, E., Beck, D., Poupet, A. and Bettachini, B. '*In vitro* vegetative propagation of *Gerbera jamesonii* Bolus'. *CR Acad. Sci.*, Paris. III. 296: 885-887, 1983.

Malaure, R. S., Barday, G., Power, J. B. and Davey, J. B. 'The production of novel plants from florets of *Chrysanthemum morifolium* using tissue culture. II. Securing natural mutations (sports)'. *J. Plant. Physiol.* 139: 14-18, 1991a.

Malaure, R. S., Barday, G., Power, J. B. and Davey, M. R. 'The production of novel plants from florets of *Chrysanthemum morifolium* using tissue culture. 1. Shoot regeneration from ray florets and somaclonal variation exhibition by the regenerated plants'. *J. Plant. Physiol.* 139: 8-13, 1991b.

Mandal, A. K. A. and Datta, S. K. 'Development of *in vitro* protocol for commercial exploitation of difficult-to-root bougainvillea cultivar'. *Ind. J. Agric. Sci.* (communicated), 2002b.

Mandal, A. K. A. and Datta, S. K. 'Introduction of gerbera cultivation in Lucknow agro-climate through tissue culture of young flower head'. *Ind. J. Biotech.* 1: 212-214, 2002a.

Mandal, A. K. A., Chakrabarty, D. and Datta, S. K. 'Tissue culture of large flowered chrysanthemum for commercial floriculture'. In: *Exploring the Chrysanthemum in India.* (Ed.). Datta, S. K. Proceedings of the *Natl. Conf. on Chrysanthemum.* EBIS, NBRI, Lucknow, pp. 57-59, 1997.

Marinangeli, P., Delmastro, S. and Curvetto, N. 'Influence of DB-cAMP, Adenosine, Forskolin and Traumatic Acid on *in vitro* bulbing of *Lilium longiflorum*'. *HortScience* 33: 151-152, 1998.

Martin, C., Carre, M. and Vernoy, R. 'La multiplication vegetative *in vitro* des vegetaux ligneux cultives: cas des rosiers'. *CR Acad. Sci.*, Paris 293: 175-177, 1981.

Matsumato, H. and Onozawa, Y. 'Development of non-chimeric mutation lines through *in vitro* culture of florets in chrysanthemum'. *Sci. Rep.* Faculty of Agriculture, Ibaraki University 37: 55-61, 1989.

McFarlane, N. J. B. 'Image-guidance for robotic harvesting of micropropagating plants'. *Comp. & Electronics in Agri.* 8: 43-56, 1993.

Mederos, S. and Enriquez, M. J. '*In vitro* propagation of Golden Times roses. Factors affecting shoot tips and axillary bud growth and morphogenesis.' *Acta. Hortic.* 212: 619-624, 1987.

Meynet, J. 'Effects of *in vitro* propagation on the further growth behaviour of some *Gerbera* varieties'. *Agronomie* 3: 839-846, 1983.

Misra, P. and Datta, S. K. 'Acclimatisation of Asiatic hybrid lilies under stress conditions after propagation through tissue culture'. *Curr. Sci.* 81: 1530-1533, 2001.

Misra, P. and Datta, S. K. '*In vitro* maintenance of F_1 hybrid'. *Curr. Sci.* 78: 383-384, 2000.

Misra, P. and Datta, S. K. '*In vitro* propagation of White marigold (*Tagetes erecta* L.) through shoot tip proliferation'. *Curr. Sci.* 77: 1138-1140, 1999.

Mityushkina, T. Y., Dolgov, S. V., Vainstein, A. and Weiss, D. 'Regeneration from leaf discs of *C. morifolium* Ramat'. *Acta. Hortic.* 420: 112-114, 1995.

Miyazaki, S. and Tashiko, Y. 'Tissue culture of *Chrysanthemum morifolium* Ramat. III. Variation in chromosome number and flower of plants regenerated from different parts of shoots *in vitro*'. *Agric. Bull.* Saga. Univ. 44: 13-31, 1978.

Morel, G. 'Producing virus free cymbidiums'. *Amer. Orchid Soc. Bull.* 29: 495-497, 1960.

Muralidhar, C. E. and Mehta, A. R. 'Clonal production of three ornamental plants through tissue culture methods'. *Plant Tissue Culture,* 5 Meet: 693-694, 1982.

Murashige, T., Serpa, M. and Jones, J. B. 'Clonal multiplication of *Gerbera* through tissue culture'. *HortScience* 9: 175-180, 1974.

Nagaraju, V., Bhowmik, G. and Parthasarthi, V. A. '*In vitro* propagation of gladiolus; optimisation of conditions for shoot proliferation and rooting'. *Int. J. Tropic. Agric.* 14: 131-139, 1998.

Nagatomi, S., Miyahira, E. and Degi, K. 'Combined effect of gamma irradiation methods and *in vitro* explants sources on mutation induction of flower colour in *C. morifolium* Ramat'. *Gamma Field Symposia* 35: 51-69, 1996.

Nhut, D. T. Micropropagation of lily (*Lilium longiflorum*) via *in vitro* stem node and pseudo-bulblet culture'. *Plant Cell Rep.* 17: 913-916, 1998.

Nhut, D. T., Le, B. V., Jaime, A., Teixeira, De Silva and Aswath, C. R. 'Thin cell layer culture system in *Lilium*: Regeneration and transformation perspectives'. *In Vitro Cell and Dev. Biol.-Plant* 37: 516-523, 2001.

Niimi, Y. 'Application of leaf segment culture to *in vitro* bulblets production of six *Lilium* species'. *Acta. Bot. Neer.* 35: 189-194, 1986.

Niimi, Y. 'Bulblet productivity of explants from scales, leaves, stems and petals of *Lilium rubellum* Baker'. *Sci. Hortic.* 22: 391-394, 1984.

Nobecourt, P. and Kofler, L. 'Culture de tissus de tige de rosier'. *CR Hebd Seanc Acad. Sci.,* Paris 221: 53-54, 1945.

Noriega, C. and Sondahl, M. R. 'Somatic embryogenesis in hybrid tea roses'. *Bio-Technology* 9: 991-993, 1991.

Oka, S., Muraoka, O., Abe, T. and Nakajima, S. 'Adventitious bud and embryoid formation in garland chrysanthemum leaf culture'. *J. Japan. Soc. Hort. Sci.* 68: 70-72, 1999.

Oliviera, P. D., Pasqual, M. and Paira, R. 'Effect of agar concentration and pH levels on *in vitro* propagation of chrysanthemum'. *Revista Ceres.* 264: 141-148, 1999.

Orlikowska, T., Nowak, E., Marasek, A. and Kucharska, D. 'Effects of growth regulators and incubation period on *in vitro* regeneration of adventitious shoots from gerbera petioles'. *Plant Cell Tiss. Org. Cult.* 59: 95-102, 1999.

Paek, K.-Y., Hahn, E.-J. and Son, S.-H. 'Application of bioreactors for large-scale micropropagation systems of plants'. *In Vitro Cell. and Dev. Biol.-Plant* 37: 149-157, 2001.

Petru, E., Jirsakova, E. and Luzny, J. 'Use of explant cultures for chrysanthemum (*C. indicum*) propagation'. *Acta Universitatis Agriculturae Brno A Facultas Agronomica* 29: 171-176, 1981.

Pierik, R. L., Steegmans, H. H. M., Verhaegh, J. A. M. and Wouters, A. N. 'Effect of cytokinin and cultivar on shoot formation of *Gerbera jamesonii in vitro*'. *Neth. J. Agr. Sci.* 30: 341-346, 1982.

Pierik, R. M. L., Jansen, J. L. M., Maasdam, A. and Binnendijk, C. M. 'Optimisation of *Gerbera* plantlet production from excised capitulum explants'. *Sci. Hortic.* (Amst.) 3: 351-357, 1975.

Pittet, H. and Moncousin, C. 'La micropropagation *in vitro* du rosier'. *Rev. Hortic. Suisse* 55: 67-70, 1982.

Podwyszynska, M. and Hempel, M. 'The factors influencing acclimatisation of *Rosa hybrida* plant multiplied *in vitro* to greenhouse conditions'. *Acta. Hortic.* 226: 639-642, 1988.

Prasad, R. N., Sharma, A. K. and Chaturvedi, H. C. 'Clonal multiplication of *Chrysanthemum morifolium* 'Otome Zakura' in long-term culture'. *Bangladesh J. Bot.* 12(1): 96-102, 1983.

Priyadarshi, S. and Sen, S. 'A revised scheme for mass propagation of Easter lily'. *Plant Cell Tiss. Org. Cult.* 30: 193-197, 1992.

Queralt, M. C., Beruto, M., Vanderschaeghe, A. and Delbergh, P. C. 'Ornamentals'. In: *Micropropagation—Technology and Application.* (Eds.). Delbergh, P. C. and Zimmerman, R. H. Dordrecht: Kluwer Academic Publishers, pp. 215-229, 1991.

Remotti, P. C. 'Primary and secondary embryogenesis from cell suspension cultures of *Gladiolus*'. *Plant Sci.* 107: 205-214, 1995.

Remotti, P. C. and Loffler, H. J. M. 'Callus induction and plant regeneration from gladiolus'. *Plant Cell Tiss. Org. Cult.* 42: 171-178, 1995.

Reynoird, J. P., Chriqui, D., Noin, M., Brown, S. and Marie, D. 'Plant regeneration from *in vitro* leaf culture of several *Gerbera* species'. *Plant Cell Tiss. Org. Cult.* 33: 203-210, 1993.

Reynoird, J. P., Chriqui, D., Noin, M., Brown, S. and Marie, D. 'High frequency bud regeneration and ploidy stability from *in vitro* leaf culture of several *Gerbera* species'. VIII Int. Congress on *Plant Tissue and Cell Culture*. Firenze, June 12-17, 1994.

Reynoird, J. P., Meynet, J., Caissard, J. C. and Chrigui, D. 'Micropropagation of Gerbera'. In: *Biotechnology in Agriculture and Forestry, High Tech Micropropagation VI.* (Ed.). Bajaj, Y. P. S. Heidelberg, Berlin, Germany: Springer Verlag, 40, 147-162, 1997.

Roest, S. and Bokelmann, G. S. 'Vegetative propagation of *Chrysanthemum morifolium* Ram. *in vitro*. *Scientia Hort.* 3: 317-330, 1975.

Rout, G. R., Debata, B. K. and Das, P. 'Somatic embryogenesis in callus cultures of *Rosa hybrida* L. cv. Landora'. *Plant Cell Tissue Org. Cult.* 27: 56-69, 1991.

Ruffoni, B. and Massabo, F. 'Tissue culture in *Gerbera jamesonii hybrida*'. *Acta Hortic.* 289: 147-148, 1991.

Sanyal, M., Dutta Gupta, S., Jana M. K. and Kundu, S. C. 1998. 'Shoot organogenesis and plant regeneration from leaf callus cultures of tuberose (*Polianthes tuberosa* L.)'. *Plant Tiss. Cult. Biotech.* 4: 81-86, 1998.

Sauer, A., Walther, F. and Peril, W. 'Different suitability for *in vitro* propagation of rose cultivars'. *Gartenbauissuscraft* 50: 133-138, 1985.

Sen, J. and Sen, S. 'Technique for a high frequency regeneration of gladiolus corm'. *Sci. Hortic.* 64: 133-138, 1995.

Seon, J. H., Kim, Y. S., Son, S. H. and Paek, K. Y. 'The fed-batch culture system using bioreactor for the bulblet production of oriental lilies'. *Acta. Hort.* 520: 53-59, 2000.

Sharma, A. K. and Chaturvedi, H. C. 'Micropropagation of *Bougainvillea* × *Buttiana* Scarlet Queen Variegated by shoot tip culture'. *Indian Experimental Biology* 26: 285-288, 1988.

Sharma, A. K., Prasad, R. N. and Chaturvedi, H. C. 'Clonal propagation of *Bougainvillea glabra* 'Magnifica' through shoot apex culture'. *Plant Cell Tiss. Org. Cult.* 1: 33-38, 1981.

Shen, T. M., Cowen, R. D. and Meyer Jr., M. M. '*In vitro* propagation of tuberose'. *HortScience* 26: 756, 1991.

Shin, K. S., Chakrabarty, D. and Paek, K. Y. 'Sprouting rate, change of carbohydrate contents and related enzymes during cold treatment of lily bulblet regenerated *in vitro*'. *Sci. Hortic.* 1823: 1-10, 2002.

Simmonds, J. A. and Cumming, B. G. 'Propagation of *Lilium* hybrids. I. Dependence of bulblet production on time of scale removal and growth substances'. *Sci. Hortic.* 5: 77-83, 1976a.

Simmonds, J. A. and Cumming, B. G. 'Propagation of *Lilium* hybrids. II. Production of plantlets from bulb-scale callus cultures for increased propagation'. *Sci. Hortic.* 5: 161-170, 1976b.

Simonsen, J. and Hildebrandt, A. C. '*In vitro* growth and differentiation of Gladiolus plants from callus cultures'. *Can. J. Bot.* 49: 473-492, 1971.

Sindhu, S. S. and Niimi, Y. 'Callus induction and plant regeneration from Asiatic hybrid lily cv. Connecticut King'. In: Natl. Symp. on *Emerging Scenario in Ornamental Horticulture in 2000 A.D. and Beyond.* Division of Floriculture and Landscaping, IARI, New Delhi, July 21-22, 1999.

Skirvin, R. M. and Chu , M. C. '*In vitro* propagation of 'Forever Yours' rose'. *HortScience* 14: 608-610, 1979b.

Skirvin, R. M. and Chu, M. C. 'Propagation of roses with tissue culture'. III *Res.* 21 :3, 1979a.

Stefaniak, B. 'Somatic embryogenesis and plant regeneration of Gladiolus (*Gladiolus* hort.)'. *Plant Cell Rep.* 13: 386-389, 1994.

Sundaram, K. and Dutta Gupta, S. 'Vision guided analysis of protocorm cluster proliferation in liquid culture of gladiolus'. In: *Exploring the Gladiolus in India.* (Ed.). Datta, S. K. Proc. *Natl. Conf. Gladiolus* held on 24-25 January 2000 at Natl. Bot. Res. Inst. Lucknow, India: C. G. and Company, pp. 192-198, 2000.

Sutter, E. and Langhans, R. W. 'Abnormalities in chrysanthemum regenerants from long term cultures'. *Ann. Bot.* 48: 559-568, 1981.

Taeb, A. G., Alderson, P. G. and Rice, R. D. 'Micropropagation of tulip: optimising shoot production from floral stem explants'. *Acta. Hortic.* 212: 677-681, 1987.

Takahashi, S., Matsubara, H., Yamagata, H. and Morimoto, T. 'Micropropagation of virus free bulblets of *Lilium longiflorum* by tank culture. I. Development of liquid culture method and large scale propagation'. *Acta. Hortic.* 319: 83-88, 1992.

Takayama, S. and Misawa, M. 'A scheme for mass propagation of *Lilium in vitro*'. *Sci. Hort.* 18: 353-362, 1983.

Tribulato, A., Remotti, P. C., Loffler, H. J. M. and Van Tuyl, J. M. 'Somatic embryogenesis and plant regeneration in *Lilium longiflorum* Thunb'. *Plant Cell Rep.* 17: 113-118, 1997.

Tsukamoto, Y. and Fujime, Y. 'The *in vitro* culture of chrysanthemum flower buds'. *J. Jap. Soc. Hort. Sci.* 40: 262-267, 1971.

Valles, M. and Boxus, P. 'Micropropagation of several *Rosa hybrida* L. cultivars'. *Acta. Hortic.* 212: 611-617, 1987b.

Valles, M. and Boxus, P. 'Regeneration from *Rosa* callus'. *Acta. Hortic.* 212: 691-696, 1987a.

Varshney, A., Dhavan, V. and Srivastava, P. S. 'A protocol for *in vitro* mass propagation of Asiatic hybrids of lily through liquid stationary culture'. *In Vitro Cell. and Dev. Biol.-Plant* 36: 383-391, 2000.

Waithaka, K. 'Micropropagation of tuberose by *in vitro* somatics organogenesis of leaf friable callus'. *Int. Congr. Plant Tissue Culture.* 6 Meet: 239, 1986.

Walter, R. J., Kamp, M. and Smith, R. H. '*In vitro* propagation of *Rosa chinensis* Jacq. var. Minima Red Cascade'. *J. Rio Grande Val. Hortic. Soc.* 33: 125-127, 1979.

Wang, Z., Heimemann, P. H., Walker, P. N. and Huser, C. 'Automated micropropagated sugarcane shoot separation by machine vision'. *Transactions of the ASAE* 42: 247-254, 1999.

Yang, X. H., Jin, D., Zhang, Y., Mu, D., Tang, X. M. and Zhu, D. W. 'Enhancement of direct shoot regeneration from internode segments of chrysanthemum by silver nitrate'. *Acta Hortic.* 404: 68-73, 1995.

Ziv, M. 'Morphogenesis of Gladiolus buds in bioreactors—implications for scaled-up propagation of geophytes'. In: *Progress in Plant Cellular and Molecular Biology.* (Eds.). Nijkamp, H. J. J., van der Plas, L. H. and van Aartrijk, J. The Netherlands: Kluwer Academic Press, pp. 119-124, 1990b.

Ziv, M. 'Morphogenetic patterns of plants micropropagated on liquid medium in shaken flask of large scale bioreactor cultures'. *Israel J. Bot.* 40: 145-153, 1991.

Ziv, M. 'The effect of growth retardants on shoot proliferation and morphogenesis in liquid cultured gladiolus plants'. *Acta. Hortic.* 280: 207-214, 1990a.

Ziv, M. 'Transplanting *in vitro* propagated *Gladiolus* plants'. *Sci. Hortic.* 11: 257-260, 1979.

Ziv, M. and Lilien-Kipnis, H. 'Bud regeneration from inflorescence explants for rapid propagation of geophytes *in vitro*'. *Plant Cell Rep.* 19: 845-850, 2000.

Ziv, M. and Lilien-Kipnis, H. In: *Handbook of Plant Cell Cultures,* vol. 5. *Ornamental Species.* (Eds.). Ammirato, P. V., Evans, D. A., Sharp, W. R. and Bajaj, Y. P. S. McGraw Hill Pub. Co., pp. 461-478, 1990.

Ziv, M., Halevy, A. H., Shilo, R. 'Organs and plantlets regeneration of Gladiolus through tissue culture'. *Ann. Bot.* 34: 671-676, 1970.

Ziv, M., Roenen, C. and Raviv, C. 'Proliferation of meristematic clusters in disposable pre-sterilised plastic bioreactors for the large scale micropropagation of plants'. *In Vitro Cell Dev. Biol.-Plant* 34: 152-158, 1998.

—oo(O)oo—

4

Micropropagation of Orchids

S. P. Das and S. K. Bhattacharjee

INTRODUCTION

ORCHIDS are the first horticultural plants cloned by tissues culture methods on a commercial scale (Goh, 1990; Griesbach, 1986). In fact, commercial application of plant tissue culture has its beginning with the micro propagation of orchids in the U.S.A. in the Seventies (Govil *et al*, 1997; Zimmerman, 1996). The technique of mericloning orchids was successfully demonstrated for the first time by Morel (Morel, 1960, 1964). The techniques were further strengthened by subsequent work by Wimber (Wimber, 1963, 1965). Since then, several workers have successfully utilised them for mass multiplication of a number of desired orchid clones. Several workers have used tissue culture techniques not only for commercial mass multiplication but also to save the orchid species from extinction (Hey and Hey, 1966). Micropropagation techniques were also instrumental for production of disease free, particularly virus free, planting materials on a large scale. Large scale multiplication of commercially important orchids through mericloning is now routinely used world wide. Standardised protocols for 83 genera were listed by Arditti and Ernst (1993). Singh (1992) reported commercial multiplication of orchids from 69 genera and their hybrids through tissue culture.

Nutritional regimes *in vitro* and the physiological age of the donor tissues greatly influence the regenerative potential of explants and morphogenetic changes involved in the regeneration, ultimately leading to the development of plantlets (Arditti and Ernst, 1993; Vij, 1995). Available literature suggests that many economically important orchid hybrids regenerate rather slowly (cf. Prakash *et al*, 1996; Arditti and Ernst, 1993). Time required for PLB or callus formation was reported to be greatly influenced by the orchid species, type of explant used and nutritional regime in the culture medium. It is now established that modification of the media by supplementing with growth regulators and growth adjuncts can lead to faster regeneration. However, one should practise caution against excessive use of growth regulators (George and Sherrington, 1984; Goh, 1990). This is due to the fact that genetic stability of the micropropagated plants is of major significance. Several researchers emphasised the higher concentration of growth regulators to be the main factor behind the genetic alterations observed in the *in vitro* cultures. Available literature suggests that use of pre-existing meristems on explant is of great importance in ensuring genetic stability of the micropropagated plants, as culture of meristematic tissues would require least addition of plant growth regulators in the media. Genetic modification of *in vitro* cultures, however, sometimes can be interesting. Aberrations like chromosome doubling may even lead to the development of new and prize-worthy cultivars (cf. Prakash *et al*, 1996; Charanasri and Wanichkul, 1978; Goh, 1990). Pierik (1987) recorded BA at concentrations lower than 0.1 mg/l induced mutation in *Phalaenopsis*. Occurrence of polyploids during *in vitro* culture of *Dendrobium* was recorded by Vajrabhaya (1977).

Most commonly used growth regulators in orchid tissue culture are Indole-3-butyric acid (IBA), Naphthaleneacetic acid (NAA), Indole-3-acetic acid (IAA) and 2-4-Dichlorphenoxyacetic acid (2,4-D), BAP, Kinetic(KN) and BA. Coconut water (CW) is the most frequently used growth adjunct. Other complex additives include banana homogenate, potato homogenate, tomato juice and peptone. Growth promoting effect of these growth adjuncts was reported to be orchid specific and, therefore, cannot be generalised; rather they need to be empirically determined for each species and hybrid (Prakash *et al*, 1996).

Juvenility of explants is another important factor that influences the regeneration response in orchid tissue culture (Fu, 1978; Vij *et al*, 1984, 1986, 1994; Vij and Pathak, 1990; Kaur and Vij, 2000). This is due to the fact that juvenile cells have less rigid cell walls and are more active physiologically and biochemically (Misra and Bhatnagar, 1995). Young seedling tissue in *Phalaenopsis* was reported to yield better PLBs than that formed by mature tissue (Tanaka *et al*, 1975). Inability of root explants of 21 orchid taxa taken from mature plants to regenerate in culture was reported by Vij (1993). *In vitro* sourced explants were reported to produce better results (Sood and Vij, 1986).

Meristem or shoot tip remains the first choice as explant for commercial orchid micropropagation. However, a number of other explants were also used successfully by different workers. These include leaf or leaf segments, roots, inflorescence and floral parts, pseudobulbs, stem or nodal segments, etc. A recently developed method, the thin section (TS) culture, utilises only a very thin cross section of tissue (Prakash *et al*, 1996).

Regeneration can follow different pathways in orchids, depending on the nutritional regime in the culture media and also varies from species to species (Arditti and Ernst, 1993). Regeneration in orchids through tissue culture can be either through direct differentiation of protocorm-like bodies (PLBs) from cultured explants and their subsequent development into plantlets, or through indirect differentiation of PLBs from explant tissues through an intermediary callus phase. It was recorded to be PLB-mediated in *Oncidium* (Kaur, 1996) and *Coelogyne* (Sharma, 1996). It was reported to be callus-mediated in *Dendrobium* and *Phaius* (Arora, 1990). Regeneration through multiple shoot bud formation was recorded in *Bletilla striata* (Yam, 1989; Vij and Dhiman, 1997), *Dendrobium chrysanthum* (Vij and Pathak, 1989), *Dendrobium transparens* (Hazarika and Sharma, 1950).

Several researchers have developed procedures for successful regeneration of large numbers of orchid species and hybrids using different explants. Some of the prominent procedures are the ones developed by Lim-Ho (1981) for *Arachnis*; Goh (1973, 1990), Loh *et al* (1975), Goh and Wong (1990) for *Aranda*; Irawati *et al* (1977), Cheah and Sagawa (1978), Ong and Chua (1980) for *Aranthera*; Churchill *et al* (1971), Ichihashi and Kako (1973), Lindemann *et al* (1970), Reinert and Mohr (1967), Scully (1967) for *Cattleya*; Morel (1960), Sagawa *et al* (1966), Wang *et al* (1981) for *Cymbidium*;

Sagawa and Shoji (1967), Arditti *et al* (1973), Kim *et al* (1970), Mosich *et al* (1974), Singh and Sagawa (1972) for *Dendrobium*; Lim-Ho *et al* (1984) for *Mokara*; Lim-Ho (1981), Lim-Ho and Lee (1987), Khaw *et al* (1978) for *Oncidium*; Bapat and Narayanaswami (1977), Lim-Ho *et al* (1984) for *Spathoglottis* (cf. Prakash *et al*, 1996).

MEDIA

A number of chemically defined media and their modifications have been used successfully for orchid micropropagation. Commonly used media are Knudson C, 1946 (KC), Vacin and Went, 1949 (VW), Murashige and Skoog, 1962 (MS), Heller's (1953), Raghavan and Torrey (1963, 1964), Ichihashi, 1979 (I) and Nitsch (1969). Some workers have also made use of other media like Hyponex, Knop, White's, Dogashima (ND), etc. However, Vacin and Went (1949), Knudson C (1946) and Murashige and Skoog (1962) are the most commonly used media.

Vacin and Went (1949) medium has been used for successful regeneration of *Phalaenopsis* (Ishi *et al*, 1988), *Dendrobium* (Sagawa and Shoji, 1967; Sobhana and Rajeevan, 1993; Sharon and Vasundhara, 1990; Pathania *et al*, 1998; Kim *et al*, 1970; Devi and Laishram, 1998; Soediono, (1983), *Aranda*, *Mokara* and *Dendrobium* (Lam and Lee, 1996), *Oncidium* (Bagde *et al*, 1996), *Spathoglottis plicata*, *Phaius wallichi* (Nagabhushana, 1982), *Oncidium*, *Dendrobium* and *Phalaenopsis* (Nuraini and Mohd. Sahib, 1992), *Bulbophyllum careyanum* (Vij *et al*, 2000), *Cattleya* (Scully, 1967), *Arachnis* (Lim-Ho, 1981), *Aranda* (Goh, 1990) and *Spathoglottis plicata* (Lim-Ho *et al*, 1984).

Murashige and Skoog (1962) or modifications of it have been used for successful regeneration of *Dendrobium* (Mujib and Jana, 1994; Devi and Laishram, 1998), *Spathoglottis plicata*, *Cymbidium aloifolium*, *Dendrobium aphyllum*, *Dendrobium moschatum* (Nayak *et al*, 1997); *Phalaenopsis* (Jimenez and Guevara, 1996); *Cymbidium ensifolium* (Chang and Chang, 1997); *Oncidium* (Bagde *et al*, 1996), *Aerides maculosum* (Kulkarni and Surwase, 1988); *Bletilla striata* (Yam and Weatherhead, 1991); *Aranda*, *Ascocenda* and *Cattleya* (Fu, 1978, 1979); *Oncidium*, *Dendrobium* and *Phalaenopsis* (Nuraini and Mohd. Sahib, 1992), *Epidendrum* (Churchill *et al*, 1973) and *Mokara*, *Spathoglottis plicata* (Lim-Ho *et al*, 1984).

Knudson C (KC) medium or its modifications were used for successful micropropagation of *Mormodes histrio* (Holters and Zimmer, 1990); *Oncidium*, *Spathoglottis unguiculata* (Prakash *et al*, 1996); *Aranda* (Goh, 1990); *Aranthera* (Irawati *et al*, 1977); *Cymbidium* (Paek *et al*, 1989); *Dendrobium* (Pathania *et al*, 1998); *Odontoglossum* (Morel, 1965); *Oncidium papilio* (Fast, 1973); *Phalaenopsis* (Latha and Seeni, 1991).

White's medium was used for micropropagation of *Aranda* Deborah (Goh, 1973) and *Vanda* Miss Joaquim (Goh, 1970). Raghavan and Torrey medium was used for micropropagation of *Arundina bambosifolia*, *Phaius tankervilleae* (Nagaraju and Parthasarthy, 1995). Heller's medium was used for successful micropropagation of *Pholidota chinensis*, *Liparis viridiflora* (Yam and Weatherhead); *Laeliocattleya* (Churchill *et al*, 1973); *Cattleya* (Churchill *et al*, 1971; Champagnat *et al*, 1970); and *Paphiopedilum* (Steward and Button, 1975). Hyponex medium was used for micropropagation of *Cymbidium* (Paek *et al*, 1989); *Cattleya* (Yoneda *et al*, 1979). Dogashima (ND) medium was used for micropropagation of *Bletilla striata* (Ishikawa *et al*, 1997). Nitsch medium was used for micropropagation of *Agrostophyllum myrianthum*, *Cymbidium longifolium*, *Phaius tankervilleae*, *Renanthera imschootiana* (Devi *et al*, 1998). Mitra *et al* (1976) medium was employed for successful regeneration of *Cymbidium mastersii*, *Dendrobium* Emma White, *Aerides multiflorum* (Vij, 1993), *Cymbidium pendulum*, *Vanda* Kasem's Delight (Vij *et al*, 1994); *Rhynchostylis retusa* (Vij *et al*, 1984; Sood and Vij, 1986). Modified Knop medium was used for successful micropropagation of *Bletilla striata*, *Dendrobium aduncum*, *Dendrobium lodigessi*, *Dendrobium transparens*, *Liparis nervosa* and *Malaxis acuminata* (Yam and Weatherhead, 1990); *Dendrobium* (Arditti *et al*, 1973). Modified Ichihashi medium was used for micropropagation of *Bletilla striata* (Vij and Dhiman, 1997).

MERISTEM OR SHOOT TIP CULTURE

Shoot meristem culture has now emerged as an important technique for rapid and mass multiplication of desired genotypes in orchids (Vij and Pathak, 1990). Sagawa and Kunisaki (1982) also suggested apical shoots or axillary buds to be the best source of explants for successful clonal propagation of orchids. However, Vij and Pathak (1990) stressed its limited utility in monopodials as the technique requires sacrifice of an entire new growth. Multiplication of monopodial orchids through shoot meristem culture poses a great threat to the very existence of the mother plant itself. More often than not, the rare elite and prized individual is killed as a result of excision of the only shoot meristem of the plant and the explant also perishes due to infection (Chaturvedi and Sharma, 1986).

Morel was the first to make use of meristematic tissues for micro-propagation of *Cymbidium* (Morel, 1960). Meristematic tissues were also used by Sagawa *et al* (1966) and Wang *et al* (1981) for micropropagation of *Cymbidium*. Apical or lateral meristem was used by Scully (1967) for regeneration and mass multiplication of *Cattleya*. They used a liquid VW medium with CW and sucrose for culture initiation and PLB multiplication and a solid medium for further development of plantlets. Sagawa and Shoji (1967) developed a method of micropropagation of *Dendrobium* using shoot tip as explant. They used a simple low salt medium (modified VW) for efficient protocorm development and vigorous plant growth. First successful regeneration of *Aranda* was obtained by Goh (1973) using meristematic tissues from shoot apices and axillary buds. Initial culturing was done on liquid White's (White, 1943) medium supplemented with peptone (2 g/l) and sucrose (30 g/l) for PLB formation. A solidified medium was used for differentiation of plantlets. Goh (1990) also developed procedures for several *Aranda* hybrids using similar method, but used VW or KC media in most of the cases. Apical meristems excised in sterile conditions and cultured in KC medium, kept at 25° C in 12 hours day, were reported to produce protocorm-like protuberance (Rosa, 1975). These, when re-cultured in fresh medium and kept at 28° C in 18 hours day and 22° C night temperature produced new protocorms after 3-4 weeks. Irawati *et al* (1977) regenerated *Aranthera* from shoot tips. A liquid KC medium supplemented with CW was used for PLB formation and a solid medium for plant development. Lim-Ho (1981) developed a very reliable procedure for micropropagation of *Arachnis* using shoot tips. A liquid VW medium supplemented with CW was used for culture initiation. A solidified VW was used for PLB multiplication and the multiplied PLBs were then grown on a VW based formulation enriched with banana homogenate, tomato juice, sucrose and CW for plantlet development. Soediono (1983) found that VW medium containing 15 per cent (by volume) coconut water was suitable for initial culture of shoot tips. Addition of 10 ppm NAA to the above led to rapid proliferation of PLBs in *Dendrobium*. He reported that addition of vitamin or 2,4-D was deleterious. He also cautioned against production of a very large number of plantlets from a single shoot tip. Sharon and Vasundhara (1990) micropropagated *Dendrobium* successfully using shoot meristem. The explants generated a profuse crop of callus mediated protocorms in coconut milk supplemented liquid medium (VW), but failed to develop further unless transferred to agar based medium. Mujib and Jana (1994) reported callus induction from apical meristem of *Dendrobium* on MS supplemented with either 2,4-D + BA or NAA + BA; protocorms were induced on MS supplemented with GA_3. Presence of sucrose (2 per cent), NA (0.1 mg/l) and charcoal induced roots. Nagaraju and Parthasarathy (1995) obtained better results on Raghavan and Torrey medium with shoot tips from *in vitro* grown plants of *Arundina bambosifolia* and *Phaius tankervilleae* on 4 basal media. Devi and Laishram (1998) obtained multiple shoots by culturing shoot tips and axillary buds of *Dendrobium* hybrids on VW medium with addition of thiamine Hcl (1.0 mg/l), Pyridoxine Hcl (1.0 mg/l), nicotinic acid (1.0 mg/l), adenine sulphate (100 mg/l), myo-inositol (100 mg/l), NAA (0.1 mg/l) and BAP (3-4 mg/l).

LEAF

Meristem culture, though widely practised, has its limitation in the fact that it requires sacrifice of an entire new growth. Attempts were therefore made to use other plant parts like leaf, root, inflorescence, etc. as explant in orchid micropropagation.

Regenerative potential of leaf tissues was demonstrated for the first time by Wimber (1965), who was able to produce PLBs from leaf tissues in *Cymbidium*. Utility of leaf segments or entire leaf in producing a large number of identical clones through direct or callus mediated organogenesis was emphasised by Arditti (1967), Churchill *et al* (1970), Harrison and Arditti (1970) and Arditti (1977). So far, regeneration ability of orchid leaves has been reported from about 60 species (cf. Vij and Pathak, 1990; Aditti 1977a, 1977b; Chaturvedi and Sharma, 1986; Mathews and Rao, 1985; Tanaka, 1987; Tanaka and Sakanishi, 1977, 1980; Tanaka *et al,* 1989; Vij *et al,* 1984, 1986; Wang, 1989).

Regeneration potential of leaf tips in *Laeliocattleya* and *Epidendrum* was reported by Churchill *et al* (1973). They found that *Epidendrum* leaf tip formed calli on MS. Plantlets were also obtained from tips of young leaves from mature *Laeliocattleya* plants cultured on Heller's medium. Chaturvedi and Sharma (1986) induced protocorm formation from excised tips of young leaves of *Vanda*. Regenerative potential of leaf tips was also recorded by Loh *et al* (1975) in *Aranda*, Allenberg (1976) in *Paphiopedilum* and Chaturvedi and Sharma (1986) in *Vanda*. Regenerative potential of leaf base was recorded by Champagnat *et al* (1970) in *Cattleya*, Seeni (1988) in *Vanda coerulea* and Vij and Pathak (1990) in *Neofinetia falcata, Satyrium nepalense, Vanda cristata* and *Vanda testacea.*

Greater regenerative potential of leaf base than leaf tip was also recorded by Fu (1978, 1979) in *Ascocenda* and *Vanda*, Abdul Karim and Hairani (1990) in *Coelogyne, Dendrobium* and *Oncidium,* Seeni and Latha (1992) in *Renanthera* and Nayak *et al* (1997) in *Acampe*. Seeni and Latha (1992) cultured young leaf explants of *Renanthera* on Mitra *et al* (1976) medium supplemented with BA, NAA and peptone. Cultured leaf bases produced large numbers of uniform plantlets.

Mathews and Rao (1985a, 1985b) reported greater response from intact leaves in *Vanda*, whereas responses from leaf tip were infrequent. A decisive role of leaf base in regeneration of leaf explants was suggested. Latha and Seeni (1991) reported the initiation of PLBs near the tip region and its quick spread in a sporadic fashion all over the sides of the leaves in *Phalaenopsis* and suggested that the entire surface of the organ was potentially meristematic and regenerative. Teng *et al* (1997), however, obtained regeneration of PLBs and subsequent plantlet development only to the extent of 6.5 per cent by using leaf explants in *Spathoglottis plicata.*

Juvenility of orchid tissues was found to be an important factor in determining the regenerative response in several orchids (cf. Kaur and Vij, 2000; Vij *et al*, 1984, 1986, 1994; Vij and Pathak, 1990; Fu, 1978). Tanaka *et al* (1975) reported better PLB development from leaf segments of *Vanda* and *Phalaenopsis* when young seedling tissue was used. Mature tissues required etiolating to induce primordia, which subsequently developed into PLBs. Kaur and Vij (2000) reported that culture initiation in *Saccolabium papillosum* leaf segments was markedly influenced by the source and juvenility of the tissues besides chemical stimulus in the nutrient pool. *In vitro* sourced tissues responded far more frequently to a wide range of stimuli in comparison to greenhouse sourced tissues.

Vij and Pathak (1990) suggested that leaf explants could be induced to become regenerative by manipulating the chemical composition of the culture media *in vitro*. They recorded that leaf explants proliferated along both abaxial and adaxial surfaces and the proliferations were traced to the epidermal or sub-epidermal cells. Loh *et al* (1975), Monorama *et al* (1984), Vij *et al* (1984) and Mathews and Rao (1985) also traced the regeneration ability of orchid leaves to the dermal layers.

Yam and Weatherhead (1991) reported that younger leaf tips produced more shoots. More number of plantlets from *Pholidota chinensis* leaf tip culture could be obtained on Heller's medium containing 1 mg BA and 0.2 mg NAA per litre. Darkness and presence of graphite induced more plants. *Liparis viridiflora* leaf tips also produced shoots more rapidly on Heller's medium containing 1 mg BA per litre. However, they reported that Ichihashi and Yamashita medium containing 1 mg BA and 0.2 mg NAA per litre was most suitable for producing the largest number of shoots from leaf tip culture of *Acampe rigida*.

Fu (1978, 1979) reported a media based on Murashige and Skoog inorganic salts and CW to be the most successful for leaf tissue culture of *Aranda, Ascocenda* and *Cattleya*. Regeneration of several orchid species using leaf explants on three media, namely, Mitra, MS and modified KC with various combinations of growth regulators was reported by Vij *et al* (1990). Rudolph *et al* (1972) observed that Orthochlorophenoxyacetic acid (OCP) enhanced water uptake and increased tissue dry weight in cultures from callus tissue derived from leaf tips of *Epidendrum* × "O" Brienianum.

ROOT

Micropropagation of orchids using root segments has its advantage in the fact that they are easily available, particularly in epiphytes, which produce plenty of aerial roots. Removal of some roots also does not affect the well-being of the mother plant much.

In 1970, Beechey suggested the possibility of propagation of orchids from aerial roots (Beechey, 1970). Goh (1970) observed that root segments of *Vanda* cultured *in vitro* developed into thin etiolated structures. Similar observations in *Epidendrum* root culture was reported by Churchill *et al* (1972). Inherent regenerative competence of orchid roots was also suggested by Scully (1971), who reported plantlet formation *in vivo* from *Phalaenopsis struitiana*. Champagnat (1971) reported that apical meristems in root tips could be induced to become regenerative and form protocorms in *Neottia*. Stewart and Button (1978) in *Epidendrum,* Kerbauy (1984a) in *Catasetum* and Kerbauy (1984b) in *Oncidium* also reported similar regeneration using root segments as explants. Holters and Zimmer (1991) also reported shoot formation from root tip cultures of *Catasetum* sp., *Clowesia* sp., *Cycnoches* sp., *Cyrtopodium* sp., *Galeandra* sp. and *Mormodes* sp. Yam and Weatherhead (1991) produced plantlets from root tips of *Bletilla striata* and *Cleisostoma fordii* cultured on modified MS medium containing 1 mg NAA and 0.2 mg BA/litre. Root tip elongation was observed with high cytokinin: auxin ratio. A low cytokinin:auxin ratio led to callus formation.

Sood and Vij (1986) reported that root segments obtained *in vivo* from well-developed *Rhynchostylis retusa* plant failed to respond in culture. However, root explants from six month old axenic cultures proliferated in some selective combinations. They suggested that the path followed by the proliferating cells, that is, via PLB or shoot bud formation, could be influenced by manipulating the growth adjuncts in the media.

Chaturvedi and Sharma (1986) cultured root explants from aseptically established plants of *Vanda* hybrid (*Vanda* TMA × *Vanda teres*) and *Rhynchostylis retusa.* Root tip explants measuring about 3 mm in length mostly responded to form PLBs in the optimum treatment containing 1 ml/l BAP, 1 ml/l IAA and 200 ml/l CH. Initially, root tips turned green and formed root hairs. In 20 per cent cultures, one or two PLBs differentiated from the cut end of root tips. The root explants with differentiated PLBs when sub-cultured in treatment containing 0.5 mg/l BAP, 1 mg/l IAA, 0.1 mg/l 2,4-D and 200 mg/l CH showed enormous proliferation of PLBs, which subsequently formed tiny plantlets when re-cultured in some media (major salts of VW medium plus 500 mg/l NH_4, NO_3, 2 mg/l Fe-Na_2-EDTA, in place of ferric tartrate, trace elements of Heller (1953) medium and 0.1 mg/l each of thiamine Hcl, pyridoxine Hcl and nicotinic acid) developed as individual plants by root development.

Latha and Seeni (1991) induced PLB formation from root explants of *Phalaenopsis* on KC medium supplemented with minor elements of Mitra *et al* (1976), 20 per cent CW, 2 per cent sucrose, 500 mg/l peptone, 500 mg/l casein hydrolysate, 1 per cent agar and factorial combinations of BA and NAA. The highest number of PLBs were formed when a combination of 10 mg/l BA and 1 mg/l NAA was employed. They also reported that the entire surface of the root explants was potentially meristematic and regenerative. On the contrary, earlier reports by Tanaka *et al* (1976) and Sanchez (1988) suggested only root tips to have regeneration potential. Presence of leaves in the explants was reported to inhibit formation of axillary shoots and of root-borne shoots (Holters and Zimmer, 1999). They reported that the root-borne shoots developed from PLBs only on isolated single roots or root systems without a shoot. Longer root sections produced more PLBs. Holters and Zimmer (1990) again reported that shoot formation was nearly 100 per cent on isolated roots of *Mormodes histrio.* Shoot formation was promoted by 0.2 mg/l potassium salt of NAA and 2 mg/l BA. Growth of protocorm-like bodies was accelerated with 0.1-0.2 mg/l potassium salt of NAA and 0.1-0.5 mg/l BA. Growth of shoots, however, was best without BA. Goh (1992) from a study on *mycorrhizal* associations in some tropical orchids found free-hanging aerial roots devoid of any fungus and, therefore, suggested their suitability as explants for micropropagation.

Vij (1993) suggested that regeneration response and rate from root explants was specific. Presence of yeast extract was suggested as obligatory for inducing proliferation in *Aerides multiflorum* and *Vanda teres.* Teng *et al* (1997) reported that in *Spathoglottis plicata* though a very good success could be achieved from nodal explants, but regeneration from root explants was occasional. Only occasionally, some root segments dissected from *in vitro* regenerated plantlets were able to produce PLBs and then plantlets. Chang and Chang (1997) induced calluses of *Cymbidium ensifolium* var. *misericors* from sections of roots of seed derived-plantlets on half strength MS medium plus 10 mg 2,4-D and 0.1 mg thidiazuran/l. Kulkarni and Surwase (1998) recorded callusing of root explants from *in vitro* germinated plants of *Aerides maculosum* on MS medium supplemented with 2 mg/l each of NAA and 2-ip and 1 g/l peptone. Successful regeneration of root explants from axenic seedlings was also reported by Stewart and Button (1978) in *Epidendrum* sp., Kerbauy (1984) in *Oncidium varicosum,* Tanaka *et al* (1976) in *Phalaenopsis* sp., Konoweiz and Janick (1984) in *Vanilla planifolia* and Vij *et al* (1997), Sood and Vij (1986) in*Rhynchostylis retusa.*

STEM OR NODAL SEGMENTS

Micropropagation of orchids using stem or nodal segments as explants were reported by several workers. Regeneration potential of uninodal explants of *Cymbidium pendulum* was examined by Vij *et al* (1994). They reported that formation of PLBs was best on basal media, shoot multiplication required BA and root development was best in the presence of activated charcoal. A new procedure for *in vitro* propagation of *Phalaenopsis* using elongated stems of benzyladenine induced young seedlings as the starting material was reported by Duan *et al* (1996). They used top nodes for cyclic propagation of new explants and the middle nodes were used for producing shoots for multiple adventitious buds. Regeneration to the extent of 98.5 per cent from nodal explants dissected from 8-month old pot grown seedlings of *Spathoglottis plicata,* on charcoal amended MS medium supplemented with various combinations of NAA and BA, was reported by Teng *et al* (1997). Jimenez and Guevara (1996) suggested that complete rooted plants regenerated from nodal segments could be transferred to greenhouses within six months of culturing *in vitro.* They cultured nodal cuttings from senescent flowering stems of commercial hybrids of *Phalaenopsis* on half strength modified MS containing 2.9 mg Ca/l supplemented with vitamins, adenine, coconut milk, activated charcoal, growth regulators, sucrose and glucose. Larger stem segments produced best results. Pathania *et al* (1998) assessed the regeneration potential of stem explants in *Dendrobium* Sonia. Both VW and KC media favoured PLB formation and subsequent development of plantlets. Rooting of the plantlets was favoured in the presence of IBA (1 mg/l) or NAA (1.8 mg/l) and Paclobutrazol (0.5 mg/l) in the media.

PSEUDOBULB AND RHIZOME

In general, *in vitro* sourced pseudobulb segments showed better response than greenhouse sourced pseudobulb explants in terms of regeneration in *Bulbophyllum careyanum* (Vij *et al,* 2000). They reported that an additional treatment with plant growth regulators was obligatory for activating meristematic activity in greenhouse sourced explants.

Chang and Chang (1997) induced calluses of *Cymbidium ensifolium* var. *misericors* from sections of pseudobulbs and rhizomes of seed derived plantlets on half strength MS. Efficient mass propagation was possible via rhizome proliferation and embryoid formation. Vij and Dhiman (1997) reported regeneration of uninodal segments of pseudobulbs of *Bletilla striata* either through development of multiple shoot bulbs or callus mediated PLB formation. Regeneration of *Dendrobium moschatum* from pseudobulb segments was carried out by Vij and Sood (1982).

INFLORESCENCE

Regenerative potential of flower stalks was demonstrated for the first time by Rotor (1949). Since his successful regeneration of *Phalaenopsis* from flower stalks, the use of inflorescence axis in mass propagation of orchids, monopodials in particular, has gained in importance. The technique, like leaf or root, has its advantages in the fact that it does not cause any damage to the mother plant and explants are easily available in sufficient number. So far, the technique has been successfully utilised in about 20 orchid species (cf. Vij *et al,* 1997; Arditti and Ernst, 1993; Arditti and Krikorian, 1996; Vij *et al,* 1986, 1991; Kaur and Vij, 1995; Kanika and Vij, 1997; Vij and Dhiman, 1997). In monopodial orchids floral primordial of 1.5 cm long or less proliferated within 20-51 days on liquid VW basic media with coconut water. However, longer inflorescences (3 cm or larger) elongated and produced normal or abnormal flowers (Intuwong and Sagawa, 1973). Khaw *et al* (1978) developed an efficient micropropagation method for *Oncidium* using inflorescence tips as explant. High percentage of PLBs from inflorescence tips of *Aranda, Mokara* and *Dendrobium* cultured in VW medium was reported by Lam and Lee (1996). Efficient micropropagation of *Spathoglottis plicata* and *Spathoglottis unguiculata* from inflorescence tip or inflorescence stalk bud culture on liquid KC medium for initiation and re-culturing on agar solidified medium for plantlet differentiation was suggested by Prakash *et al* (1996).

The regenerative potential of inflorescence segments was suggested to vary according to their position on the donor axis and chemical stimulus in the nutrient pool (Vij *et al,* 1997). They observed that in *Rhynchostylis gigantea* explants from the proximal region with well differentiated buds failed to regenerate, whereas explants from the distal region, with more or less, undifferentiated buds followed PLB or callus mediated regeneration along their cut ends. They recorded that proliferation of PLBs or calluses was along the extra nodal regions. Lin (1986) in *Doritaenopsis* and *Phalaenopsis* and Kanika and Vij (1997) in *Cleisostoma* observed similar proliferation of inter-nodal sections. The distal-most region of the inflorescence axis, however, was reported to follow only PLB mediated regeneration (Vij *et al,* 1997). Similar proliferation was also recorded by Lin (1986) in *Phalaenopsis.* Higher regenerative potential of younger inflorescence explants was also recorded by Nuraini and Mohd. Shahib (1992). They observed higher shoot production from younger flower scapes.

Perianth segments were also used for micropropagation of orchids. Regenerative potential of perianth segments was demonstrated by Kins and Kako (1984) in *Cymbidium.* Vij and Sharma (1996) in *Vanda cristata* reported that juvenility of tissues was an important factor in influencing the regenerative competence of *Vanda cristata* perianth segments. Cell proliferation was better achieved in labellum-sourced explants than those from normal petals.

BUD

Kim *et al* (1970) reported formation of PLBs from bud explants as well as from leaf axils of *Dendrobium.* Apical or axillary bud explants were used by Teo *et al* (1973) for standardisation of an efficient micropropagation procedure for *Vanda.* They used liquid VW medium supplemented with CW and sucrose for initiation and multiplication of PLBs. Plantlet differentiation was obtained on Phytagel solidified VW medium. *In vitro* plantlet development from apical buds within 2-4 months in *Arachnis, Aranda, Cattleya, Dendrobium, Renanthera* and *Vanda* was reported by Teo and Teo (1974). Vajrabhaya (1978) was able to produce PLBs from dormant buds on old pseudobulbs of *Cattleya.* Lateral buds on flower stalks of *Phalaenopsis* were regenerated by Fu (1978) on VW medium. Initial culturing was done in liquid media to induce PLB formation, whereas plantlets from PLBs were regenerated on solid media. Lim-Ho (1981) developed an apical bud culture procedure for effective mass multiplication of *Oncidium.* Young flower buds of *Spathoglottis plicata* were used by Lim-Ho *et al* (1984) to produce PLBs on MS medium supplemented with IAA, kinetin, adenine and sucrose. A growth regulator-free VW medium added with CW and banana homogenate was used for plantlet formation and growth. Regeneration of plantlets from immature buds of *Cymbidium* was reported to yield somaclonal variants (Laneri, 1990). Jimenez and Guevera (1996) recorded that regenerative buds in *Phalaenopsis* rapidly produced leaves and roots when cultured on half strength modified MS. However, flower buds developed into inflorescence with nodal segments. To initiate vegetative growth, they were required to bere-cultured. Lam and Lee (1996) reported that culturing of young inflorescence buds (1-3 mm) resulted in best yield of PLBs in *Aranda, Mokara* and *Dendrobium.* Dormant buds obtained from inflorescence stalks also gave very good regeneration. Nayak *et al* (1997) induced high frequency shoot proliferation from shoot segments of *Cymbidium aloifolium, Dendrobium aphyllum* and *Dendrobium moschatum* on MS medium containing TDX (thidiazuron). Regenerated shoots were rooted on MS medium containing 10.8 m IBA. Regeneration of *Dendrobium* hybrids from axillary buds through multiple shoots induction was recorded by Devi and Laishram (1998).

THIN SECTION CULTURE

Prakash *et al* (1996) developed a thin section (TS) culture system for a number of commercially important orchids to improve the regeneration efficiency without sacrificing the genetic stability of the plants produced. The method used cross sections of tissue (0.6 mm thick). Their experiments with hybrids of *Aranda, Cattleya, Dendrobium, Mokara, Oncidium, Spathoglottis* and *Vanda* indicated that the method could be used as an efficient micropropagation method for commercial operation. Explants were prepared by transverse sectioning (0.6 mm) of actively growing plant parts such as shoot tips, young leaves and inflorescence or developing PLBs. The thin section showed 40-60 per cent regeneration. As the number of thin sections from each explant can be as high as 10-20, the actual regeneration is increased manifold. They also reported that survival of TS explants and production of PLBs were found to be dependent on the amount of CW in the medium. Maximum survival was recorded at 20 per cent CW. In *Ananda* Deborah, they recorded that TS explants taken from young inflorescence produced PLBs much more rapidly than those obtained from shoot tips. Further, they suggested that TS culture alone or in combination with ribavirin treatment proved to be an effective procedure for elimination of viruses from infected plants in the production of virus-free orchids. Linn *et al* (1993) demonstrated that higher percentage of virus-free plantlets could be obtained by thin section cultures of infected plantlets and PLBs along with chemotherapy of ribavirin. Regenerates from thin section culture without ribavirin were also free from CYMV and ORSV.

GROWTH REGULATORS

The effects of combinations of auxins and cytokinins vary from orchid to orchid, growth regulators used, their concentrations and ratios (cf. Kumaria and Tandon, 2000; Arditti and Pridgeon, 1997). Higher concentration BA was reported to increase the frequency and number of PLBs from flower stalk cultures (Tanaka and Sakanishi, 1978; Ernst, 1985). Inhibitory action of NA on PLB formation in inter-nodal segment cultures of flower stalks was reported by Lin (1986). A combination of NAA either with KN or BAP resulted in better PLB production in *Luisia trichorhiza, Satyrium nepalense, Vanda cristata* and *Vanda testacea* (Vij an Pathak, 1990). Beneficial effect of NAA when used with BAP in leaf tissue culture of *Vanda* was reported by Mathews and Rao (1985) and Vij *et al* (1986).

A combination of 0.44 M BA and 5.37 M NAA was reported to induce the greatest regeneration of *Spathoglottis plicata* from leaf and stem nodal explants (Teng *et al*, 1997). Nayak *et al* (1997) reported that high frequency shoot proliferation from shoot segments of *Cymbidium* and *Dendrobium* on media supplemented with BA or TDZ, NAA and BA were suggested to influence shoot elongation; whereas rooting of the shoots was obtained with supplementation of IBA. Beneficial effect of BA, NAA, Kinetin and 2,4-D in inflorescence tip culture of *Aranda, Mokara* and *Dendrobium* was reported by Lam and Lee (1996). Mujib and Jana (1994) reported that supplementation of 2,4-D + BA or NAA + BA in media resulted in callus induction from apical meristem of *Dendrobium,* whereas supplementation with GA_3 + BA induced protocorm formation. They also recorded the influence of NAA in root induction. Efficient mass propagation of *Cymbidium ensifolium* from sections of pseudobulbs, rhizome and roots in the presence of 2,4-D and thidiazuron was reported by Chang and Chang (1997). Supplementation of IBA or NAA + Paclobutrazol in *Dendrobium* culture using stem explants favoured rooting of the regenerated plantlets (Pathania *et al,* 1998). Rooting in *Dendrobium* was reported to be favoured by NAA (Sharon and Vasundhara, 1990). Vij *et al* (1994) suggested that shoot multiplication in *Cymbidium pendulum* regenerated from uninodal explants required BA in the basal media. Acceleration in growth of PLBs of *Mormodes histrio* by BA and potassium salt of NAA was recorded by Holters and Zimmer (1990). They, however, reported that growth of shots was best without BA in the media. Presence of NAA and BAP in the media was reported to assist multiple shoot formation in *Dendrobium* culture from shoot tips and axillary buds (Devi and Laishram, 1998). Influence of IBA and kinetin on better shoot production in *Dendrobium moniliforme* culture was reported by Lin *et al* (1993). They recorded that IBA at 0.1 mg/l was best for producing many tall and rooted shoots.

Ratio of cytokinin and auxin in the culture media is an important factor. Yam and Weatherhead (1991) observed that in root tip culture of *Bletilla striata,* callus induction was at low cytokinin:auxin ratio. When ratio was high, root tips elongated. Importance of correct auxin:cytokinin ratio for optimum propagation of orchids was also stressed by Meyer (1988). It was suggested that although high cytokinin concentration in the nutrient media resulted in rapid multiplication, it could also lead to abnormal growth and flowering, such as bushy growth and small flowers. Paek *et al* (1989) reported that in rhizome cultures of seed derived *Cymbidium* species, better rhizome growth was obtained under a low Kinetin:NAA ratio, whereas a high ratio resulted in good shoot formation and growth. *In vitro* flowering from rhizomes of *Cymbidium niveomarginatum* and *Cymbidium ensifolium* × *Cymbidium kanran* cv. Dogui could be obtained by supplementing 3 mg and 0.3 mg/litre of kinetin and NAA, respectively, in the media.

Paek and Teung (1991) examined the effect of plant growth regulators on rhizome growth and shoot formation in *Cymbidium forrestii.* All tested auxins were found to stimulate rhizome growth with 2 mg/l NAA being the most effective. All auxins, however, suppressed normal shoot formation and retarded leaf formation and development. Rhizome growth was reduced by cytokinins, but shoot formation was enhanced by BA. Addition of 0.1 mg/l BA and 10 mg/l NAA in MS media proved to be most effective in shoot induction from apical flower bud culture of *Cymbidium goeringii.* However for production of plantlets with shoots as well as roots, optimal doses for BA and NAA were 0.1 mg/l and 1 mg/l, respectively (Shimasaki and Kemota, 1991).

BAP at the rate of 1 mg/l was reported to promote callus mediated PLB formation in pseudobulb uninodal segments culture of *Bletilla striata* whereas, at higher dose (2 mg/l), BAP promoted regeneration through multiple shoot bud formation (Vij and Dhiman, 1997). They, however, recorded detrimental effect of auxin (NAA) on regeneration frequency. A combined treatment of BAP and NAA, however, was essential for root development in pseudobulb derived shoots of *Dendrobium moschatum* (Vij and Sood, 1982). Beneficial effect of IBA and Kinetin in good shoot production in *Dendrobium moniliforme* was reported by Lim *et al* (1993). Vij and Sharma (1996) observed positive influence of BAP and NAA on regenerative potential of *Vanda cristata* perianth segments.

Significance of various growth regulators in transforming floral buds into vegetative buds was demonstrated in several orchids such as *Dendrobium crepidatum* and *Dendrobium pierardii* (Vij *et al*, 1991), *Oncidium* (Lim-Ho and Lee, 1987) and *Rhynchostylis retusa* (Kaur and Vij, 1995). Ichihashi (1992) however, reported that regeneration in *Phalaenopsis* floral bud was possible without any growth hormones. Vij *et al* (1997) emphasised that organogenesis of inflorescence segments of *Rhynchostylis gigantea* was directly proportional to BAP level in the medium. It had the most pronounced effect at 10 mg /l dose. The benign effect of BAP or KN in inducing callus and/or multiple shoots has also been reported by Vij and Dhiman (1997) in *Malaxis acuminata*, Lin (1986) in *Phalaenopsis,* Kanika and Vij (1997) in *Cleisostoma micranthum.*

TABLE 1
Protocols Perfected by Indian Research Group

Orchid Species/Hybrid	Explants	Contact Scientist
Ascocenda 50th State Beauty	Leaf	Prof. S. P. Vij Botany Department Panjab University Chandigarh
Cattleya 'Almakee'	Leaf	Prof. S. P. Vij
Dendrobium densiflorum Wall	Stem node	Prof. S. P. Vij
Dendrobium 'Emma White'	Stem node, leaf	Prof. S. P. Vij
Dendrobium 'Snowfire'	Leaf	Prof. S. P. Vij
Dendrobium 'Sonia Jo'	Stem, node, leaf	Prof. S. P. Vij
Dendrobium 'Sonia NR 28'	Leaf	Prof. S. P. Vij
Oncidium 'Gower Ramsey'	Stem, node, leaf	Prof. S. P. Vij
Papilionanthe teres (Roxb.) Schltr.	Leaf	Prof. S. P. Vij
Phaius tankervilliae (L. Herit) Bl.	Leaf	Prof. S. P. Vij
Rhynchostylis retusa Bl.	Leaf	Prof. S. P. Vij
Vanda coerulea Griff. ex. Lindle.	Leaf	Prof. S. P. Vij
Vanda 'Kasem's Delight' Tom Boykin	Leaf, root	Prof. S. P. Vij

Source: Govil and Gupta (2001).

Exogenous supply of IAA was reported to increase peroxidase activity (Kumaria *et al,* 1990; Dendsay, 1989). Peroxidase activity in plants is associated with plant defence against pathogens. Krsnik-Rasol (1991) suggested to consider them as chief developmental markers in tissue culture studies. Kumaria and Tandon (2000) also reported that auxin application increased activities of both peroxidase and polypenol oxidase on fresh weight basis with the growth and development of protocorms in most of the treated tissues. Treatment with IAA, NAA and GA_3 increased IAA oxidase activity. The KN and GA_3 treated protocorms showed a decline in the total phenolic content. They also suggested that growth regulators in the media might have acted in a manner similar to endophytic fungi, to bring about physiological changes during protocorm development.

Auxins (IAA/IBA/NAA/2,4-D) were reported to be detrimental for culture initiation in greenhouse sourced explants, but the same could be successfully used for the purpose in case of *in vitro* sourced explants (Vij *et al*, 2000). They found IBA to be more beneficial, replacement of IAA in the media with IBA resulted in 100 per cent shoot bud regeneration.

COMPLEX ADDITIVES

Ishii *et al* (1998) reported that addition of coconut water (200 ml/l) to the media was beneficial for callus induction in regeneration of *Phalaenopsis* through somatic embryogenesis. Supplementation of CW in the media was reported to favour better development of leafy shoots in *Dendrobium* micropropagation using shoot meristem (Sharon and Vasundhara, 1990). Beneficial effect of coconut water was observed by Fu (1978, 1979) in leaf tissue culture of *Aranda, Ascocenda*

TABLE 2
Protocols Transferred by Indian Research Group

Orchid Species/Hybrid	Explants	Contact Scientist	Transferred to
Spathoglottis plicata	Seeds, protocorm, young seedlings	Dr. G. Bhowmik Guwahati University Guwahati	TERI, New Delhi
Vanilla planifolia	Root and shoot meristem tip	Dr. V. J. Philip University of Calicut Calicut	A. V. Thomas & Co., Cochin
Dendrobium (3 sp.)	Seeds	Dr. C. R. Raju COSTFORD Mavelikkara, Kerala	Private entrepreneurs
Phalaenopsis (3 sp.)	Meristems	Dr. C. R. Raju COSTFORD Mavelikkara, Kerala	Private entrepreneurs

Source: Govil and Gupta (2001).

and *Cattleya*. Fu (1978) also observed beneficial effect of coconut water and tomato juice in regeneration of *Phalaenopsis* from lateral buds. A treatment of CW (10 per cent) alone or in combination with BAP was reported to induce multiple shoot buds besides activating additional meristematic loci in the explants (Vij *et al,* 2000). Beneficial effect of CW was also reported by Kerbauy (1984), Tanaka *et al* (1988) and Sanchez (1988). Promotory effect of CW and banana pulp in plant development from protocorm was reported by Lopez *et al* (1994). Intuwong and Sagawa (1973) reported addition of CW, potato extract and homogenised banana to the media was essential for plantlet differentiation from floral primordia culture in three monopodial orchids. Beneficial effect of CW plus inositol and casein hydrolysate in profuse PLB formation was recorded by Hass-Von Schmude (1983). Similar effect of thiamine and peptone was recorded by Sanchez (1988) Survival of TS explants and production of PLBs were reported to be dependent on the amount of CW in the medium, maximum survival being observed at 20 per cent (Prakash *et al,* 1996).

Multiple shoot induction in *Dendrobium* shoot tip and axillary bud culture was reported to be influenced by thiamine Hcl, Pyridoxine Hcl, Nicotinic acid, Adenine sulphate and myo-inositol addition in the media (Devi and Laishram, 1998). Growth promotory effect of Biostimin (extract of active substances from *Aloe* sp.) on *in vitro* orchid culture was suggested by Kukulezanka (1985). He recorded that addition of Biostimin markedly increased fresh and dry weight of culture. Like BA it increased the number of protocorms and like auxin it stimulated the formation of leaf, shoots and roots. Beneficial effect of Fe EDTA in shoot and root growth and in reducing leaf browning was reported by Paek *et al* (1993).

Vij *et al* (1984, 1986) reported that yeast extract was essential for regeneration of *Aerides multiflorum, Papilionanthe teres* and *Satyrium nepalense,* whereas peptone was essential for regeneration of *Rhynchostylis* and *Vanda.*

TABLE 3
Commercial Farms Engaged in Orchid Tissue Culture

Commercial Tissue Culture Unit	Location	Orchid Species/Hybrids Micropropagated
Hafi Biotech	Kalamassery, Kerala	*Dendrobium, Phalaenopsis* and tropical orchids
A. V. Thomas & Co.	Kochy, Kerala	*Phalaenopsis, Vanilla*
Beena Nursery Pvt. Ltd.	Kochy, Kerala	Orchids
Bush, Boake Allen Ltd.	Chennai, Tamil Nadu	*Vanilla*
COSTFORD promoted unit	Thiruvanthapuram, Kerala	*Phalaenopsis, Dendrobium*
Harrisons Malayalam Ltd.	Ernakulum, Kerala	*Vanilla*
IAHS American Hybrid Seeds	Bangalore, Karnataka	*Vanilla* and other orchids
Kenibreed Plants	Kalimpong, West Bengal	*Cymbidium* and *Phalaenopsis*
Phytoclone Laboratories	Kottayam, Kerala	*Dendrobium*
Raj Agrobase Pvt. Ltd.	Pune, Maharashtra	Orchids
Rallis India	Bangalore, Karnataka	Orchids
SIV Industries Ltd.	Coimbatore, Tamil Nadu	Orchids
SPIC (Southern Petro-chemicals Ind. Corp. Ltd.)	Coimbatore, Tamil Nadu	*Vanilla* and other orchids
Ganesh Villa	Kalimpong, West Bengal	*Cymbidium*
U. C. Pradhan Orchid Labs.	Kalimpong, West Bengal	*Cattleya* and other orchids
Universal Nursery	Kalimpong, West Bengal	Orchid species
The Waysid Gardens	Tadong, Gangtok, Sikkim	*Cymbidium* and other orchids
Himalayan Flora & Biotech	Gangtok, Sikkim	*Cattleya* and other orchids
Hidden Forest	Gangtok, Sikkim	*Cymbidium* and other orchids

Source: Govil and Gupta (2001).

CULTURE CONDITION

Though a wide range of procedures adopting varied cultural conditions were developed for micropropagation of orchids, but in general, for initial PLB formation and multiplication an agitated liquid medium was found to be more effective. On the other hand, a solid medium was preferred for differentiation and plantlet development (Prakash *et al*, 1996). A liquid media for culture initiation and a solidified media for differentiation and plantlet development was used by Lim-Ho (1981) for regeneration of *Arachnis,* Goh (1973) for *Aranda*, Irawati *et al* (1977) for *Aranthera*, Scully (1967) for *Cattleya* and Teo *et al* (1973) for *Vanda.*

Prakash *et al* (1996) suggested that an incubation temperature of 25 ± 2° C and 16 hour photoperiod proved satisfactory for micropropagation of most of the commercially important orchid genera. Lim-Ho (1981) regenerated *Arachnis* from shoot tip and axillary bud by maintaining the cultures at 26° C with 16-hour photoperiod. Irawati *et al* (1977) and Cheah and Sagawa (1978),

TABLE 4
Successful Regeneration of Orchids Using Different Explants and Media

Orchid Species/Hybrid	Media	Explant	Reference
Acampe	—	Leaf	Nayak *et al* (1997)
Acampe rigida	Ichihashi & Yashimata	Leaf Tip	Yam & Weatherhead (1991)
Aerides maculosum	MS	Root	Kulkarni & Surwase (1998)
Aerides multiflorum	MS	Leaf segment	Vij & Pathak (1990)
Agrostophyllum myrianthum	Nitsch. Agar, water	Encapsulated protocorm	Devi *et al* (1998)
Arachnis	—	Apical bud	Teo and Teo (1974)
Arachnis hookeriana, *Arachnis* 'Capama' Merah	VW	Shoot tip, axillary bud	Lim-Ho (1981)
Aranda	VW	Inflorescence tip	Lam and Lee (1996)
Aranda	MS inorganic salt based	Leaf	Fu (1978, 1979)
Aranda	—	Leaf	Loh *et al* (1975)
Aranda	—	Apical bud	Teo and Teo (1974)
Aranda	—	Leaf	Loh *et al* (1975)
Aranda	—	Inflorescence	Goh & Wong (1990)
Aranda Deborah	White's	Shoot apices, axillary buds	Goh (1973)
Aranda Deborah	White's	Shoot tip, axillary bud	Goh (1973)
Aranda Deborah	Modified VW	Shoot tip	Prakash *et al* (1995)
Aranda hybrids	VW, KC	Shoot apices axillary buds	Goh (1990)
Aranthera	KC	Shoot tip	Irawati *et al* (1977)
Arundina bambosifolia	Raghavan & Torrey	Shoot tip	Nagaraju & Parthasarathy (1995)
Arundina bambusifolia		Shoot tip, Stem discs	Mitra (1971)
Ascocenda	MS inorganic salt based	Leaf	Fu (1978, 1979)
Ascofinetia	Liquid VW	Inflorescence	Intuwong & Sagawa (1973)
Bletilla striata	Dogashima (ND)	Vitrified embryos	Ishikawa *et al* (1997)
Bletilla striata	Modified MS	Root tip	Yam & Weatherhead (1991)
Bletilla striata	Modified Knop's	Node	Yam & Weatherhead (1990)

Continued...

...Continued

Orchid Species/Hybrid	Media	Explant	Reference
Bletilla striata	Modified Ichihashi	Pseudobulb segment	Vij and Dhiman (1997)
Bletilla striata	Modified MS	Root tip	Yam and Weatherhead (1991b)
Bulbophyllum careyanum	VW	Pseudobulb segment	Vij *et al* (2000)
C. fordii	Modified MS	Root tip	Yam & Weatherhead (1991)
Catasetum hybrid		Root tip	Kerbauy (1984a)
Catasetum sp.		Root	Holters & Zimmer (1991)
Cattleya	MS inorganic salt based	Leaf	Fu (1978, 1979)
Cattleya	Hyponex		Yoneda *et al* (1979)
Cattleya	—	Leaf	Champagnat *et al* (1970)
Cattleya	—	Apical bud	Teo and Teo (1974)
Cattleya	VW	Apical or lateral meristem	Scully (1967)
Cattleya	Liquid Heller's	Shoot tip, leaf tip	Churchill *et al* (1971)
Cattleya	—	Lateral bud	Reinert & Mohr (1967)
Cattleya	Heller's	Leaf base	Champagnat *et al* (1970)
Cattleya	VW	Shoots	Scully (1967)
Cleisostoma		Internodal sections	Kanika & Vij (1997)
Cleisostoma fordii	Modified MS	Root tip	Yam & Weatherhead (1991a)
Clowesia sp.	—	Root tip	Holters & Zimmer (1991)
Coelogyne	—	Leaf	Abdul Karim & Hairani (1990)
Cynoches sp.	—	Root tip	Holters & Zimmer (1991)
Cymbidium ensifolium	Half strength MS	Pseudobulb rhizome, root	Chang & Chang (1997)
Cymbidium longifolium	Nitsch, Agar water	Encapsulated protocorm	Devi *et al* (1998)
Cymbidium	—	Perianth segment	Kim & Kako (1984)
Cymbidium	—	Leaf	Wimber (1965)
Cymbidium	Modified Tsuchiya	Shoot tip	Wimber (1963)
Cymbidium aloifolium	MS	Shoot segments	Nayak *et al* (1997)

Continued...

...Continued

Orchid Species/Hybrid	Media	Explant	Reference
Cymbidium goeringii	MS	Apical flower buds	Shimasaki & Kemoto (1991)
Cymbidium hybrid	—	Immature flower bud	Laneri (1990)
Cymbidium hybrid	Liquid VW	Shoots	Sagawa *et al* (1966)
Cymbidium sp.	Hyponex, KC, MS	Rhizome of seed derived plant	Paek *et al* (1989)
Cypripedium macranthos var. *taiwanianum*	TT	Immature seed derived callus	Tomita & Tomita (1997)
Cyropodium punctatum	—	Root	Sanchez (1988)
Cyropodium sp.	—	Root tip	Holters and Zimmer (1991)
Dendrobium	MS	Apical meristem	Mujib & Jana (1994)
Dendrobium sp.	VW		Sobhana & Rajeevan (1993)
Dendrobium	VW	Inflorescence tip	Lam & Lee (1996)
Dendrobium moschatum	MS	Shoot segments	Nayak *et al* (1997)
Dendrobium	VW	Shoot meristem	Sharon & Vasundhara (1990)
Dendrobium cv. Sonia	VW, KC	Stem	Pathania *et al* (1998)
Dendrobium	VW	Bud, leaf axil	Kim *et al* (1970)
Dendrobium	VW	Shoot tip	Soediono (1983)
Dendrobium	VW, modified MS	Flower scape	Nuraini & Mohd. Sahib (1992)
Dendrobium	—	Leaf	Abdul Karim & Hairani (1990)
Dendrobium	—	Apical bud	Teo & Teo (1974)
Dendrobium	VW modified	Shoot tip	Sagawa & Shoji (1967)
Dendrobium	VW	Shoot tip	Kim *et al* (1970)
Dendrobium	Modified Knop's	Nodal segment	Arditti *et al* (1973)
Dendrobium	Modified VW	Flower stalk	Singh & Sagawat (1972)
Dendrobium moniliforme	—		Lin *et al* (1993)
Dendrobium aduncum	Modified Knop's	Node	Yam & Weatherhead (1990)
Dendrobium aphyllum	MS	Shoot segments	Nayak *et al* (1997)
Dendrobium crepidatum	—	Floral bud	Vij *et al* (1991)
Dendrobium hybrid	KC	Stem	Pathania *et al* (1998)
Dendrobium hybrids	VW, MS	Shoot tips, axillary buds	Devi & Laishram (1998)

Continued...

...Continued

Orchid Species/Hybrid	Media	Explant	Reference
Dendrobium Jonnie Ostehault	VW	Shoot tip	Sharon & Vasundhara (1990)
Dendrobium loddigessi	Modified Knop's	Node	Yam & Weatherhead (1990)
Dendrobium moschatum	—	Psuedobulb segment	Vij & Sood (1982)
Dendrobium pierardii	—	Floral bud	Vij *et al* (1991)
Dendrobium transparens	Modified Knop's	Node	Yam & Weatherhead (1990)
Dendrobium wardianum	—	Encapsulated PLBs	Sharma *et al* (1992)
Doritaenopsis	—	Internodal sections	Lin (1986)
Epidendrum	MS	Leaf tip	Churchill *et al* (1973)
Epidendrum	Modified liquid MS	Leaf tip, root tip	Churchill *et al* (1970, 1973)
Epidendrum × O'Brieniannum	—	Root tip	Steward & Button (1978)
Epidendrum × O'Brieniannum	—	Leaf tip	Rudolph *et al* (1972)
Galeandra sp.	—	Root tip	Holters & Zimmer (1991)
Laeliocattleya	Heller's	Leaf tip	Churchill *et al* (1973)
Liparis nervosa	Modified Knop	Node	Yam & Weatherhead (1990)
Liparis viridiflora	Heller's	Leaf tip	Yam & Weatherhead (1991)
Luisia trichorhiza	MS	Leaf segment	Vij & Pathak (1990)
Malaxis acuminata	Modified Knop	Node	Yam & Weatherhead (1990)
Mokara	VW	Inflorescence tip	Lam & Lee (1996)
Mokara	MS	Flower buds	Lim-Ho *et al* (1984)
Mormodes histrio		Root	Holters & Zimmer (1990)
Mormodes histrio	KC	Pseudobulb, nodal discs	Holters & Zimmer (1990)
Moromodes sp.	—	Root tip	Holters & Zimmer (1991)
Neofinetia falcata		Leaf	Vij & Patkhak (1990)
Neostylis	VW	Inflorescence	Intuwong & Sagawa (1973)
Oncidium	VW, MS	Shoot tip	Bagde *et al* (1996)
Oncidium	VW, Modified MS	Flower scape	Nuraini & Mohd. Sahib (1992)
Oncidium	—	Floral bud	Lim-Ho & Lee (1987)
Oncidium	—	Leaf	Abdul Karim & Hairani (1990)

Continued...

...*Continued*

Orchid Species/Hybrid	Media	Explant	Reference
Oncidium	—	Apical bud	Lim-Ho (1981)
Oncidium	KC	Inflorescence or shoot tip	Prakash *et al* (1996)
Oncidium hybrid	—	Inflorescence tip	Khaw *et al* (1978)
Oncidium papilio	Modified KC	Flower stalk	Fast (1973)
Oncidium varicosum	—	Root tip	Kerbauy (1984b)
Paphiopedilum	—	Leaf	Allenberg (1976)
Phaius tankervilleae	Raghavan & Torrey	Shoot tip	Nagaraju & Parthasarathy, (1995)
Phaius tankervileae	Nitsch, Agar water	Encapsulated protocorm	Devi *et al* (1998)
Phaius wallichii	VW		Nagabhushana (1982)
Phalaenopsis	VW		Ishii *et al* (1998)
Phalaenopsis hybrid	Half strength modified MS	Nodal cuttings	Jimenez & Guevara (1996)
Phalaenopsis	—	Stem	Duan *et al* (1996)
Phalaenopsis	VW	Lateral buds of flower stalk	Fu (1978)
Phalaenopsis	VW, Modified MS	Flower scape	Nuraini & Mohd. Sahib (1992)
Phalaenopsis	—	Flower stalk	Rotor (1949)
Phalaenopsis	—	Internodal sections	Lin (1986)
Phalaenopsis	—	Floral bud	Ichihashi (1992)
Phalaenopsis	—	Leaf	Latha & Seeni (1991)
Phalaenopsis	—	Leaf	Tanaka *et al* (1975)
Phalaenopsis	Modified VW	Flower stalk	Intuwong *et al* (1972)
Phalaenopsis	—	Root tip	Yoneda & Momose (1988)
Phalaenopsis amabilis	—	Root tip	Tanaka *et al* (1976)
Phalaenopsis hybrid	Modified KC	Leaf and root	Latha & Seeni (1991)
Pholidota chinensis	Heller's	Leaf tip	Yam & Weatherhead (1991)
Pholidota chinensis	Heller's	Leaf tip	Yam & Weatherhead (1991b)
Renanthera	—	Leaf	Seeni & Latha (1992)
Renanthera	—	Apical bud	Teo and Teo (1974)
Renanthera imschootiana	Nitsch, Agar water	Encapsulated protocorm	Devi *et al* (1998)

Continued...

...Continued

Orchid Species/Hybrid	Media	Explant	Reference
Rhynchostylis	—	Leaf	Chaturvedi & Sharma (1986)
Rhynchostylis gigantea	—	Inflorescence segments	Vij *et al* (1997)
Rhynchostylis retusa	—	Floral bud	Kaur & Vij (1995)
Rhynchostylis retusa	Modified VW	Root tip, leaf tip	Chaturvedi & Sharma (1986)
Rhynchostylis retusa	Mitra *et al*	Leaf segment	Vij *et al* (1984)
Rhynchostylis retusa	VW	Inflorescence segments	Kaur & Vij (1995)
Rhynchostylis retusa	Mitra *et al*	Root tip	Sood & Vij (1986)
Saccolabium pappillosum		Leaf	Kaur & Vij (2000)
Satyrium nepalense		Leaf	Vij & Pathak (1990)
Spathoglottis plicata	MS half strength	Leaf, stem	Teng *et al* (1997)
Spathoglottis plicata	MS	Encapsulated protocorms	Nayek *et al* (1997)
Spathoglottis plicata	VW		Nagabhushana (1982)
Spathoglottis plicata	MS, VW		Lim-Ho *et al* (1984)
Spathoglottis plicata, S. unguiculata	KC	Inflorescence, stalk bud, inflorescence tip	Prakash *et al* (1996)
Vanda		Leaf	Mathews & Rao (1985a, b)
Vanda		Leaf	Fu (1978, 1979)
Vanda		Leaf	Chaturvedi & Sharma (1986)
Vanda		Apical bud	Teo & Teo (1974)
Vanda		Leaf	Tanaka *et al* (1975)
Vanda	VW	Apical or axillary buds	Teo *et al* (1973)
Vanda	VW	Shoot tip	Kunisaki *et al* (1972)
Vanda	VW	Stem sections	Sagawa & Sehgal (1967)
Vanda	Modified liquid VW	Shoot tip	Teo *et al* (1973)
Vanda coerulea		Leaf	Seeni (1988)
Vanda cristata		Perianth segment	Vij & Sharma (1996)

Continued...

...Continued

Orchid Species/Hybrid	Media	Explant	Reference
Vanda cristata		Leaf	Vij & Pathak (1990)
Vanda hybrid (*Vanda* TMA × *V. teres)*	Modified VW	Leaf-tips, root tips	Chaturvedi & Sharma (1986)
Vanda 'Kasem's Delight'	Mitra *et al*	Young leaves	Vij *et al* (1994)
Vanda 'Miss Joaquim'	White's	Shoot tip, axillary bud, roots	Goh (1970)
Vanda testacea		Leaf	Vij & Pathak (1990)
Vanilla planifolia		Root tip	Philip & Nainar (1986, 1988)
Vanilla planifolia	Modified MS	Nodal stem segments	Kononowicz & Janik (1984)

however, recommended 24-hour photoperiod at 24-26° C for micropropagation of *Aranthera*. Rosa (1975) suggested that culture condition of 25° C and 12 hour day was ideal for inducing protocorm like protuberances from apical meristems. An environmental regime of 18 hours day, 28° C and 22° C night temperature was, however, suitable for production of new protocorms from them.

Tanaka *et al* (1998) reported that red light promoted leaf growth but decreased chlorophyll content in *in vitro* grown plantlets of *Cymbidium* cv. Sleeping Beauty 'Golden Bird'. However, blue was found to result in the reverse effect. They also reported growth enhancement by CO_2 enrichment. Heo *et al* (1996) reported that *Cymbidium* plantlets grown *in vitro* produced significantly greater fresh weight, dry weight, leaf area per plant and photosynthetic rate, when kept under conditions of high CO_2 and high light intensities. They reported that under such conditions, plants could be grown photo-autotropically. Lopez *et al* (1994) recorded a light intensity of 1500 lux resulted in best *in vitro* plant development in *Cattleya aurantiaca, Encyclia belizensis* subsp. p*arviflora, Oncidium leucochilum* and *Stanhopea intermedia*. Hew *et al* (1995) obtained increased growth in an *in vitro* propagated *Mokara* by using a novel method of CO_2 enrichment in an optimised photoautotrophic open system of culture. The optimisation process for the open system involved the manipulation of external CO_2 concentration (0.03 per cent, 1 per cent and 10 per cent), sucrose requirement, light intensities (80 and 200 $mol^{-2} s^{-1}$) and the venting of headspace ethylene from culture vessels.

Advantage of suspension medium over a semi-solid medium in orchid embryo culture was suggested by Singh and Prakash (1985). They suggested that suspension medium favoured better germination due to better aeration provided by suspension medium. Though a liquid media was good for culture initiation but a solidified medium for plantlet differentiation and further development seemed essential. Sharon and Vasundhara (1990) reported that shoot meristem of *Dendrobium* though it produced a profuse crop of callus mediated protocorms of liquid media, failed to develop further unless transferred to agar based media.

TABLE 5
Growth Regulating Chemicals Used in Orchid Micropropagation

Orchid Species/Hybrid	Growth Regulators	Reference
Spathoglottis plicata	BA, NAA	Teng *et al* (1997)
Dendrobium	BA, NAA, 2,4-D, GA_3	Mujib & Jana (1994)
Dendrobium sp.	BA, NAA	Sobhana & Rajeevan (1993)
Cypridedium macranthos var. *taiwanianum*	NAA, BAP	Tomita & Tomita (1997)
Spathoglottis plicata	NAA, BA	Nayak *et al* (1997)
Aranda	BA, NAA, 2,4-D, KN	Lam & Lee (1996)
Mokara	BA, NAA, 2,4-D, KN	Lam & Lee (1996)
Dendrobium	BA, NAA, 2,4-D, KN	Lam & Lee (1996)
Cymbidium aloifolium	BA, TDZ, NAA, IBA	Nayak *et al* (1997)
Dendrobium aphyllum	BA, TDZ, NAA, IBA	Nayak *et al* (1997)
Dendrobium moschatum	BA, TDZ, NAA, IBA	Nayak *et al* (1997)
Cymbidium ensifolium	2,4-D, Thidiazuron	Chang & Chang (1997)
Dendrobium cv Sonia	IBA, NAA, Paclobutrazol	Pathania *et al* (1998)
Agrostophyllum myrianthum	IAA	Devi *et al* (1998)
Cymbidium longifolium	IAA	Devi *et al* (1998)
Phaius tankervilliae	IAA	Devi *et al* (1998)
Renanthera imschootiana	IAA	Devi *et al* (1998)
Dendrobium hybrids	NAA, BAP	Devi & Laishram (1998)
Aerides maculosum	NA, 2-ip	Kulkarni & Surwase (1998)
Dendrobium		Kim *et al* (1970)
Dendrobium moniliforme	IBA, Kinetin	Lin *et al* (1993)
Bletilla striata	NAA, BA	Yam & Weatherhead (1991)
Cleisostoma fordii	NAA, BA	Yam & Weatherhead (1991)
Pholidota chinensis	NAA, BA	Yam & Weatherhead (1991)
Acampe rigida	NAA, BA	Yam & Weatherhead (1991)
Liparis viridiflora	BA	Yam & Weatherhead (1991)
Dendrobium	NAA	Soediono (1983)
Mormodes histrio	NAA, BA	Holters & Zimmer (1990)
Mormodos histrio	BA, NAA	Holters & Zimmer (1990)
Cymbidium sp.	KN, NAA	Paek *et al* (1989)
Cymbidium goeringii	BA, NAA	Shimasaki & Kemoto (1991)
Bletilla striata	BAP	Vij & Dhiman (1997)
Dendrobium moschatum	BAP, NAA	Vij & Sood (1982)

Continued...

...Continued

Orchid Species/Hybrid	Growth Regulators	Reference
Rhynchostylis gigantea		Vij *et al* (1997)
Dendrobium wardianum		Sharma *et al* (1992)
Cattleya	BA, GA	Yoneda *et al* (1979)
Onycidium	BA	Nuraini & Mohd. Sahib (1992)
Dendrobium	BA	Nuraini & Mohd. Sahib (1992)
Phalaenopsis	BA	Nuraini & Mohd. Sahib (1992)
Bulbophyllum careyanum	BAP, KN, NAA, IAA, BA, 2,4-D	Vij *et al* (2000)
Mokara	IAA, KN	Lim-Ho *et al* (1984)
Oncidium	BA	Prakash *et al* (1996)
Arachnis hookeriana	NAA	Lim-Ho (1981)
Spathoglottis plicata	IAA, KN	Lim-Ho *et al* (1984)
Spathoglottis plicata, *S. unguiculata*	BA	Prakash *et al* (1996)

Hew *et al* (1990) suggested use of polypropylene membrane or filter paper supported by glass wool in the liquid medium for culture of orchid callus. The practice simplifies the culture process and provides oxygen and nutrition to the cultures without agitation. Addition of glass wool as support for the plantlets in liquid media was also observed to enhance growth of *Phalaenopsis* seedlings (Ernst, 1975). Adelberg *et al* (1992) compared micropropagation of *Brassolaelio cattleya* and a *Laeliocattleya* hybrid on solid media gelled with agar to that on the media covered by polypropylene membrane (Celgard.). They recorded that proliferation on the membrane resulted in more harvestable plants than with agar medium. Membrane cultures did not require routine transfer to fresh medium, adding water alone to membrane cultures after three to five months was shown to accelerate growth. They reported that membrane method was time saving and showed 100 per cent acclimatisation in the greenhouse when shifted after eight months in culture.

Suitability of gellan gum over agar as gelling agent in *in vitro* culture of *Phalaenopsis* was suggested by Ishii *et al* (1998). Use of gellan gum as gelling agent was also made by Tomita and Tomita (1997) for plant regeneration from seed derived callus tissues of *Cypripedium macranthos* var. *taiwanianum*. Oddie *et al* (1994) reported better germination of *Elythranthera brunonis* and *Diuris longifolia* on media gelled with gelrite than medium gelled with agar. Phytagel (2.5 mg/l) solidified VW media was found effective for plantlet differentiation in *Vanda* (Teo *et al*, 1973).

Tanaka *et al* (1992) compared *in vitro* growth in different disposable culture bags, culture packs made from fluorocarbon polymer film, culture bags from polyethylene film with that in test tube. They suggested better suitability of the disposable bags based on their observation of a

TABLE 6
Complex Additives Used in Orchid Micropropagation

Orchid Species/Hybrid	Complex Additives	Reference
Phalaenopsis	Coconut water	Ishii *et al* (1998)
Phalaenopsis hybrid	Vitamins, adenine, coconut milk	Jimenez & Guevara (1996)
Oncidium	Coconut milk, casein hydrolysate	Bagde *et al* (1996)
Dendrobium	Coconut milk	Sharon & Vasundhara (1990)
Dendrobium hybrids	Thiamine Hcl, Pyridoxine Hcl, Nicotinic acid, adenine sulphate, myoinositol	Devi & Laishram (1998)
Dendrobium	Coconut water	Kim *et al* (1970)
Aranda	Coconut water	Fu (1978, 1979)
Ascocenda	Coconut water	Fu (1978, 1979)
Cattleya	Coconut water	Fu (1978, 1979)
Phalaenopsis	Coconut water, tomato juice	Fu (1978)
Dendrobium	Coconut water	Soediono (1983)
Bulbophyllum careyanum	Coconut water	Viz *et al* (2000)
Cattleya	Coconut water	Scully (1967)
Oncidium	Coconut water	Prakash *et al* (1996)
Arachnis hookeriana, *Arachnis* 'Capama' Merah	CW, banana homogenate, tomato juice	Lim-Ho (1981)
Aranda Deborah	Peptone	Goh (1973)
Aranthera	CW	Irawati *et al* (1977)
Spathoglottis plicata	CW, banana homogenate	Lim-Ho *et al* (1984)
Spathoglottis plicata, *S. unguiculata*	CW	Prakash *et al* (1996)
Vanda	CW	Teo *et al* (1973)

greatest number of leaves, shoot fresh weight, number of roots and root fresh weight of *Phalaenopsis* cultured in the disposable bags. In another study also they observed larger and more vigorous growth of *Cymbidium* and *Phalaenopsis* cultured in culture packs made from Neoflon Pfa or high density polyethylene film in comparison to flasks. Better performance of the culture bags was suggested due to higher light transmittance and better gas permeability.

Ionising radiation was also reported to have positive effect on *in vitro* culture of orchids. Kozlowska (1979) reported stimulatory effect of gamma rays on growth and development of protocorms and corresponding increased activities of gibberellin, auxin and cytokinin like substances.

Darkness and presence of graphite in the media were reported to induce more plantlets in leaf tip cultures of *Pholidota, Acampe* and *Liparis* (Tam and Weatherhead, 1991). Promotory effect

TABLE 7
Universities and Institutions Engaged in Orchid Tissue Culture Research in India

National Research Centre for Orchids
ICAR
Pakyong 737 106 (Sikkim)

Orchid Laboratory
Department of Botany
Punjab University
Chandigarh 160 014

Centre of Advanced Study
(Cell and Chromosome Research)
Department of Botany,
University of Calcutta
Kolkata (West Bengal)

Plant Biotechnology Division
Tropical Botanic Garden and Research Institute,
Palode, Thiruvanthapuram 696 562 (Kerala)

Division of Ornamental Crops
Indian Institute of Horticultural Research
Hassarghatta
Bangalore 560 089 (Karnataka)

Department of Agricultural Biotechnology/
Plant Breeding & Genetics
Assam Agricultural University
Jorhat 785 013 (Assam)

Tissue Culture Laboratory
Department of Botany
St. Anthony's College
Shillong 793 001 (Meghalaya)

Department of Pomology and Floriculture
College of Horticulture
Kerala Agricultural University
Trichur 680 656 (Kerala)

Division of Biotechnology
Institute of Himalayan Bioresource Technology,
Palampur 176 061 (Himachal Pradesh)

Department of Botany
School of Life Sciences
North-Eastern Hill University
Shillong 793 022 (Meghalaya)

Tissue Culture Laboratory
National Botanical Research Institute
Lucknow (Uttar Pradesh)

Department of Life Sciences
Manipur University
Imphal 795 003 (Manipur)

State Forest Research Institute
Van Vihar, Post Box No. 159
Itanagar 791 111 (Arunachal Pradesh)

Department of Biotechnology
Guwahati University
Guwahati 781 014 (Assam)

Orchid Laboratory
Department of Botany
Karnataka University
Dharwad 580 003 (Karnataka)

Plant Biotechnology Laboratory
R. J. College
Ghatkopar (W), Mumbai 400 086
(Maharashtra)

Biotechnology Laboratory
Division of Plant Breeding and Genetics
College of Agriculture
Central Agriculture University
Imphal 795 004 (Manipur)

Regional Plant Resource Centre
Nayapalli, Bhubaneshwar 751 005 (Orissa)

Department of Floriculture & Landscaping
Dr. Y. S. Parmar University of Horticulture and Forestry
Nauni 173 230 (Himachal Pradesh)

Continued...

...Continued

Department of Life Sciences University of Mumbai, Vidyanagari Santacruz (E), Mumbai 400 098 (Maharashtra)	Department of Botany Faculty of Science, M. S. University of Baroda Vadodara 390 002 (Gujarat)
Department of Life Sciences Shivaji University Kolhapur 416 064 (Maharashtra	Cytogenetics Laboratory Utkal University Bhubaneshwar 751 004 (Orissa)

of activated charcoal on rhizome growth *Cymbidium* culture was observed by Pack and Yeung (1991). Presence of activated charcoal in the basal media favoured best root development in plantlets of *Cymbidium pendulum* (Vij *et al*, 1994). Vij *et al* (2000) recorded that presence of AC in the media replaced callus development and led to direct shoot buds development. NAA was also found effective only in the presence of AC in the media. Growth enhancing effect of charcoal *Phalaenopsis* seedling was also recorded by Ernst (1975).

SUCROSE

Some of the commercially most useful *in vitro* procedures used sucrose at the rate of 20 g/l in the media. They were by Scully (1967) in *Cattleya;* Morel (1960), Sagawa *et al* (1966) and Wang *et al* (1981) in *Cymbidium;* and Teo *et al* (1973) in *Vanda.* A dose of 30 g/l of sucrose was used by Goh (1973) for PLB formation from shoot apices and axillary buds in *Aranda* Deborah. Fu (1978) also carried out plantlet production from PLBs induced from lateral buds of *Phalaenopsis* in the presence of 20 g/l sucrose in the media.

Presence of 40 g/l sucrose in the media, however, was found effective for callus induction in *Phalaenopsis*. In the absence of sucrose, PLBs were formed from calluses (Ishii *et al*, 1998). Root induction in *Dendrobium* apical meristem culture in the presence of 2 per cent sucrose was reported by Mujib and Jana (1994).

Heo *et al* (1996) suggested that presence of sucrose in the media could be avoided in *in vitro* culture of *Cymbidium* if maintained under enriched CO_2 and high photosynthetic photon flux density (PPFD). Intuwong and Sagawa (1973) reported a solid sucrose free medium was essential for plantlet formation from floral primordial in monopodials.

COMMERCIAL PROTOCOL

Among commercially important orchids, *Dendrobium* accounts for about 80 per cent of the total micropropagated tropical orchids, followed by *Mokara* and *Aranda* (10 per cent), *Oncidium* (5 per cent) and *Vanda*, *Cattleya*, etc (5 per cent). Among the temperate orchids, *Cymbidium* (50 per cent) and *Phalaenopsis* (40 per cent) constitute the bulk of total micropropagated temperate orchids (cf. Prakash *et al*, 1996; Debargh and Zimmerman, 1991).

Though several authors have developed a large number of procedures, only a few of them are suitable for direct adaptation for commercial purposes. Prakash *et al* (1996) reviewed the methods available for micropropagation of commercially important orchids. They suggested that a commercially suitable procedure should involve meristematic tissues, a simple media, least possible growth regulators and growth adjuncts supplementation, high and fast multiplication. They listed that methods proposed by Lim-Ho (1981) for *Arachnis*, Goh (1973, 1990), Loh *et al* (1975), Goh and Wong (1990) for *Aranda;* Irawati *et al* (1977), Cheah and Sagawa (1978) for *Aranthera;* Scully (1967) for *Cattleya*; Morel (1960), Sagawa *et al* (1966), Wang *et al* (1981) for *Cymbidium*; Sagawa and Shoji (1967) for *Dendrobium*; Lim-Ho *et al* (1984) for *Mokara;* Lim-Ho (1981), Khaw *et al* (1978) for *Oncidium;* and Teo *et al*, (1973) for *Vanda* may directly be adopted for commercial purposes.

REFERENCES

Abdul Karim, A. G. and Hairani, H. 'Leaf cutting of some Malaysian orchids'. In: Proc. *Int. Conf. and Exhibition on Orchids and Ornamental Plants.* (Ed.). Nair, H. Petaling Jeya, Malaysia: The Orchid Soc. Malaysia, 1990.

Adelberg, J., Desamero, N., Hale, A. and Young, R. 'Orchid micropropagation on polypropylene membranes'. *American Orchid Soc. Bull.* 61(7): 688-695, 1992.

Allenberg, H. 'Notizen zur keimung, meristemkultur and regeneration von Erdorchideen'. *Die Orchidee* 27: 28-31, 1976.

Arditti, J. 'Factors affecting the germination of orchid seeds'. *Bot. Rev.* 33: 1-97, 1967.

Arditti, J. 'Clonal propagation of orchids by means of tissue culture—a manual'. In: *Orchid Biology: Reviews and Literature.* (Ed.). Arditti, J. Ithaca and London: Cornell University Press, pp. 203-293, 1977.

Arditti, J. 'Clonal propagation of orchids by means of leaf culture *in vitro*'. *Orchid Rev.* 85(1006): 102-103, 1977a.

Arditti, J. 'Orchid seed germination and seedling culture'. In: *Orchid Biology: Reviews and Perspectives,* vol. 2. (Ed.). Arditti, J. Ithaca and London: Cornell University Press, pp. 244-370, 1982.

Arditti, J. and Ernst, R. 'Physiology of germinating orchid seeds'. In: *Orchid Biology: Reviews and Perspectives,* vol. 3. (Ed.). Arditti, J. Ithaca and London: Cornell University Press, pp. 177-322, 1984.

Arditti, J. and Ernst, R. In: *Micropropagation of Orchids.* New York, U.S.A.: John Wiley and Sons, Inc., 1993.

Arditti, J. and Kirkorian, A. D. 'Orchid micropropagation: the small path from laboratory to commercialisation and account of several unappreciated investigators'. *Bot. J. Linnean Soc.* 122: 183-241, 1996.

Arditti, J. and Pridgeon, A. M. (Eds.). In: *Orchid Biology: Reviews and Perspectives,* VII. Dordrecht, The Netherlands, Kluwer Academic Publishers, 1997.

Arditti, J., Healey, P. L. and Ernst, R. 'The role of mycorrhiza in nutrient uptake of orchids. II. Extra-cellular hydrolysis oligasaccharides by asymbiotic seedlings'. *Am. Orchid. Soc. Bull.* 47: 503-510, 1972.

Arditti, J., Michaud, J. D. and Oliva, A. P. 'Practical germination of North American and related orchids. I. *Epipactis atrorubens, E. gigantea* and *E. helleborine*'. *Am. Orchid Soc. Bull.* 51: 162-171, 1982a.

Arditti, J. M. A., Clements, G., Fast, G., Hadley, G., Nishimura, G. and Ernst, R. 'Orchid seed germination and seeding culture—a manual'. In: *Orchid Biology, Reviews and Perspectives,* vol. II. (Ed.). Arditti, J. Ithaca, New York, U.S.A.: Cornell University Press, pp. 243-370, 1982b.

Arditti, J., Michaud, J. D. and Oliva, A. P. 'Seed germination of North American orchids. I. Native California and related species of *Calypso, Epipactis, Goodyera, Piperia* and *Plantanthera*'. *Bot. Gaz.* 142: 442-453, 1981.

Arditti, J., Mosich, S. K. and Ball, E. A. 'Dendrobium node cultures: a new means of clonal propagation'. *Aust. Orchid Rev.* 38: 175-179, 1973.

Arora, S. K. Asymbiotic germination and micropropagation of some commercially viable and endangered orchids of India. Ph.D. Thesis. Panjab University, Chandigarh, India, 1990.

Bagde, Prasant; Sharon, Madhuri; Badge, P. and Sharon, M. '*In vitro* regeneration of *Oncidium* Gower Ramsey by high frequency protocorm-like bodies proliferation'. *Indian J. Pl. Physiology* 2(1): 10-14, 1997.

Barroso, J., Feveroiro, P., Oliveira, M. M. and Pais, M. S. S. '*In vitro* seed germination, differentiation production of mini tubers from *Ophrys lutea* Cav., *Ophrys fusca* Link. and *Ophrys speculum* Link'. *Scientia Hort.* 42: 329-337, 1990.

Beechey, N. 'Propagation of orchid aerial roots'. *Am. Orchid Soc. Bull.* 39: 1085-1088, 1970.

Bergman, F. J. *Amer. Hort.* 51: 20-23, 1972.

Bhaskar, J. and Rajeevan, P. K. 'Embryo culture of *Vanda* 'John Club.' *South Indian Horticulture* 44(1-2): 36-38, 1996.

Boris, H. 'Samenvermeehrung und unzucht europien Erdorchideen'. In: Proc. *2nd European Orchid Congr.*, pp. 74-78, 1969.

Burgeff, H. '*Samenkeimeng der Orchideen G. Fisher, Jena,* 1936.

Champagnat, M. 'Recherches sur la multiplication ve'getative de Neottia nidus-avis Rich.' *Ann. Sci. Natl. Bot. et Biol Ve'g.* 12: 209-247, 1971.

Champagnat, M., Morel, G. M. and Mounetou, B. 'La multiplication vegetative des Cattleya. A partirde de juenes feveilles cultivees aseptigusment *in vitro*'. *Ann. Sci. Nat. Botan. Biol. Vegetable* 11: 87-114, 1970.

Champagnat, M., Morel, G. M. and Gambade, G. 'Particularities morphologiques at pouvoir de regeneration de *Cymbidium virescens* cultive *in vitro*'. *Memoires Soc. Bot. France* 115: 235-244, 1968.

Chang, C. and Chang, W. C. 'Plant regeneration from callus culture of *Cymbidium ensifolium* var. *misericors*'. *Pl. Cell Reports* 17(4): 251-255, 1997.

Chaturvedi, H. C. and Sharma, A. K. 'Mericloning of orchids through culture of tips of leaves and roots'. In: *Biology, Conservation and Culture of Orchids.* (Ed.). Vij, S. P. New Delhi, India: Affiliated East West Press, pp. 469-472, 1986.

Cheah, K. T. and Sagawa, Y. '*In vitro* propagation of *Aranda* Wendy Scott and *Aranthera* James Storie'. *HortScience* 13: 660-662, 1978.

Chung, J. D., Chun, C. K. and Choi, S. K. 'Asymbiotic germination of *Cymbidium ensifolium.* II. Effect of several supplements to the medium, pH values and light and/or dark culture periods on rhizome'. *J. Kor. Soc. Hort. Sci.* 2: 186-192, 1985.

Churchill, M. E., Ball, E. A. and Arditti, J. 'Tissue culture of orchids. II. Methods for root tips. *Am. Orchid Soc. Bull.* 41: 726-730, 1972.

Churchill, M. E., Arditti, J. and Ball, E. A. 'Clonal propagation of orchids from leaf tips'. *Am. Orchid Soc. Bull.* 40: 109-113, 1971.

Churchill, M. E., Ball, E. A. and Arditti, J. 'Production of orchid plants from seedling leaf tips'. *Orchid Digest* 34: 271, 1970.

Churchill, M. E., Ball, E. A. and Arditti, J. 'Tissue culture of orchids. 1. Methods for leaf tips'. *New Physiologist* 72(1): 161-166, 1973.

Currah, R. S., Zelmer, C. D., Hambleton, S. and Richardson, K. A. 'Fungi from orchid mycorrhizas'. In: *Orchid Biology—Reviews and Perspectives,* vol. VII. (Eds.). Arditti, J. and Pridgeon, A. M. Dordrecht, The Netherlands: Kluwer Academic Publishers, pp. 117-170, 1997.

Curtis, J. T. 'Germination and seedling development in five species of *Cypripedium* L'. *Amer. J. Bot.* 30: 199-206, 1943.

Curtis, J. T. and Nichol, M. A. 'Culture of proliferating orchid embryos *in vitro*'. *Bull. Torrey Bot. Club* 75: 358-373, 1948.

Curtis, J. T. and Spoerl, E. 'Studies on the nitrogen nutrition of orchid embryos. II. Comparative utilisation of nitrate and ammonium nitrate'. *Amer. Orchid Soc. Bull.* 17: 111-114, 1948.

Dendsay, J. P. S. 'Activation of peroxidase by auxins in cotyledons of *Vigna radiata* L'. *Indian J. Exptl. Biol.* 27: 360-362, 1989.

DePauw, M. A., Remphrey, W. R. and Palmer, C. E. 'The cytokinin preference for *in vitro* germination and protocorm growth of *Cypripedium candidum*'. *Ann. Bot.* 75(3): 267-275, 1995.

Devi, J., Nath, M., Devi, M. and Deka, P. C. 'Effect of different media on germination and growth of some north-east Indian species of *Dendrobium*'. *J. Orchid Soc. India* 4(1-2): 45-49, 1990.

Devi, J., Ray, B. K., Chetia, S. and Deka, P. C. 'Regeneration of low temperature stored encapsulated protocorms of orchids'. *J. Orchid Soc. India* 12(1-2): 39-41, 1998.

Devi, S. Y. and Laishram, J. M. '*In vitro* propagation of *Dendrobium* hybrids through shoot tip and axillary bud culture'. *J. Orchid Soc. India* 12(1-2): 79-81, 1998.

Dodds, J. H. and Roberts, L. W. *Experiments in Plant Tissue Culture.* (Eds.). Dodds, J. H. and Roberts, L. W. U.S.A.: Cambridge University Press, 1987.

Downie, D. G. 'Notes on the germination of some British orchids'. *Trans. Proc. Bot. Soc. Edinb.* 33: 94-103, 1941.

Downie, D. G. 'The germination of *Listera ovata* L'. *Trans. Proc. Bot. Soc. Edinb.* 55: 126-130, 1949.

Downie, D. G. 'On the germination and growth of *Goodyera repens*'. *Trans. Proc. Bot. Soc. Edinb.* 33: 36-51, 1940.

Duan, J. X., Chen, H. and Yazawa, S. '*In vitro* propagation of *Phalaenopsis* via culture of cytokinin induced nodes'. *J. Plant Growth Regulation* 15(3): 133-137, 1996.

Ernst, R. 'The use of activated charcoal in asymbiotic seedling culture of *Paphiopedilum*'. *Amer. Orchid Soc. Bull.* 43: 35-38, 1974.

Ernst, R. 'Studies in asymbiotic culture of orchids'. *Am. Orchid Soc. Bull.* 44: 12-18, 1975.

Ernst, R. 'Charcoal or glass wool in asymbiotic culture of orchids'. In: Proc. *8th World Orchid Conf.* Palmengorten, Frankfurt, Germany, pp. 379-383, 1976.

Ernst, R. 'Seed and clonal propagation of *Phalaenopsis*'. In: *5th ASEAN Orchid Cong.* (Ed.). Rao, A. N. Singapore, pp. 31-41, 1985.

Fast, G. *Die Orchidee* 24: 240-246, 1973.

Fast, G. 'Die Vermehrung von Oncidium papilio duch Triebspitzen-kultur and Besprechung einiger Nahrmedian'. *Die Orchidee* 24: 240-246, 1973.

Fonnesbech, M. 'Growth hormones and propagation of *Cymbidium in vitro*'. *Physiol. Pl.* 27: 310-316, 1972.

Fu Fan May Lay. 'Studies on the tissue culture of orchids. I. Clonal propagation of *Phalaenopsis* by lateral buds from flower stems'. *Orchid Review* 86(1024): 308-310, 1978.

Fu Fan May Lay. 'Studies on the tissue culture of orchids. 2. Clonal propagation of *Aranda, Ascocenda, Cattleya* by leaf tissue culture'. *Orchid Review* 87(1037): 343-346, 1979.

Fu, F. M. L. 'Clonal propagation of *Aranda, Ascocenda, Cattleya* by leaf tissue culture'. *Gardens Bull.* 31(2): 132-138, 1978.

Gaya Suner, T. '*In vitro* cultivation of orchids'. *Horticultura Revista de Hortalizas Flores y Plantas Ornamtales* 108: 99-100, 1995.

Goh, C. J. 'Tissue culture of *Vanda* Miss Joaquim'. *J. Singapore Natl. Acad. Sci.* 2: 31-33, 1970.

Goh, C. J. 'Meristem culture of *Aranda* Debora'. *Malayan Orchid Rev.* 11: 10-14, 1973.

Goh, C. J. 'Orchids—monopodials'. In: *Handbook of Plant Cell Culture.* vol. 5. *Ornamental Species.* (Eds.). Ammirato, P. V., Evans, D. A., Sharp, W. R. and Bajaj, Y. P. S. New York, U.S.A.: McGraw Hill, pp. 598-637, 1990.

Goh, C. J. and Wong, P. F. 'Micropropagation of the monopodial orchid hybrid *Aranda* Deborah using inflorescence explants'. *Scientia Hortic.* 44: 315-321, 1990.

Goh, C. J., Sim, A. A. and Lim, G. 'Mycorrhizal associations in some tropical orchids'. *Lindleyana* 7(1): 13-17, 1992.

Govil, S., Gupta, S. C. and Govil, S. 'Commercialisation of plant tissue culture in India'. *Plant Cell Tissue and Organ Culture* 51(1): 65-73, 1997.

Govil, S. and Gupta, S. C. 'Plant tissue culture research and its commercialisation in India'. In: *Orchids, Science and Commerce.* (Eds.). Pathak, P. *et al.*, pp. 485-508, 2001.

Graeflinger, B. 'Repicagem Precoce de Orquideos sobre'. *Musceneas Orquidea* 12: 131-134, 1950.

Haas-Von Schumude, N. F. 'Klonale Massenvermehrung von Phalaenopsis'. *Die Orchidee* 34: 242-247, 1983.

Hadley, G. 'The interaction of kinetin, auxin and other factors in the development of north temperate orchids'. *New Phytol.* 69: 549-555, 1970.

Hadley, G. 'Cellulose as a carbon source for orchid mycorrhiza'. *New Phytol.* 68: 933-939, 1969.

Harrison, C. R. and Arditti, J. 'Growing orchids from seeds'. *Orchid Dig.* 34: 199-204, 1970.

Harvais, G. 'Notes on the biology of some native orchids of Thunder Bay. Their endophytes and symbionts'. *Can. J. Bot.* 52: 451-460, 1974.

Harvais, G. 'An improved culture for growing the orchid *Cypripedium reginae* axenically'. *Can. J. Bot.* 60: 2547-2555, 1982.

Harvais, G. 'The development and growth requirements of *Dactylorhiza purpurella* in asymbiotic cultures'. *Can. J. Bot.* 50: 451-460, 1972.

Hazarika, R. B. and Sarma, C. M. '*In vitro* germination and regeneration of *Dendrobium transparens* Lindl'. *J. Orchid Soc. India* 9(1-2): 51-54, 1995.

Hegarty, C. P. 'Observations on the germination of orchid seed'. *Am. Orchid Soc. Bull.* 27(7): 457-464, 1955.

Heller, R. 'Recherches sur la nutrition minerale des tissue vegetaux cultives *in vitro*'. *Ann. Sci. Natl. Biol. Veg.* 14: 1-223, 1953.

Hew, C. S., Hin, S. E., Yong, J. W. H., Gouk, S. S. and Tanaka, M. '*In vitro* CO_2 enrichment of CAM orchid plantlets'. *J. Hort. Sci.* 70(5): 721-736, 1995.

Hew, C. S., Chan, Y. S., Lee, Y. K. and Chia, T. F. 'Culture of orchid tissue on polypropylene membrane'. In: *Malayan Orchid Review* 24: 78-81, 1990.

Hew, C. S., Clifford, P. E. and Young, J. W. H. 'Aspects of carbon partitioning in tropical orchids'. *J. Orchid Soc. India* 10(1-2): 53-80, 1996.

Hey, G. L. and Hey, M. G. 'Raising rare orchids from seeds'. In: Proc. *5th World Orchid Conf.* (Ed.). De Garmo, L. R., pp. 35-38, 1966.

Holters, J. and Zimmer, K. 'Shoot regeneration from root tips or orchids in *in vitro* propagation of other orchids using the method developed for *Mormodes histrio*'. *Gartenbauwisenschaft* 56(3): 114-117, 1991.

Holters, J. and Zimmer, K. 'Shoot regeneration from root tips of orchids *in vitro*. II. Effect of shoot parts on root system and root length and width on regenerative ability of *Mormodes histrio* roots'. *Gartenbauwisenschaft* 55(4): 182-186, 1990.

Hoty, G. M. 'Seed germination made simple'. *Am. Orchid Soc. Bull.* 29: 597-599, 1960.

Ichihashi, S. 'Micropropagation of *Phalaenopsis* through the culture of lateral buds from young flower stalks'. *Lindleyana* 7: 208-215, 1992.

Ichihashi, S. 'Studies on the media for orchid seed germination. IV. Influence of the characteristics of some culture media on the growth of orchid seedlings'. *J. Japanese Soc. Hort. Sci.* 48(3): 345-352, 1979.

Intuwong, O. and Sagawa, Y. 'Clonal propagation of Sarcanthine orchids by aseptic culture of inflorescence'. *Am. Orchid Soc. Bull.* 42: 209-215, 1973.

Intuwong, O., Kunisaki, J. T. and Sagawa, Y. 'Vegetative propagation of *Phalaenopsis* by flower stalk cuttings'. *Na Okika O'Hawaii* 1: 13-18, 1972.

Irawati, Harjadi, S. S., Suseno, H. and Idris, S. 'Tissue culture of *Aranthera* James Storie'. *Orchid Rev.* 85: 138-142, 1977.

Ishii, Y., Takamura, T., Goi, M. and Tanaka, M. 'Callus induction and somatic embryogenesis of *Phalaenopsis*'. *Plant Cell Reports* 17(6-7): 445-450, 1998.

Israel, H. W. 'Production of *Dendrobium* seedlings by asceptic culture of excised ovularies'. *Am. Orchid Soc. Bull.* 32: 441-443, 1963.

Jimenez, V. M. and Guevara, E. 'Propagation *in vitro* of *Phalaenopsis* (Orchidaceae) through culture of sections from floral axes after flower stem senescence'. *Agronomia Costaricense* 20(1): 76-79, 1996.

Kanika and Vij, S. P. 'Root to shoot transformation in *Vanda coerulea* Griff. possibilities *in vitro*'. In: *Proc. Natl. Seminar on Developmental Biology and Commercialisation of Orchids*, Gangtok, Sikkim, 1997.

Kano, K. 'Studies on the media for orchid seed germination'. *Memoirs Faculty Agri. Kagawa Univ.* 20: 1-68, 1965.

Kaur, P. and Vij, S. P. 'Morphogenetic response of *Rhynchostylis retusa* Bl. inflorescence segments; a study *in vitro*'. *J. Orchid Soc. India* 9(1-2): 85-90, 1995.

Kaur, S., Sarma, C. M. and Kaur, S. 'Selection of best medium for *in vitro* propagation of *Dendrobium lindle* Yi Steud'. *Adv. Plant Sci.* 10(1): 1-5, 1997.

Kaur, S. and Vij, S. P. 'Micropropagation of *Saccolabium papillosum* lindl. through leaf segments'. *J. Orchid Soc. India* 14(1-2): 67-74, 2000.

Kerbauy, G. B. 'Regeneration of protocorm-like bodies through *in vitro* culture of root tips of *Catasetum* (Orchidaceae)'. *Z. Pflanzen Physiol.* 113: 287-291, 1984a.

Kerbauy, G. B. '*In vitro* flowering of *Oncidium varicosum* mericlones (Orchidaceae)'. *Plant Sci. Lett.* 35: 73-75, 1984b.

Kerbauy, G. B. and Handro. W. 'Culture of orchid embryos in liquid medium'. *Orchid Rev.* 89: 316-318, 1981.

Khaw, C. H., Wong, H. T. and Nair, H. 'Hormones in the nutrition of orchid tissues in mericloning'. *Malayan Orchid Rev.* 13: 60-65, 1978.

Kim, K. K. and Kakao, S. 'Studies on clonal propagation in the *Cymbidium* floral organ culture *in vitro*'. *J. Kor. Soc. Hort. Sci.* 25: 65-71, 1984.

Kim, K. K., Kunisaki, J. T. and Sagawa, Y. 'Shoot tip culture of *Dendrobium*'. *Am. Orchid Soc. Bull.* 39: 1077-1080, 1970.

Knudson, L. 'Non-symbiotic germination of orchid seeds'. *Bot. Gaz.* 73: 1-25, 1922.

Knudson, L. 'A new nutrient solution for germination of orchid seed'. *Am. Orchid Soc. Bull.* 15: 214-217, 1946.

Knudson, L. 'Germination of seeds of *Vanilla*'. *Am. J. Bot.* 37: 241-247, 1950.

Konoweiz, H. and Janick, J. '*In vitro* propagation of *Vanilla planifolia*'. *HortScience* 19: 58-59, 1984.

Kotomori, S. and Murashige, T. 'Some aspects of aseptic propagation of orchids'. *Am. Orchid Soc. Bull.* 34: 2184-2189, 1965.

Kozlowska, Kalizz, J. 'The influence of ionising radiation on biological activity of endogenous growth regulators in orchids (Cymbidium) in tissue culture'. *Acta Horticulturae* 91: 261-267, 1979.

Krsnik, Rasd, M. 'Peroxidase as a developmental marker in plant tissue culture'. *Int. J. Dev. Biol.* 35(3): 259-263, 1991.

Kukulczanka, K. 'Effect of Biostimin on the growth of meristematic tissue and protocorm formation of some orchids in *in vitro* culture'. *Acta Horticulturae* 167: 273-278, 1985.

Kulkarni, A. R. and Surwase, B. S. '*In vitro* propagation of *Aerides maculosum* Lindl.'. *J. Orchid Soc. India* 12(1-2): 47-51, 1998.

Kumaria, S., Chrungoo, N. K. and Tandon, P. 'Activities of some oxidative enzymes in axenic cultures of protocorms of *Cymbidium giganteum* Wall. as influenced by different growth regulators'. *J. Orchid Soc. India* 4(1-2): 37-44, 1990.

Kunisaki, J. T., Kim, H. H. and Sagawa, Y. 'Shoot tip culture of *Vanda*'. *Am. Orchid Soc. Bull.* 41:435-439, 1972.

Lam Chan, L. T. and Lee, S. M. 'Inflorescence culture of some tropical orchid varieties'. *Singapore J. Primary Industries* 24: 25-29, 1996.

Laneri, U. 'A somaclonal variant in *Cymbidium*'. *Acta Horticulturae* 280: 451-453, 1990.

Latha, P. G. and Seeni, S. 'Rapid multiplication of *Phalaenopsis* from leaf root tissues cultured *in vitro*'. *J. Orchid Soc. India* 5(1-2): 1-8, 1991.

Lawrence, D. and Arditti, J. 'A new medium for the germination of orchid seeds'. *Am. Orchid Soc. Bull.* 33: 766-768, 1964.

Letham, D. S. 'Regulators of cell division in plant tissues. XX. The cytokinins of coconut milk'. *Physiol. Pl.* 32: 66-70, 1974.

Lim, S. T., Wong, S. M. and Goh, C. J. 'Elimination of *Cymbidium* mosaic virus and *Odontoglossum* ring spot virus from orchids by meristem culture'. *Ann. App. Biol.* 122: 92-124, 1993.

Lim-Ho, C. L. 'Experimental findings of the tissue culture of orchid hybrids of the Singapore Botanic Gardens'. *Gard. Bull.* (Singapore) 34: 148-160, 1981.

Lim-Ho, C. L. and Teo-Lee, G. C. 'Clonal propagation of *Oncidium* from dormant buds of flower stalk'. *Malayan Orchid Rev.* 21: 48-52, 1987.

Lim-Ho, C. L., Teo-Lee, G. C. and Phua, L. K. 'Clonal propagation of orchids from flower buds'. In: Proc. *5th ASEAN Orchid Cong.* (Ed.). Rao, A. N. Singapore, pp. 98-101, 1984.

Lin, C. C. '*In vitro* culture of flower stalks internodes of *Phalaenopsis* and *Doritaenopsis*'. *Lindleyana* 1: 158-163, 1986.

Lin, Y. J., Chen, C. C., Yeh, F. T., Chiu, N. Y. and Tsay, H. S. 'Tissue culture of *Bletilla formosana*. I. The influence of seed maturity and pre-treatment on seed germination and seedling development'. *J. Agric. Res. China* 43(1): 40-50, 1994.

Lindel, R. W. 'Notes on germinating *Cypripedium* seed'. *Am. Orchid Soc. Bull.* 22: 195-197, 1953a.

Lindel, R. W. 'Notes on germinating *Cypripedium* seed'. *Am. Orchid Soc. Bull.* 22: 580-582, 1953b.

Linden, B. 'Two new methods of pre-treatment of seeds of northern orchids to improve germination in axenic culture'. *Annales Botanici Femici* 29(4): 305-313, 1992.

Loh, C. S., Rao, A. N. and Goh, C. J. 'Clonal propagation from leaves in the orchid *Aranda*'. *J. Singapore Natl. Acad. Sci.* 4: 97-99, 1975.

Lopez Villalobos, A., Mejia Munoz, J. and Sosa Moss, C. 'Tissue culture of orchids: *Cattleya, Encyclia, Oncidium* and *Stanhopea*. National Congr. *Ornamental Horticulture*, Chapingo, Mexico, 20-25 Mon. 1994. *Revista ChapingoSerie Horticultura* 1(1): 58-62, 1994.

Malmgren, S. 'Large scale asymbiotic propagation of *Cypripedium calceolus*—plant physiology from a surgeon's point of view'. *Botanic Gardens Micropropagation News* 1(5): 59-63, 1992.

Mathews, V. H. and Rao, P. S. '*In vitro* culture of *Vanda* hybrid (*Vanda* TMA × *Vanda* Miss Joaquim). I. Studies of protocorm explants'. Proc. *Indian Natl. Sci. Acad.* 51(1): 96-103, 1985a.

Mathews, V. H. and Rao, P. S. '*In vitro* culture of *Vanda* hybrid (*Vanda* TMA × *Vanda* Miss Joaquim). II. Studies on seedling explants'. Proc. *Indian Natl. Sci. Acad.* 51(4): 496-504, 1985b.

McIntyre, D. K., Veitch, G. J. and Wrigley, J. W. 'Australian terrestrial orchids from seeds. II. Improvements in techniques and further successes'. *Am. Orchid Soc. Bull.* 43(1): 52-53, 1974.

Mead, J. W. and Bullard, C. 'Effect of vitamins and nitrogen source on asymbiotic germination and development of *Orchis laxiflora* and *Ophrys sphaegodes*'. *New Phytol.* 74: 33-40, 1975.

Mead, J. W. and Bullard, C. 'Vitamins and nitrogen requirement of *Orchis laxiflora* Lank'. *New Phytol.* 83: 129-136, 1979.

Meyer, B. '*In vitro* multiplication of ornamental plants in the horticultural cooperative 'Blutenfreude" in Luckenwalde'. *Gartenbau* 35(6): 185-186, 1988.

Meyer, J. R. and Pelloux, A. 'Acao estimulante de does normonios de crescimento vegetal sobre germinacae de sementes a desenvlvimento de seedling de orquideas em mieo assimbiotico'. *O. Biologica* 14(4): 151-161, 1948.

Mitra, G. C. 'Studies of seeds, shoot tips and stem discs of an orchid grown in aseptic culture'. *Indian J. Exptl. Biol.* 9: 79-85, 1971.

Mitra, G. C. 'Some aspects of asymbiotic nutrition of orchid embryos'. *J. Orchid Soc. India* 1(1-2): 91-103, 1987.

Mitra, G. C., Prasad, R. N. and Roychowdhury, A. 'Inorganic salts and differentiation of protocorms in seed callus of an orchid and correlated changes in its free amino acid content'. *Indian J. Exp. Biol.* 14: 350-351, 1976.

Mitra, G. C. '*In vitro* culture of orchid seeds for obtaining seedlings'. In: *Biology, Conservation and Culture of Orchids.* (Ed.). Vij, S. P. New Delhi, India: Affiliated East West Press Pvt. Ltd., pp. 401-412, 1986.

Miyoshi, K. and Mii, M. 'Photohormone pre-treatment for the enhancement of seed germination and protocorm formation by the terrestrial orchid, *Calanthe discolor* (Orchidaceae), in asymbiotic culture'. *Scientia Hort.* 63(3-4): 263-267, 1995.

Monorama, P., Rao, A. N., Goh, C. J. and Loh, C. S. 'Leaf callus development and organogenesis in *Aranda* and *Dendrobium*'. Proc. 5th ASEAN Orchid Congress Seminar. (Ed.). Rao, A. N., pp. 102-109, 1984.

Moore, D. 'On growing orchids from seeds'. *Gardeners Chron.* 35: 549, 1849.

Morel, G. M. 'Producing virus-free *Cymbidiums*'. *Am. Orchid Soc. Bull. 29*: 495-497, 1960.

Morel, G. M. 'A new means of clonal propagation of orchids'. *Am. Orchid Soc. Bull.* 31: 473-477, 1964.

Mujib, A. and Jana, B. K. 'Clonal propagation of *Dendrobium* 'Madame Pompadour' through apical meristem culture'. *Adv. Pl. Sci.* 7(2): 340-346, 1994.

Murashige, T. and Skoog, F. 'A revised medium for rapid growth and bioassays with tobacco tissue culture'. *Physiol. Pl.* 15: 473-497, 1962.

Muriat, F. 'Action de l'acide nicotinique sur la germination at al development des embryos de *Cattleya*'. *Compt. Rend. Acad. Sci. Paris* 229: 1355-1357, 1949.

Nagaraju, V. and Parthasarthy, V. A. '*In vitro* propagation of Phaius and Bamboo orchid by shoot tip culture'. *Annals of Plant Physiology* 9(2): 102-104, 1995.

Nakamura, S. J. 'Nutritional conditions required for the non-symbiotic culture of an achlorophyllous orchid *Galeola septentrialis*'. *New Physiol.* 90: 701-715, 1982.

Nath, M., Debi, J., Borthakur, B., Sharma, J. and Deka, P. C. 'Embryo culture of *Rhynchostylis retusa* and *Vanda coerulea*'. *J. Orchid Soc. India* 5(1-2): 97-101, 1991.

Nayak, N. R., Patnaik, S. and Rath, S. P. 'Direct shoot regeneration from foliar explants of epiphytic orchid *Acampe praemorsa* (Roxb.) Blatter and McCann'. *Pl. Cell Rep.* 16(8): 583-586, 1997.

Nayak, N. R., Rath, S. P. and Patnaik, S. N. '*In vitro* propagation of *Spathoglottis plicata* BL. (Orchidaceae) using artificial seeds'. *Pl. Sci.* 10(1): 742, 1997.

Nayak, N. R., Rath, S. P. and Patnaik, Satyanarayana. '*In vitro* propagation of three epiphytic orchids, *Cymbidium aloifolium* (L.) Sw., *Dendrobium aphyllum* (Roxb.) Fisch. and *Dendrobium moschatum* (Buch-Han) Sw. Hmough Thidiazuron-induced high frequency shoot proliferation'. *Scientia Horticulturae* 71(3-4): 243-250, 1997.

Nimoto, D. H. and Sagawa, Y. 'Ovule development in *Dendrobium*'. *Am. Orchid Soc. Bull.* 30: 813-819, 1961.

Nuraini, T., Mohd.-Shaib, J., Wan Mohamad, W. O. (Ed.), Rosli Mohamad (Ed.) and Khoo Khay Chong. 'Micropropagation of orchids using scape nodes as the explant material'. *Acta Horticulturae* 292: 169-172, 1990.

Oddie, R. L. A., Dixon, K. W. and Mccomb, J. A. 'Influence of substrate on symbiotic and asymbiotic *in vitro* germination and seedling growth of two terrestrial orchids'. *Lindleyana* 9(3): 183-189, 1994.

Paek, K. Y. and Yeung, E. C. 'The effects of l-naphthaleneacetic acid and N6-benzyladenine on the growth of *Cymbidium forrestii in vitro*'. *Pl. Cell Tissue Organ Culture* 24(2): 65-71, 1991.

Paek, K. Y., Shim, G. B. and Kim, J. J. 'Exploitation of temperate Cymbidiums and establishment of a micropropagation system. I. Asymbiotic germination of temperate Cymbidiums and the effect of media and growth regulators on organogenesis'. *J. Korean Soc. Hort. Sci.* 30(3): 234-247, 1989.

Paek, K. Y., Park, J. D. and Shin, G. B. 'Growth response of several terrestrial orchids to Fe-EDTA level *in vitro*'. *Journal of the Korean Society for Horticultural Science* 34(4): 308-314, 1993.

Pages, P. D. 'Banana homogenate, coconut water, peptone and auxins as nutrient supplements in the *in vitro* culture of *Dendrobium* and *Phalaenopsis* ovules'. College Leguna, Philippines, *SERCA Bull.* No. 2, 1971.

Pathak, P., Mahant, K. C. and Gupta, A. '*In vitro* propagation as an aid to conservation of Indian orchids: seed culture'. In: *Orchids, Science and Commerce.* (Eds.). Pathak P. *et al.* Dehradun, India: Bishen Singh Mahendra Pal Singh, pp. 319-362, 2001.

Pathania, N. S., Sehgal, O. P., Paul, D. and Ditta, B. S. 'Studies on micropropagation of *Dendrobium* cv. Sonia'. *J. Orchid Soc. India* 12(1-2): 35-38, 1998.

Philip, V. J. and Nainar, S. A. Z. *Ann. Bot.* 61: 143-199, 1998.

Philip, V. J. and Nainar. *J. Plant Physiol.* 122: 211-215, 1986.

Pierik, R. L. M. and Steegmans, H. H. M. 'The effect of 6 benzylaminopurine on growth and development of *Cattleya* seedlings grown from unripe seeds'. *Z. Pfanzenphysiol.* 68: 228-234, 1972.

Prakash, L., Lee, C. L. and Goh, C. J. '*In vitro* propagation of commercial orchids: an assessment of current methodologies and development of a novel approach—thin section culture'. *J. Orchid Soc. India* 10(1-2): 31-41, 1996.

Prakash, L., Loh, C. S. and Goh, C. J. 'An *in vitro* method for rapid regeneration of a monopodial orchid hybrid *Aranda* Deborah using thin section culture'. *Plant Cell Rep.* 14: 510-514, 1995.

Prasad, R. N. and Mitra, G. C. 'Nutrient requirements for germination of seeds and development of protocorm and seedlings of *Cymbidium* in aseptic culture'. *Indian J. Exp. Biol.* 13: 123-126, 1975.

Raghavan, V. and Torrey, J. G. 'Inorganic nitrogen nutrition of the embryos of orchid *Cattleya*'. *Am. J. Bot.* 51: 264-274, 1964.

Raghavan, V. and Torrey, J. G. 'Inorganic nitrogen nutrition of the embryos of the orchid *Cattleya*'. *Am. J. Bot.* 50: 617, 1963.

Rao, A. N. and Avadhani, P. N. 'Some aspects of *in vitro* culture of *Vanda* seeds'. In: Proc. *4th World Orchid Conf.* Singapore: Straits Times Press, pp. 194-202, 1964.

Reinert, R. A. and Mohr, H. C. 'Propagation of *Cattleya* by tissue culture of lateral bud meristems'. Proc. *Am. Soc. Hort. Sci.* 91: 644-671, 1967.

Rosa M. Dalla and Laneri, U. 'Modification of nutrient solution for germination and growth *in vitro* of some cultivated orchids and for the vegetative propagation of *Cymbidium* cultivars'. *Am. Orchid Soc. Bull.* 46(9): 813-820, 1977.

Rosa, M. 'More rapid vegetative propagation of orchids by culture *in vitro*'. *Italia Agricola* 112(2): 120-125, 1975.

Rotor, G. 'A method for vegetative propagation of *Phalaenopsis* species and hybrids'. *Am. Orchid Soc. Bull.* 18: 738-739, 1949.

Rudolph, M. J., Ball, E. A. and Arditti, J. 'Tissue culture of orchids. III. Does Orthochlorophenoxyacetic acid select for or induce anthocyanin production'. *Am. Orchid Soc. Bull.* 41(12): 1074-1078, 1972.

Sagawa, Y. and Sehgal, O. P. 'Aseptic stem propagation of *Vanda* × Miss Joaquim'. *Bull. Pacific Orchid Soc.*, Hawaii 25: 17-18, 1967.

Sagawa, Y. and Shoji, T. 'Clonal propagation of *Dendrobium* through meristem culture'. *Am. Orchid Soc. Bull.* 36: 856-889, 1967.

Sagawa, Y., Shoji, T. and Shoji, T. 'Clonal propagation of Cymbidiums through shoot meristem culture'. *Am. Orchid Soc. Bull.* 35: 118-122, 1966.

Sagawa, Y. and Kunisaki, J. T. 'Clonal propagation of orchids by tissue culture'. *Plant Tissue Culture* 683: 1982.

Salisbury, E. J. In: *The Reproductive Capacity of Plants.* London, U.K.: G. Bell and Sons, Ltd., 1942.

Sanchez, M. I. 'Micropropagation of *Cyrtopodium* (Orchidaceae) through root tip culture'. *Lindleyana* 3: 93-96, 1988.

Scully, R. M. 'Orchid plant sales and breeding'. Proc. *6th World Orchid Conf.*, Sydney, 1971.

Scully, R. M. Jr. 'Aspects of meristem culture in the *Cattleya* alliance'. *Am. Orchid Soc. Bull.* 36: 103-108, 1967.

Seeni, S. 'Micropropagation of blue *Vanda* using leaf bases'. In: Proc. Natl. Seminar *Current Research Trends in Indian Orchids with Special Reference to Tissue Culture Technology.* Shillong: NEHU, 1988.

Seeni, S. and Latha, P. G. 'Foliar regeneration of the endangered Red Vanda, *Renanthera imschootiana* Rolf (Orchidaceae)'. *Plant Cell Tissue Organ Culture* 29(3): 167-172, 1992.

Sharma, A. P., Tandon, P. and Kumar, A. 'Regeneration of *Dendrobium wardianum* Worner (Orchidaceae) from synthetic seeds'. *Indian J. Exp. Biol.* 30: 747-748, 1992.

Sharma, J. In: *Orchids of India: Commercialisation and Conservation.* Delhi, India: Daya Publishing House, 1996.

Sharma, M. and Vij, S. P. 'Micropropagation of *Eulophia dabia* (D. Don) Hochr'. *Orchid News* 2(1): 11, 1986.

Sharma, S. K. and Tandon, P. 'Influence of growth regulators on asymbiotic germination and early seedling development of *Coelogyne punctulata* Lindl'. In: *Biology, Conservation and Culture of Orchids.* (Ed.). Vij, S. P. New Delhi, India: Affiliated East West Press Pvt. Ltd., pp. 441-451, 1986.

Sharma, S. K., Tandon, P. and Mishra, R. R. 'Vitamins as related to axenic seed germination and seedling growth of *Cymbidium elegans* Lindl. *Coelogyne punctulata* Lindl'. *J. Orchid Soc. India* 5(1-2): 25-28, 1991.

Sharon, M. and Vasundhara, G. 'Micropropagation of *Dendrobium* Joannie Ostenhault'. *J. Orchid Soc. India* 4(1-2): 145-148, 1990.

Shimasaki, K. and Kemoto, S. 'Rhizome induction and plant regeneration of *Cymbidium goeringii* from flower bud cultures *in vitro*'. *Plant Cell Tissue Organ Culture* 25(1); 49-52, 1991.

Shobana, A. and Rajeevan, P. K. '*In vitro* multiple shoot production in *Dendrobium* influenced by cytokinins'. *J. Ornamental Hort.* 1(2): 1-5, 1993.

Singh, F. '*In vitro* propagation of orchids: state of the art'. *J. Orchid Soc. India* 6(1-2): 11-14, 1992.

Singh, F. and Prakash, D. 'Suspension culture technique for the culture of orchid embryos'. *Gartenbanuissenschaft* 50(5): 236-238, 1985.

Singh, F. 'Micropropagation of orchids'. In: *Advances in Horticulture.* 12. *Ornamental Plants.* (Eds.). Chadha, K. L. and Bhattacharjee, S. K. New Delhi, India: Malhotra Publishing House, pp. 583-602, 1995.

Singh, H. and Sagawa, Y. 'Vegetative propagation of *Dendrobium* by flower stalk cuttings'. *Na Okika O Hawaii-Hawaii Orchid J.* 1: 19, 1972.

Soediono, N. 'Use of coconut water, NAA, 2,4-D and vitamins in shoot tip cultures of *Dendrobium* cv. Jacqueline Thomas 'White'. *Orchid Rev.* 91(1073): 86-87, 1983.

Sood, A. and Vij, S. P. '*In vitro* root segment culture of *Rhynchostylis retusa*'. In: *Biology, Conservation and Culture or Orchids.* (Ed.). Vij, S. P. New Delhi, India: Affiliated East West Press, pp. 463-468, 1986.

Stewart, J. and Button, J. 'Development of callus from *Epidendrum* root tips in culture *in vitro*'. *Am. Orchid Soc. Bull.* 47: 606-612, 1978.

Stoutamire, W. 'Terrestrial orchid seedlings'. In: *Orchids: A Scientific Survey.* (Ed.). Withner, C. L. New York, U.S.A.: Wiley Interscience, pp. 101-128, 1974.

Strauss, M. S. and Reisinger, D. M. 'Effect of NAA on seed germination'. *Am. Orchid Soc. Bull.* 45: 722-723, 1976.

Suryoninoto, M. and Sumaryo, L. 'Totipotency of the pollen tetrads of *Dendrobium* Tommy White'. In: Proc. *4th Asian Orchid Conf.* Manila: Council of Agriculture and Research Development, pp. 185-187, 1985.

Tanaka, M. 'Studies on the clonal propagation of *Phalaenopsis* through *in vitro* culture'. *Mem. Fac. Agric. Kagasa University* 49: 1-85, 1987.

Tanaka, M. and Sakanishi, Y. 'Clonal propagation of *Phalaenopsis* by leaf tissue culture'. In: Proc. *9th World Orchid Conf.* (Ed.). Sukshom Kashemsanta, M. R. Thailand: Amarin Press, pp. 215-221, 1977.

Tanaka, M. and Aakanishi, Y. 'Factors affecting the growth of *in vitro* cultured lateral buds from *Phalaenopsis* flower stalks. *Scientia Hort.* 8: 169-178, 1978.

Tanaka, M., Hasegawa, A. and Goi, M. 'Studies on the clonal propagation of monopodial orchids by tissue culture. I. Formation of protocorm-like bodies from leaf tissue of *Phalaenopsis* and *Vanda*'. *J. Japanese Soc. for Hort. Sci.* 44(1): 47-58, 1975.

Tanaka, M., Jayakawa, J., Nishimura, M. and Goi, N. 'Micropropagation of *Phalaenopsis* by leaf segment culture'. In: Proc. *3rd Asia Pacific Orchid Conf.* (Ed.). Shooter, R. Adelaide, South Australia: Lane Bros., Master Printers Pty. Ltd., pp. 6-13, 1989.

Tanaka, M., Kumara, M. and Goi, M. 'Optimal conditions for shoot production from *Phalaenopsis* flower stalk cuttings *in vitro*'. *Scientia Hort.* 35: 117-126, 1988.

Tanaka, M., Senda, Y. and Hasegawa, A. 'Plantlet formation in *Phalaenopsis*'. *Am. Orchid Soc. Bull.* 45: 1022-1024, 1976.

Tanaka, M., Ikada, M., Fukai, S., Goi, M., Hayashi, M. (Ed.) and Goto, E. 'Effect of different films used for film culture vessels on plantlet development of *Phalaenopsis* and *Cymbidium*'. *Acta Horticulturae* 319: 225-230, 1992.

Tanaka, M., Takamura, T., Watanabe, H., Endo, M., Yanagi, T. and Okamoto, K. '*In vitro* growth of *Cymbidium* plantlets cultured under superbright red and blue light-emitting diodes (LEDs)'. *Journal of Horticultural Science and Biotechnology* 73(1): 39-44, 1998.

Teng. W. L., Nicholson, L. and Teng, M. C. 'Micropropagation of *Spathoglottis plicata*'. *Plant Cell Reports* 16(12): 831-835, 1997.

Teng Whei Lan; Nicholson, L., Yu, L. Y. Y., Teng, W. L., Altman, A. (Ed.) and Ziv, M. 'Clonal propagation of *Spathoglottis plicata* from young plants'. *Acta Horticulturae* 447: 193-197, 1997.

Teo, C. K. H. and Teo, F. S. 'Clonal propagation of orchids by tissue culture'. *Planter* 50(580): 213-216, 1974.

Teo, C. K. H., Kunisaki, J. T. and Sagawa, Y. 'Clonal propagation of strap leafed *Vanda* by shoot tip culture'. *Am. Orchid Soc. Bull.* 42: 402-405, 1973.

Tomita, M. and Tomita, M. 'Plant regeneration from immature seed-derived callus of *Cypripedium macranthos* Swartz var. *taiwanianum* (Masamune) F. Maekwa'. *Breeding Sci.* 47(3): 279-281, 287, 1997.

Vacin, E. F. and Went, F. W. 'Some pH changes in nutrient solution'. *Bot. Gaz.* 110: 605-913, 1949.

Vajrabhaya, M. 'Tissue culture of dormant buds from *Cattleya* back bulbs'. *Orchid Rev.* 86: 256-257, 1978.

Vajrabhaya, M. and Vajrabhaya, T. 'Tissue culture of *Rhynchostylis gigantea,* a monopodial orchid'. *Am. Orchid Soc. Bull.* 39: 907-910, 1970.

Vajrabhaya, M., Supaokit, S. and Vajrabhaya, T. 'A simple medium for orchid seedlings'. In: Proc. *Nagoya Int. Orchid Show.* (Eds.). Ichihashi, S. and Nagata, N. Nagoya, Japan: The Organising Committee, NIOC 94, pp. 82-85, 1994.

Vajrabhaya, T. 'Variations in clonal propagation'. In: *Orchid Biology: Reviews and Perspectives.* vol. I. (Ed.). Arditti, J. Ithaca, New York, U.S.A.: Cornell University Press, pp. 176-201, 1977.

Valmayor, H. L. and Sagawa, Y. 'Ovule culture in some orchids'. *Am. Orchid Soc. Bull.* 36: 766-769, 1967.

Van Standen, J. and Drewes, S. E. 'Identification of Zeatin and Zeatin riboside in coconut milk'. *Physiol Pl.* 34: 106-109, 1975.

Vejsadova, H. and Mala, M. 'Seed germination of some endangered terrestrial orchids under aseptic conditions'. *Acta Pruhoniciana* 63: 77-84, 1996.

Vij, S. P. and Pathak, P. 'Asymbiotic germination of the saprophytic orchid, *Cymbidium macrorhizon:* a study *in vitro*'. *J. Orchid Soc. India* 2(1-2), 25-32, 1988a.

Vij, S. P. and Pathak, P. 'Green pod culture of *Pholidota articulate* Lindl'. In: Proc. *National Seminar on Current Research Trends in Indian Orchids with Special Reference to Tissue Culture Technology.* (Eds.). Vij, S. P. and Khullar, S. P. Chandigarh, India: The Orchid Society of India, 1988b.

Vij, S. P., Sood, A. and Plaha, K. K. '*In vitro* seed germination of some epiphytic orchids'. In: *Contemporary Trends in Plant Sciences.* (Ed.). Verma, S. C. 'New Delhi, India: Kalyani Publishers, pp. 473-481, 1981.

Vij, S. P., Sharma, M. and Shekhar, N. 'Asymbiotic seed generation of therapeutically important orchids: Genus Habenaria Willd., a study *in vitro*'. In: *Indigenous Medicinal Plants Including Microbes and Fungi.* (Ed.). Kaushik P. New Delhi, India: Today and Tomorrow's Printers and Publishers, pp. 71-77, 1988c.

Vij, S. P., Shekhar, N. and Sushma. '*In vitro* asymbiotic seed germination in *Pachystoma senile* Reichb.f (Orchidaceae)'. In: National Seminar on *Reproductive Biology of Plants Including Endangered Species.* Meerut, India, 1985.

Vij, S. P., Pathak, P. and Kaur, H. 'Split-pod culture'. In: National Seminar on *A Decade of Orchid Research and Development.* (Ed.). Vij, S. P. Chandigarh, India: The Orchid Society of India, 1994.

Vij, S. P. (Ed.). *Biology, Conservation and Culture of Orchids.* New Delhi, India: Affiliated East West Press, 1986.

Vij, S. P. 'Regeneration response of orchid roots: a study *in vitro*'. *J. Orchid Soc. India* 7: 61-72, 1993.

Vij, S. P. 'Genetic resources in orchids'. In: *Advances in Horticulture.* 12. *Ornamental Plants.* (Eds). Chadda, K. L. and Bhattacharjee, S. K. New Delhi, India: Malhotra Publishing House, pp. 153-181, 1995.

Vij, S. P. and Dhiman, A. '*In vitro* inflorescence segment culture in *Malaxis acuminata* D. Don'. In: Proc. Natl. Sem. *Developmental Biology and Commercialisation of Orchids.* Gangtok, Sikkim, 1997.

Vij, S. P. and Pathak, P. 'Micropropagation of *Dendrobium* chrysanthemum through pseudobulb segment'. *J. Orchid Soc. India* 3: 25-28, 1989.

Vij, S. P. and Pathak, P. 'Micropropagation of orchids through leaf segments'. *J. Orchid Soc. India* 4(1-2): 69-88, 1990.

Vij, S. P. and Sood, A. '*In vitro* pseudobulb segment culture, a means for rapid clonal propagation of *Dendrobium moschatum* (Orchidaceae)'. In: Proc. Natl. Sem. *Development and Comparative Aspects of Plant Structure and Function.* Hyderabad, 1982.

Vij, S. P., Pathak, P. and Sharma, M. 'On the regeneration potential of *Rhynchostylis retusa* root segments—a study *in vitro*'. *J. Orchid Soc. India* 1: 71-74, 1987.

Vij, S. P., Sharma, V. and Kaur, S. 'Foliar explant and orchid micropropagation: *Vanda* Kasem Delight 'TomBoykin'. *J. Orchid Soc. India* 8(1-2): 79-83, 1994.

Vij, S. P., Sood, A. and Plaha, K. K. 'Propagation of *Rhynchostylis retusa* Bl. (Orchidaceae) by direct organogenesis from leaf segments culture'. *Bot. Gaz.* 145: 210-214, 1984.

Vij, S. P., Sood, A. and Plaha, K. K. '*In vitro* breakdown of apical dormancy and development of vegetative shoots from inflorescence segments in *Saccolabium calceolane* Lindl'. In: *Biology, Conservation and Culture of Orchids.* (Ed.). Vij, S. P. New Delhi, India: Affiliated East West Press, pp. 473-478, 1986.

Vij, S. P., Sood, A. and Sharma, M. '*In vitro* leaf segment culture of *Vanda testcea* (Lindl.) Reichb. f. (*V. parviflora* Lindl.) (Orchidaceae)'. *Curr. Sci.* 55(21): 1100-1101, 1986.

Vij, S. P., Sood, A. and Sharma, M. 'Morphogenetic response of floral buds of *Dendrobium:* A study *in vitro*'. In: Proc. Natl. Sem. *Biology Improvement, Propagation and Commercialisation of Indian Orchids.* Bangalore, pp. 40-41, 1991.

Vij, S. P. and Dhiman, A. 'Regenerative competence of *Bletilla striata* pseudobulb segments: a study *in vitro*'. *J. Orchid Soc. India* 11(1-2): 93-97, 1997.

Vij, S. P. and Kaur, S. 'Micropropagation of therapeutically important orchids: *Malaxis acuminata* D. Don'. *J. Orchid Soc.* India, 1998.

Vij, S. P. and Pathak, P. 'Micropropagation of orchids through leaf segments'. *J. Orchid Soc. India* 4(1-2): 69-88, 1990.

Vij, S. P. and Sharma, V. 'Regenerative competence of *Vanda cristata* perianth segments: a study *in vitro*'. *J. Orchid Soc. India* 10(1-2): 25-29, 1996.

Vij, S. P., Kher, A. and Pathak, A. 'Regeneration competence of *Bulbophyllum careyanum* (Hook.) Spreng. pseudobulb segments'. *J. Orchid Soc. India* 14(1-2): 47-55, 2000.

Waes, J. M. and Debergh, P. C. '*In vitro* germination of some western European orchids'. *Physiologia Plantarum* 67(2): 253-261, 1986.

Waes, J. M. Van; Debergh, P. C. and Van Waes, J. M. 'Adaptation of tetrazolium method for testing the seed viability and scanning electron microscopy study of some western European orchids'. *Physiologia Plantarum* 66(3): 435-442, 1986.

Wang, H. 'Rapid clonal propagation of *Phalaenopsis* by tissue culture'. *Acta Hort. Sin.* 16(1): 73-77, 1989.

Wang, X., Chen, J. C., Liu, G. Y., Gui, M. X. and Bas, C. H. 'Clonal propaganda of orchids by means of tissue culture'. *Acta Physiol. Sin.* 10: 391-396, 1981.

Weatherhead, M. A., Zee, S. Y. and Baretto, G. 'Some observations on the early stages of development of *Eulophia yushuiana* Men'. *Hong Kong Natl. Hist. Sci.* 117: 85-90, 1986.

Weatherhead, M. A. and Harberd, D. J. 'Micropropagation of *Pleione*'. *Hort. Res.* 20: 87-89, 1980.

Weatherhead, M. A., Burdon, J. and Henshaw, G. G. 'Some effects of charcoal as an additive to plant tissue culture media'. *Z. Pflanzenphysiol.* 89: 141-147, 1978.

Weatherhead, M. A., Burdon, J. and Henshaw, G. G. 'Effect of activated charcoal as an additive to plant tissue culture media, II'. *Z. Pflanzenphysiol.* 94: 399-405, 1979.

White, P. R. *A Handbook of Plant Tissue Culture.* Lancaster, Penn., U.S.A.: Jaques Cattel Press, 1943.

Wimber, D. E. 'Clonal multiplication of *Cymbidium* though tissue culture of shoot meristem'. *Am. Orchid Soc. Bull.* 32: 105-107, 1963.

Wimber, D. E. 'Additional observations on clonal multiplication of *Cymbidium* through tissue culture of the shoot meristem'. *Cymbidium Soc. News* 20: 7-10, 1965.

Withner, C. L. 'Orchid physiology'. In: *Orchids: A Scientific Survey.* (Ed.). Withner, C. L. London, U.K.: Ronald Press, pp. 315-360, 1959.

Withner, C. L. Effect of plant hormones and other compounds on the growth or orchids'. *Am. Orchid Soc. Bull.* 20: 276-278, 1951.

Yam, T. W. 'Conservation, ecology and propagation of wild orchids of Hong Kong. Ph.D. Diss. University of Hong Kong, Department of Botany, Hong Kong, 1989.

Yam, T. W. and Weatherhead, M. A. 'Leaf tip culture of several native orchids of Hong Kong'. *Lindleyana* 6: 147-150, 1991a.

Yam, T. W., Arditti, J. and Weatherhead, M. A. 'The use of darkening agents in seed germination and tissue culture media for orchids. A review'. *J. Orchid Soc. India* 3: 35-39, 1989.

Yam, T. W. and Weatherhead, M. A. 'Germination and seedling development of some Hong Kong orchids'. *Lindleyana* 3: 156-160, 1988.

Yam, T. W. and Weatherhead, M. A. 'Nodal culture of some native orchids of Hong Kong'. *Lindleyana* 5(4): 218-223, 1990.

Yam, T. W. and Weatherhead, M. A. 'Root tip cultures of several native orchids of Hong Kong'. *Lindleyana* 6(3): 151-153, 1991b.

Yates, R. and Curtis, J. T. 'The effect of sucrose and other factors on the shoot ratio of orchid seedlings'. *Am. J. Bot.* 36: 390-396, 1949.

Yoneda, K. and Momose, H. *Bull. Coll. Agric. Vet. Nihon. Univ.* 9: 191-196, 1988.

Yoneda, K., Sakamoto, T. and Sasaki, H. 'Studies on mericlones of orchids. I. Propagation of *Cattleya* protocorms by means of meristem culture'. *Bull. College of Agric. Vet. Medicine, Nihon University* 36: 64-73, 1979.

Zimmerman, R. H. 'Commercial application of tissue culture to horticultural corps in the United States'. *J. Korean Soc. Hort. Sci.* 37(3): 486-490, 1996.

—oo(O)oo—

5

GREENHOUSES AND GLASSHOUSES FOR ORNAMENTAL CROPS UNDER DIFFERENT CLIMATIC CONDITIONS

PITAM CHANDRA AND M. J. GUPTA

INTRODUCTION

ANYTHING in the landscape that does not have to be mowed is termed an ornamental plant (*Land and Landscape Digest*). Obviously, this covers a wide variety of plant material including trees, shrubs, flowers, vines, ground covers, etc., all or any of which can play an important functional and beautifying role in a particular landscape. Ornamentals, by definition, do not bear edible fruits, nuts or berries. Ornamental crops should primarily be seen in the light of cash crops. Though ornamental crops have increasingly become important as an export commodity, they are at the same time, becoming important consumer items on local markets—due largely to the expansion of many densely populated urban centres in the country, where the availability of cut flowers and ornamental plants is being appreciated more and more for the family and social environment, as natural greenery becomes more rare and remote. Cultivation of ornamental plants at a given location in the open is constrained by local climatic conditions.

There is a technology, which has developed over the last two centuries, which promises control over all the climatic parameters for commercial ornamental crop production. The obvious reference is to greenhouse technology. A greenhouse is a framed or inflated structure covered with a transparent or a translucent material such that crops could be grown in it under at least partially controlled environmental conditions and which is large enough to permit a person to enter into it to carry out crop cultural practices (Dalrymple, 1973). Glass had been the most popular covering material for greenhouses until the development of plastic films in the 1950s. Therefore, greenhouses were generally called glasshouses. However, the widespread usage of plastic films and sheets for greenhouse glazing gave rise to such terms as polyhouse, polycarbonate house and fibreglass house. All these plant-growing structures are essentially greenhouses.

A glasshouse, due to the heavy weight of glass, requires a stronger and heavier frame. A greenhouse glazed with plastic materials, however, is structurally much lighter. Resultantly, a plastic greenhouse construction cost can be about an order of magnitude lower than that of a glasshouse. In terms of economics of greenhouse operations, therefore, plastic greenhouses could be more attractive. Plastic greenhouses permit better environmental control due to their inherent air tightness. Glass has superior solar transmission properties over long life and, therefore, glass may still be the right choice for places having lower sunshine levels. While full advantage is taken of the available sunshine for crop production in a greenhouse by way of selecting proper covering material, the enclosure provides the opportunity to control the other environmental parameters such as temperature, carbon dioxide, humidity, and nature of the root medium also. As a result, greenhouse crop productivity is largely independent of outdoor environmental conditions and the availability of ornamental crops could be maintained all the year round. The possibilities offered by protected cultivation of ornamental crops in India could represent one of the favourable elements in the general development of our nation.

The prospects of climatic control permit the raising of plants anywhere at any time of the year. Further, crop productivity is at the maximum on a per unit area, per unit volume and per unit input bases. Microclimate control also implies superior quality of produce, free from pathogens, insect damage and chemical residues. These basic advantages translate into the following specific benefits (Chandra, 1994).

- Crop cultivation under inclement climatic conditions.
- Certain crops cultivated all the year round to meet the market demands.
- High value and high quality crops grown for export markets.
- Income from small landholdings increases several-fold.
- Successful nurseries from seeds or by vegetative propagation prepared as and when necessary.
- More self-employment opportunities for educated youth on the farm.
- Manipulation of microclimate and insect proof features of the greenhouse for plant breeding and, thus, the evolution of new varieties and production of seeds.

HISTORY OF GREENHOUSE DEVELOPMENT

Agriculture is as old as human civilisation. In the beginning, the human population was small and Mother Nature produced enough. Therefore, man only needed to gather food and consume it. As the human population grew and food preferences developed, the practices of crop cultivation began. The concepts of seed, tillage, harvesting and storage took shape. Appropriate tools were developed; need for irrigation, weeding and plant protection was realised and relevant practices were developed. It did not take very long to realise that human food needed to include a variety of natural products to meet the body requirements for healthy survival. Early human settlements were confined to those regions where year-round food availability, either from nature or from cultivation, was ensured. When man moved to harsher climates, the food habits and cultivation practices got modified. Development of greenhouse technology is an example of mankind's efforts to ensure its survival under natural adversities. The technology also indicated the immense potential for the conservation and cultivation of ornamental plants for aesthetic purposes.

Plato in the fourth century BC indicated in his *Phadeon* that plants were grown under protection. The Roman emperor, Tiberius Caesar, who ruled during AD 1437 was advised by his doctors to eat one cucumber every day. Consequently, efforts were made to grow cucumbers the year round. Movable beds were placed in the open sun on favourable days. On cold days, the beds were covered with transparent stones, such as specularia, to protect the plants (Pliny, trans. H. Rackhan, 1961).

In one of the publications, which appeared in London in 1597, use of insulated beds containing well-manured soil and covered with mats, oil painted cloth or straw was demonstrated for early forcing of cucumbers. Claude Mollet, gardener to Louis XIII, who ruled from 1610 to 1643, was probably the first to mention the use of frames that were heated with manure and covered with glass panes. It appears that before glass frames came into existence, bell jar coverings and lanterns without bottoms were used to cover plants. Stoves for heating the frames appeared in the late 17th century. Hot water and steam for keeping greenhouses warm became popular in the late 19th century.

Several references of vegetable forcing in England and France are available from the early 18th century. A Swedish horticulturist, Kalm, who visited England in 1748, described some gardens where the beds were covered with glass frames, which could be taken off at will. Beans, peas, cabbage, leeks, chives, radishes, lettuce, asparagus and spinach were available in London in the months of May that year. Early vegetables from forcing frames in France became available around 1780. Literature on forcing of vegetables began appearing in the late 18th century. Initially, the forcing techniques were limited to the private estates and the produce was available only to the wealthy. It was only after the mid-19th century that the greenhouse products such as grapes, melons, peaches and strawberries became available to the general public. By the end of the 19th century, commercial greenhouse cultivation was well established. Greenhouse technology in its present form is more than two centuries old. The eighteenth century witnessed rapid development in greenhouse technology in Europe. The technology then reached the North American shores across the Atlantic. Development of transparent plastic triggered a new phase of greenhouse development after the second world war. More than 90 per cent of all new greenhouses constructed in the world now use plastic films and sheets as glazing material. The status of greenhouse use worldwide is summarised in Table 1 (Jensen and Malter, 1994).

GREENHOUSE EFFECT

A greenhouse is covered with a transparent or a translucent material such as glass or plastic. Depending upon its transparency, the greenhouse cover admits a major fraction of sunlight. The crop, floor and other objects in the greenhouse absorb the sunlight admitted to the greenhouse. These objects in the greenhouse in turn emit long wave thermal radiation for which the cover material has lower transparency. As a result, the solar energy is trapped in the greenhouse raising its temperature. This phenomenon is generally known as *greenhouse effect*. It is this natural rise in the greenhouse air temperature that is utilised under cold climates to grow successful crops. The same natural phenomenon during summers, however, requires greenhouse cooling to maintain favourable temperatures (Chandra, 2000).

Greenhouse applications for ornamentals in India can be divided into two categories. The first category of greenhouse use is in the natural ornamentals growing areas to prevent damage from weather aberrations and other natural agents. This category of greenhouses does not have any environmental control systems except the provision of adequate ventilation. The second category of greenhouses for ornamentals is that which helps to extend the growing season or permit off-season production by way of controlling micro-climatic parameters to a significant extent. The microclimate refers to light, temperature, humidity, carbon dioxide level and nature of root medium. The extent of micro-climatic control in a given situation depends on the local climate and the crop

TABLE 1
Greenhouse Area in Different Countries

No.	Country	Area in ha.	No.	Country	Area in ha.
1.	Algeria	5,000	15.	Israel	2,200
2.	Australia	600	16.	Italy	18,500
3.	Belgium	2,400	17.	Japan	42,000
4.	Bulgaria	1,350	18.	Jordan	450
5.	Canada	400	19.	Morocco	3,000
6.	Chile	1,600	20.	The Netherlands	9,600
7.	China	48,000	21.	Poland	1,500
8.	Colombia	2,600	22.	Portugal	2,500
9.	Egypt	1,000	23.	Romania	3,500
10.	England	3,500	24.	Spain	25,000
11.	France	5,800	25.	South Korea	21,000
12.	Greece	4,240	26.	Turkey	9,800
13.	Hungary	5,500	27.	U. A.R.	55
14.	India*	500	28.	U. S. A.	4,250

* Estimated for year 2000.

requirements. General attributes of different greenhouses are presented in the following sections.

Category I (Naturally Ventilated)

The greenhouses of Category I have a frame of steel, aluminium, bamboo or wood, which are covered with flexible plastic films or net or a combination of plastic film and net. These greenhouses serve the purpose of protecting the plant material from untimely rain, frost, hail, insects and diseases. In cases where the entire frame is covered with plastic film, provision is made for adequate natural ventilation through doors, windows or collapsible sides. The size of the frame and the choice of the covering materials depend upon the location and the plant material. If slight shading and protection from insects is all that is expected, then the structure could well be covered with plastic nets of suitable perforations. A 30 mesh net is claimed to keep aphids away. The net houses also protect the plants from occasional hail. However, if protection from rain is also desired, then the roof can be made impervious and the sides covered with net. Such greenhouses maintain adequate ventilation naturally. Irrigation in these greenhouses could be through micro sprayers, which would maintain favourable humidity levels in the greenhouses. The Category I greenhouses could have any shape (Figure 1) and size, that is, single span, multispan, gothic arch, gable, Quonset, saw tooth wave ranging from as small as 20 sq. metres floor area to as large as several hectares.

The Category I greenhouses could be easily modified to permit temperature rise during cold periods by attaching an additional plastic film to the netted frame area. The plastic cover can be manually or mechanically collected on one side during periods of ventilation.

Category II (Active Environmental Control Greenhouses)

Category II greenhouses are used when it is required to achieve a significant level of environmental control. Like Category I greenhouses, these greenhouses could also have different shapes and sizes. The choice of covering material and environmental control systems would require

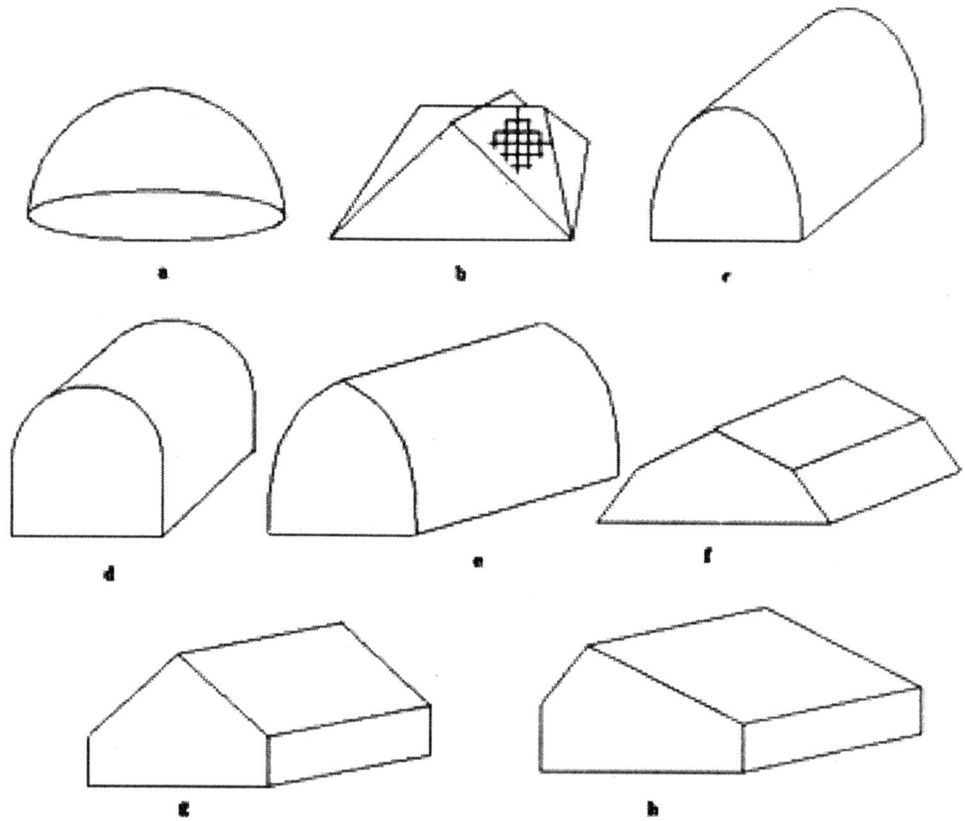

Figure 1. Architectural forms of greenhouses. Structures are arranged in descending order of transmission of solar energy. (a). Spherical dome; (b) Hyperbolic paraboloid; (c). Quonset; (d). Modified quonset or elliptical; (e). Gothic arch; (f). Mansard roof; (g). Gable even span; (h). Gable uneven span.

consideration of crop and agro-climatic regions. As mentioned earlier, complete environmental control is seldom attempted in view of economic considerations. It is adequate environmental control that is sought. For example, a Category II greenhouse in the north Indian plains will definitely have a well designed cooling system (Figures 2 and 3, A-G) and it may not have any active heating system (Figure 4-H). This is based on the premise that the loss in productivity during the short cold spell would be far less than the cost of the heating system. A similar trade off exists in other situations also. Photoperiodic control is attempted only in the case of chrysanthemums because of its essentiality. Some relevant aspects of greenhouse planning and environmental control are presented in the following sections.

SITE SELECTION

The proposed site should not be located on the downstream of an industry because of possible effects of pollutants on greenhouse crops. The location should be fairly shadow-free with sufficient windbreaks on all sides to minimise wind damage. Greenhouses could be located on hill slopes. In the Northern Hemisphere, however, a greenhouse should be located on the south-facing slope for better illumination and protection from northerly winds.

Figure 2. Different equipment for greenhouse environment control. A. Cooling pad; B. Exhaust fan; C. Desert cooler.

There should be sufficient room for the expansion of the facility in the future. In areas where snow is expected, the trees for windbreaks should be at least 30 metres away from the greenhouse for checking snowdrifts. Consideration of governmental policies on agricultural development in the particular area is important to the success of a greenhouse facility. If a big greenhouse complex is planned, the availability of labourers also becomes important. Unlike traditional agriculture, the labour requirement in a greenhouse operation is sustained the year round. This provides stability to the labour, which is beneficial to all.

GREENHOUSE ORIENTATION

As far as sunlight availability is concerned, in the case of a free-standing greenhouse, there is more sunlight available in winter in an east-west oriented greenhouse as compared to a north-south oriented one. The advantage keeps increasing as we move away from the equator. More sunlight during winter in an east west oriented greenhouse helps in reducing winter heating costs as well.

In case of multi-span greenhouses, however, shadow of structural components persists in some areas of the greenhouse if the gutters are oriented east west. This leads to loss of production in the shaded areas. On the other hand, if the gutters are oriented in the north-south direction, the shadow moves during the day and, therefore, the production in a specific area does not suffer.

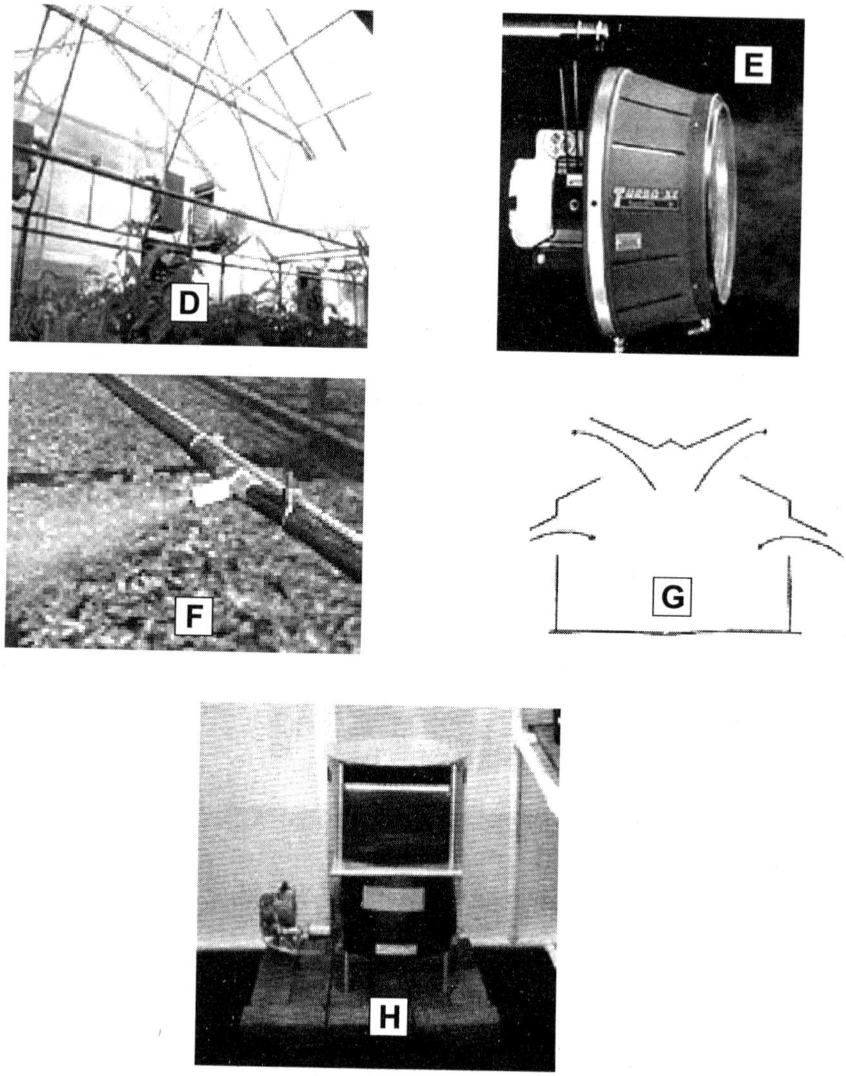

Figure 3. Different equipment for greenhouse environment control. D. Fan-tub ventilation system; E. Humidifier; F. Fogging system; G. Butterfly vents for better natural ventilation; H. Heater.

Figure 4. Multispan greenhouse.

Wind direction, if known at the location, also influences greenhouse orientation. In a naturally ventilated greenhouse, the ventilators should open on the leeward side. A free-standing greenhouse should have its long axis perpendicular to the wind direction.

PLAN LAYOUT

A greenhouse facility includes space for storage of inputs, sales, work, etc. in addition to production space. A general layout of the entire facility should be prepared keeping the following factors in view:

The workspace, also called head house, should be located to the north of the production area if possible, to avoid shading due to it.

1. The ways to the sales area and the storage/receiving area should be separate.
2. Customers should not be made to pass through the production area.
3. The wind breaks should be atleast 30 metres on the north and west from the facility.

One should also consider whether to build a single multispan greenhouse or a number of single span units. While a multispan facility permits economy in construction, heating and operation, a battery of single span greenhouses permit convenient expansion/contraction of the business. Further, a group of single span greenhouses permits the production of different plant materials by

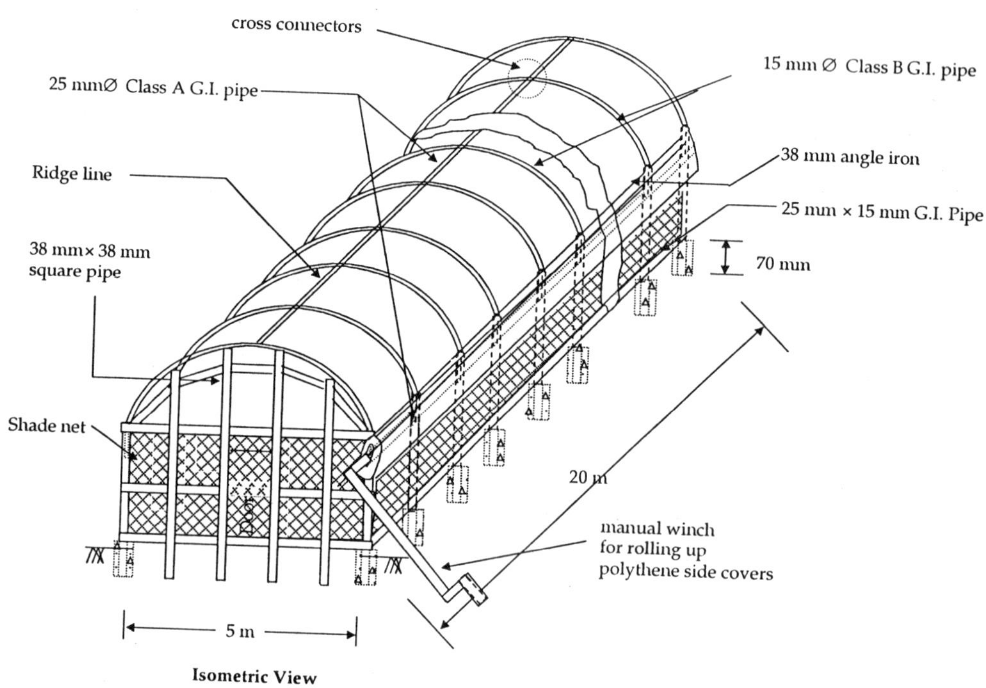

Figure 5. Naturally ventilated greenhouse.

maintaining different microclimates in different units. The economics of operation of several single span units give a lot of flexibility. Therefore, a multispan facility should be preferred if only one or similar type of plant material is to be grown.

A head house layout should be such that there is minimum material handling and cross traffic. Use of benches in the production area could be considered if pot plant operation is planned. Movable benches give high floor use factor. For vegetables, bedding plants and flower production, ground production systems are used.

GREENHOUSE STRUCTURES FOR DIFFERENT CLIMATIC CONDITIONS

Over the past two centuries, a lot of developments have taken place worldwide. Practically all possible shapes and sizes of greenhouse structures have been constructed. Various materials for construction and glazing have been tried. Efforts, however, continue to be made in terms of optimising the structural designs for newer locations and applications. Energy efficiency and environmental safety concerns require more work. A structural material, which does not cast shadows and an

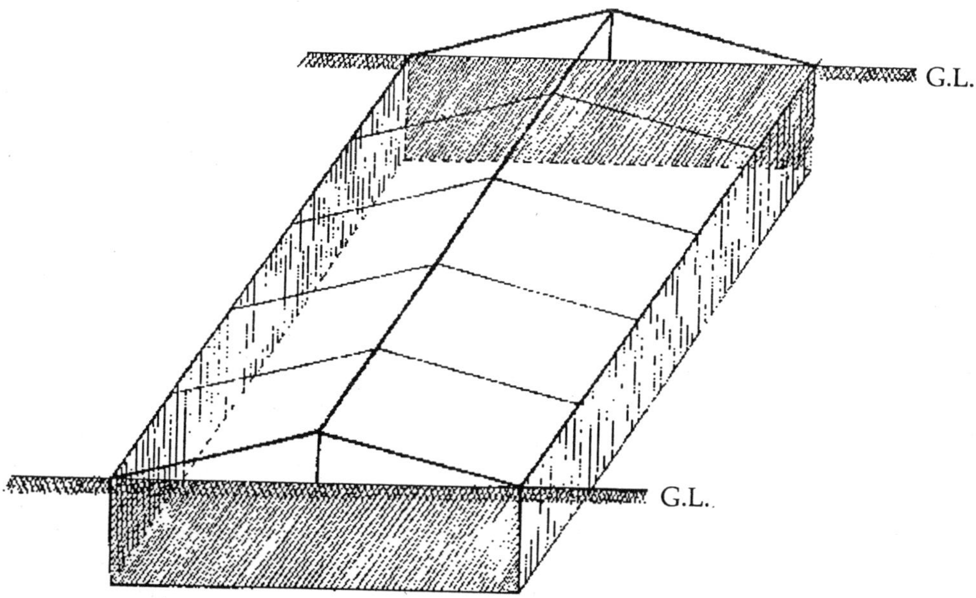

Figure 6. Trench greenhouse for Ladakh region.

intelligent glazing material are still not available. Application of artificial intelligence effectively to manage greenhouse activities would require modifications in greenhouse structures.

Greenhouse design activities in India are in a nascent stage. A few potentially useful greenhouse designs have been selected and methods of their fabrication have been developed (Figures 4-6) (Chandra, 2000). The Bureau of Indian Standards has only recently released a standard on greenhouse designs. There is a need to encourage a few groups of greenhouse designers to scrutinise the existing greenhouse designs from different parts of the world for their applicability under Indian conditions. Any material and component substitution required under specific situations should be studied both from the structural safety as well as the functional requirement points of view. It is conceivable that a catalogue of approved designs to meet the Indian requirements is prepared which could keep getting updated periodically. Linkages with industries such as the Steel Authority of India Ltd. (SAIL), Indian Petrochemicals Ltd. (IPCL), etc. need to be forged to synchronise the structural designs with material developments. Greenhouses for many specific locations and crops need to be designed. There is not much success in designing a suitable greenhouse for hot desert areas and cyclone prone coastal areas. With a view to reduce operational costs of greenhouses, the greenhouse designs should not only be energy conserving but as much use of renewable energy sources should be made as possible. Similarly, greenhouse designs should permit optimum use of water and other agro-inputs.

GREENHOUSE ENVIRONMENTAL CONTROL SYSTEMS

The control of greenhouse environment means the control of temperature, light, air composition and nature of the root medium. Obviously, a control over all these parameters makes a greenhouse completely controlled. However, in general, greenhouses with partial environmental control are more common and economical. In the following sections, an effort has been made to present general details of controlling the environmental parameters.

VENTILATION AND COOLING

A greenhouse is ventilated for either reducing the very high greenhouse air temperatures, for replenishing carbon dioxide supply or for moderating the relative humidity in the greenhouse. Air temperatures above 35° C are generally not suited to greenhouse crops. It is quite possible to bring greenhouse air temperature below this upper limit during spring and autumn seasons by providing adequate ventilation for the greenhouse. The ventilation in a greenhouse could be either natural or forced. In natural ventilation, air becomes less dense as it gets heated and, thus, rises. This chimney effect could be used to move the warm air out and cool air in. Winds also permit creation of additional natural ventilation in the greenhouses. In case of small greenhouses (less than 6 metres wide), natural ventilation could be quite effective during the spring and autumn seasons. However, fan ventilation is essential to have precise control over air temperature, humidity and carbon dioxide levels. A fan ventilation system requires that the capacity for the fans be determined and proper fans be selected to fulfil the requirements. The locations for mounting the fans and the inlet window should also be determined. One of the recommendations for summer ventilation is to provide for two air changes per minute under tropical conditions. The fans selected for the ventilation should be able to deliver 7-9 cubic metres per second of airflow per horsepower rating. The selection of fans for forced ventilation is made with the help of fan specification tables; a pressure drop of 3 mm of water gauge is allowed while selecting the fans for a given airflow rate. In order to obtain several stages of cooling, a larger number of smaller fans is selected as against a few large diameter fans. All fans should be equipped with louvres and fan guards for safety. Location and size of air inlets for the ventilation system are equally important. Fans are usually mounted on one end of the house and air inlets on the opposite end. Air inlets are designed to give an apparent mixing velocity of 3.5 m/sec.

Cooling Systems

While ventilation may be used for cooling during autumn and spring seasons, other methods have to be employed for cooling during summers.

Roof Shading

Applying opaque coatings directly to the glazing or by placing wood or aluminium laths over the glazing can reduce the amount of solar radiant energy entering the greenhouse. Commercial shading compounds or mixtures prepared with paint pigments are preferred for this purpose. White compounds are preferred, for they reflect a maximum amount of sunlight (83 per cent) as against green (43 per cent), blue or purple (25 per cent).

Lath Shades

Lath shades constructed from wood or aluminium are more flexible than shading compounds but their cost is considerably higher. Shading compounds are slightly less effective than lath covers for approximately the same degree of reduction in radiant energy. Although most of the solar radiant energy is reflected by the shading compound, some of it is absorbed and transmitted to the inside by conduction. Lath shades provide more cooling than shading compounds because air circulates between the laths and the glazing. Another method of shading is to install curtains of various cloth materials inside the greenhouse. Cloth shades are used most effectively for short periods reducing the solar radiant flux density by about 50 per cent, which will lower leaf temperature by 6° C.

Evaporative Cooling (EC)

The degree of cooling obtained from an evaporative system is directly related to the wet-bulb depression that occurs with a given set of climatic conditions. EC systems are most effective in areas where a consistently low relative humidity exists. Generally, the period of lowest humidity occurs during the hottest part of the day when the greatest degree of cooling is required and evaporative cooling is more effective. Relative humidity increases as night temperature declines and little or no cooling is required. The efficiency of EC is at its lowest during the night. Exhaust fans or overhead ventilators are usually sufficient to control greenhouse temperature after sunset.

HUMIDITY CONTROL

A greenhouse is a closed space in which plants transpire and evaporation takes place from the floor. Some of this moisture, added to the greenhouse air, is taken away by the air leaving the greenhouse due to ventilation and/or leakage. Sensible heat inputs modify the relative humidity to some extent. In order to maintain desirable relative humidity levels in greenhouses, efforts are made to use humidification or dehumidification. For most crops the acceptable range of relative humidity is between 50 to 80 per cent. However, for plant propagation work, relative humidity up to 90 per cent may be desirable. Humidification in summer can be achieved in conjunction with greenhouse cooling by employing appropriate evaporative cooling methods such as fan-pad and fogging systems. Sometimes during winter when sensible heat is being added to raise the greenhouse air temperature during nights, the relative humidity level might fall below the acceptable limit. In that situation, humidifiers might need to be operated to circumvent the problem. Dehumidification is often a problem not amenable to simple solutions. During the rainy season, the ambient relative humidity is high along with that of the greenhouse. In this situation, ventilation cannot lower the humidity of the greenhouse air. But, when the ambient relative humidity is lower, then ventilation could be practised to reduce the greenhouse relative humidity. Chemical dehumidification systems are technically feasible but expensive at present. Use of refrigeration systems (cooling coils) has also been made for dehumidification, but only at a smaller level.

GREENHOUSE HEATING

There are essentially three main categories of efforts needed to maintain desirable greenhouse temperatures during winter:

1. Design of energy efficient greenhouses with passive solar heating components.
2. Design of active heating systems based on renewable energy sources such as solar and biogas.
3. Design of active heating systems based on conventional fuels.

While the conventional fuel-based heating systems are many and dependable, the other two categories are still evolving.

CONTROLS FOR GREENHOUSES

Precise control of different parameters of greenhouse environment is necessary to optimise energy inputs and, thereby, maximise the economic returns. The objective of environmental control is to maximise plant growth. However, the maximisation of plant growth with simultaneous optimisation of inputs is needed to maximise profits.

Automatic controllers are designed to energise the environment control system in stages so that the rate of change of greenhouse environment parameters is in phase with the natural forcing functions. Experience has shown that the most commercially available high voltage bimetal thermostats are unsuitable for greenhouse use. Pneumatic thermostats are good for greenhouse use. Thermostats are becoming popular now because of their high response. The thermostatic sensors must be aspirated to give accurate control. The sensor should be shielded from solar and sky radiation and should be, preferably, located just below the top of the plants. If properly arranged, one sensing unit could control all the environmental equipment. Microprocessor based microcontrollers are now being envisioned for efficient greenhouse management. Development of microcontrollers of this type have recently been reported by Dogra and Chandra (2002).

RECENT TRENDS

Greenhouse technology today is undergoing substantial modernisation in the wake of developments in electronics and information technology. The input costs need to be optimised for maximum benefits. There is an added parameter of environmental safety. The development of microprocessor-based controllers coupled with appropriate decision-support systems are being envisioned for efficient greenhouse management. Speaking plant approach is being utilised increasingly for determining the nutritional, irrigation and plant protection requirements. Plant environment interactions are being reinvestigated to determine the appropriate environment parameters for more efficient control strategies. Closer monitoring of plants in a greenhouse with appropriate GPS and tillage processing techniques should result in more precise controls. Spatial and temporal variability in greenhouse crop production activities are being attempted to be eliminated. Greenhouse produce is generally healthier and safer for human consumption. Even then efforts are being made to enhance these parameters of quality and safety. Organic farming in greenhouses is a practice, which could not only revolutionise the availability of 'green' food in cities but also greatly enhance the export potential of organic foods. Greenhouses use plastic materials for glazing as well as for other components in irrigation, nutrient application, etc. Strategies are being contemplated for safer disposal of plastic materials. Besides the development of 'green' plastic is being pursued vigorously.

REFERENCES

Chandra, P. *Design and Development of Greenhouse for Floriculture in India.* Paper presented at APO Seminar on *Strategies for Development of Commercial Floriculture*, May 2-6, 2000.

Chandra, Pitam. 'Status of greenhouse development in India'. Proc. XXIX Annual Convention of Indian Society of Agricultural Engineers held at Gujarat Agricultural University, Junagadh, February 10-12, 1994.

Chandra, Pitam and Singh, J. K. *Instruction Manual for Establishing a 4 m × 24 m Plastic Greenhouse.* New Delhi, India: IARI, Division of Agricultural Engineering, 1988.

Dalrymple, D. G. 'A global review of greenhouse food production'. Foreign Agricultural Economic Report No. 89. Economic Research Service, USDA, Washington, D.C., U.S.A., 1973.

Dogra, A. K., and Chandra, Pitam. 'Development of an intelligent controller for greenhouse management'. Paper presented at *Recent Trends in Automation of Agriculture and Agro-based Industries* held at CEDTI, Aurangabad, February 26-27, 2002.

Jensen, M. H, and Malter, A. J. *Protected Agriculture: Global Review.* World Bank Technical Paper No. 253, 1994.

Land and Landscape Digest. www.turfgrassod.org/lawninstitute/digest10.

N.C.P.A.H. *Greenhouse Designs and Environmental Control.* Government of India, New Delhi, India. National Committee on Plasticulture Application in Horticulture, Department of Agriculture and Co-operation, Ministry of Agriculture, 2001.

Pliny. *Natural History.* (trans. by H. Rackhan). vol. V. book XIX, No. XXII: 463, 1961.

—oo(O)oo—

6

Greenhouse Design

A. S. Jadhav, M. T. Patil and P. V. Patil

INTRODUCTION

AGRICULTURE in India has made excellent progress after independence. Better varieties have been bred, improved agronomic and engineering practices have been developed and more effective measures to control insect and pest attack have been devised due to which productivity of foodgrains has increased during the last 20 years. The Green Revolution and the White Revolution both have taken place during the last two decades. But the scenario of other crops such as vegetables, flowers and fruits has not been as rosy as was the case with other crops.

A greenhouse is primarily a plant growing structure covered with transparent material such as glass or polyethylene to admit light necessary for plant growth, suitable to increase the quality and yield of flowers and vegetables. A basic important point for designing a greenhouse has been that the structure should be able to admit the maximum possible amount of sunlight during the winter season. The greenhouse designs are available from simple structures to complete machinated structures. The major consideration in the design of the greenhouse has been the design of the shape of the roof. These designs are more convenient for plastic roofs than glass cover. The design should be such that the angle of incidence of solar radiation is never greater than 40 degrees.

The structural design of the greenhouse must provide safety from wind and snow damage while permitting maximum light transmission to the crops. Thus, opaque framing members should be of minimum size consistent with providing adequate strength to resist anticipated loads over the expected life of the structure. The various loads should be considered to produce structurally sound and economically useful greenhouses.

DESIGN LOAD CALCULATIONS

It is essential to consider all loads while erecting the greenhouse. The design of the greenhouse structure is mainly governed by the dead load, live load, wind load and snow load. The combination of dead load, live load plus snow load and of dead load, live load plus wind load are combinations that usually control the design; but for extreme shapes and climatic conditions, the designer or architect must consider suitable combinations.

- Dead loads are the weights of all materials (such as frame made of GI pipe or wooden structure, weight of polyethylene sheet, etc.) used in the construction of the greenhouse and structures such as the floor, roof, frame and covering, etc.
- Live loads are the weights superimposed by use (in greenhouses they would include hanging baskets, shelves attached with vine ropes, etc.), but not wind, snow and dead load.
- Snow loads are vertical loads applied to the horizontal projection of the building roof.
- Wind loads are those caused by winds blowing from any horizontal direction, that is, load due to wind velocity.

(a.). Snow Load Calculations

This is based on the climatological data of the place on the snow load map. The map indicates isolines of ground snow load in pounds/sq. ft. or kg/sq. metres.

A greenhouse is not used for human stay. Hence, it is classified as a C type building, where there is negligible risk to human life such as crop storage buildings, farm shops, machinery sheds or stores, etc. Therefore, greenhouses should be designed for 25 years recurrence interval load. The minimum snow loads for free standing, ridge and furrow, pitched or curved roofs are determined by multiplying the ground snow loads by the coefficient for snow loads. Design snow loads may be found by the formula:

$$L = 0.6\,(\cos\theta)\,(S)$$

where L = Design snow load for the projected area (kg/sq m).

0.6 = A dimensionless factor which estimates the proportion of snow pack on the ground which occurs as a snow load on a flat roof.

θ = The roof slope with the horizontal (degrees).

S = The snow pack on the ground by map value (kg/sq m).

The snow loads may be decreased or increased as per the conditions, for example:

- Decreased load when a roof slope exceeds 30° and snow slides off.
- Decreased load when roof has clear exposure in wind swept areas.
- Increased load when a non-uniform accumulation occurs on pitched or curved.
- Increased loads when valleys are formed by multiple roofs.

(b). Wind Load Calculations

Similar to the snow load, the climatological data of the place provides us with wind maps for different recurrent intervals. For greenhouse design, 25 years recurrence interval is generally sufficient. While selecting map values, care must be taken for unusual exposures causing channelling or uplift, such as ocean promontories, mountains or gorges and for elevations where wind records or experience indicates illustrated wind speeds to be inadequate.

The effective wind pressure is determined by

$$q = 0.00466\, V^2 K$$

where q = Basic velocity pressure in kg/sq. m.

V= Map value of wind velocity for extreme km in km/h. Figure 1 gives 'q' at 9.14 m height.

K = Building height correction factor = $(h/9.144)^{2/7}$.

where 'h' is the height of the building in metres (Figure 2).

The final wind load factor for a greenhouse is calculated

$$W = q * C$$

where C = Dimensionless shape coefficient.

The values for most common shapes (Figure 3) are presented in Tables 1, 2 and 3. These values for 'W' at various locations may be used for design of main frames (trusses, columns, rigid frames and other main members) and anchorage. The value of the shape coefficients for buildings with both sides open are as follows:

(a). For roof slopes of 30° or more

Windward side, $C_2 = +0.8$

Leeward side, $C'_2 = -0.8$

(b). For roof slopes less than 30°

Windward side, $C_3 = +0.6$

Leeward side, $C'_3 = -0.6$

Once the load (W) has been calculated, the purlins, rafters, roof panels or fastenings for covering, whether a structural part of the framing or covering should be designed for 1.25 W. acting outward (suction) normal to surface consider, whereas wall girts, studs and wall panelling or coverings should be designed for pressures whether inward or outward equal to the design wind pressure 'q'.

DESIGN CONDITIONS

The design members of greenhouses must be strong enough to carry the following loads:

- Dead load.
- Live load: The minimum value of this is taken as 50 kg/sq m in non-snow zones. In cooler areas, the snow load may be added to it.
- Wind load: The minimum value of this is taken as 100 kg/sq m on vertically projecting area below 30 ft or 9.14 metre height.

In designing the members, the loads may be considered to act on one of the following combinations.

- Dead load plus live load.
- Dead load plus wind load plus half live load.
- In addition to these roof bars, purlins and rafters should be carrying a minimum 50 kg concentrated load at the centre of any span.

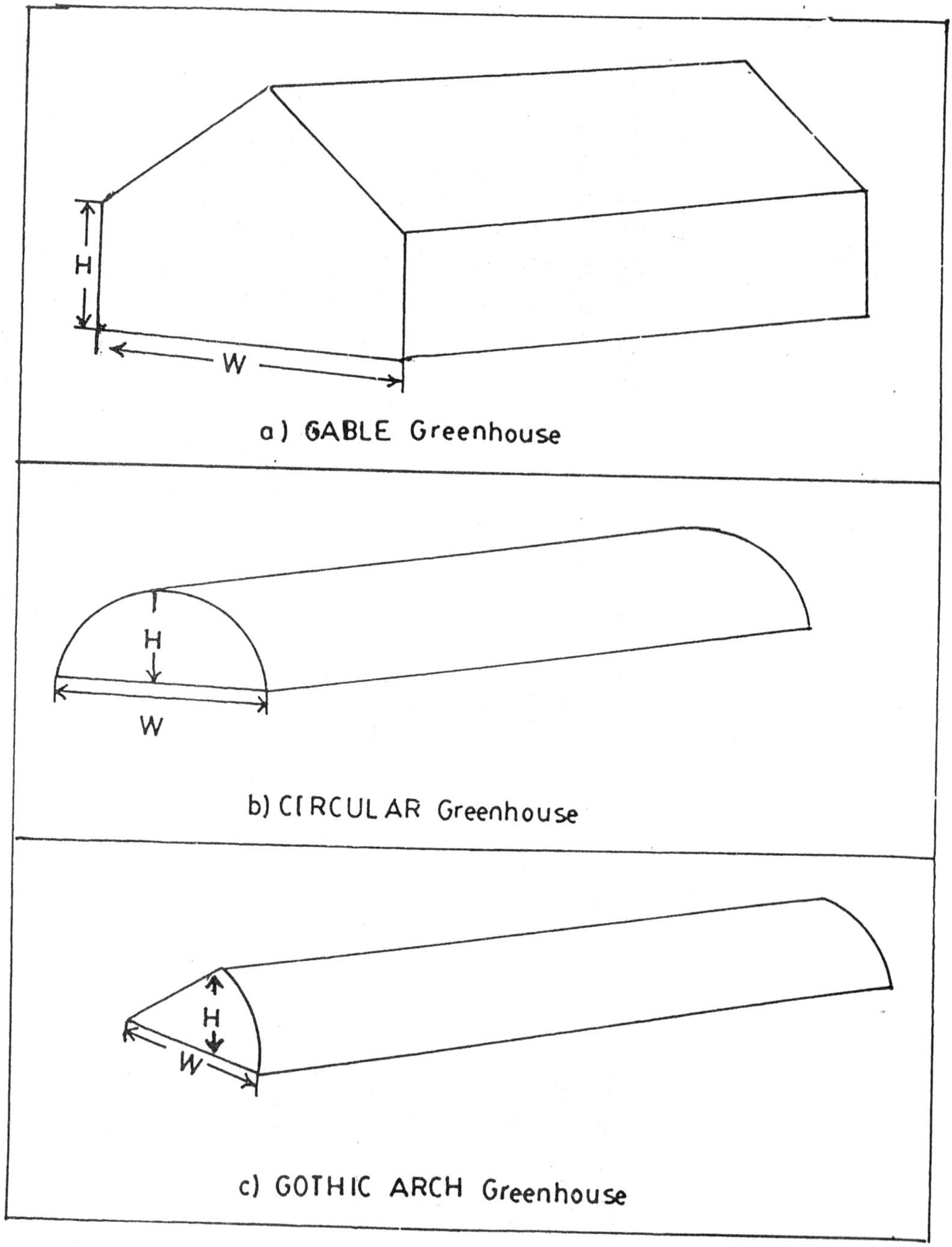
H
W
a) GABLE Greenhouse
H
W
b) CIRCULAR Greenhouse
H
W
c) GOTHIC ARCH Greenhouse

Figure 1.

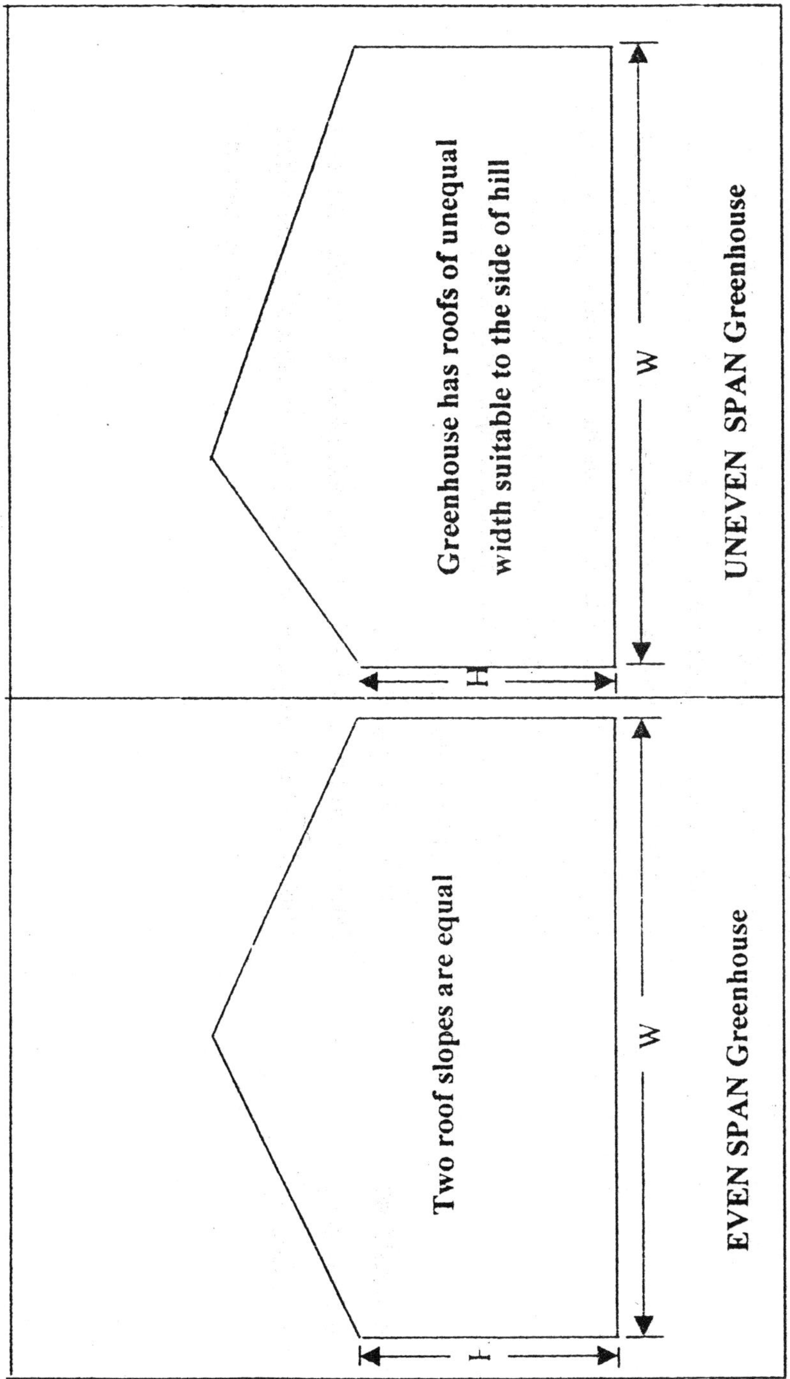

Figure 2. COMMON GREENHOUSE SHAPES

TABLE 1
Shape Coefficients, C, for External Wind Loads on Single Span Gable Buildings—Totally Enclosed
(For Designing Trusses, Columns, Rigid Frames and for Other Main Members)

H/W	Windward Wall Coeff, C_1	Windward Roof Coeff, C_2 — Roof Slope 1:12	2:12	3:12	4:12	5:12	6:12	7:12
0.10	0.70	–0.34	–0.24	–0.13	–0.03	0.05	0.12	0.19
0.15	0.70	–0.51	–0.35	–0.20	–0.05	0.05	0.12	0.19
0.20	0.70	–0.60	–0.47	–0.27	–0.06	0.05	0.12	0.19
0.25	0.70	–0.60	–0.59	–0.34	–0.08	0.05	0.12	0.19
0.30	0.70	–0.60	–0.60	–0.41	–0.18	0.01	0.08	0.16
0.35	0.70	–0.60	–0.60	–0.47	–0.26	–0.07	0.05	0.12
0.40	0.70	–0.60	0.60	–0.53	–0.33	–0.15	0.01	0.09
0.45	0.70	–0.60	–0.60	–0.57	–0.39	–0.22	–0.06	0.05
0.50	0.70	–0.60	–0.60	–0.60	–0.44	–0.29	–0.14	0.00
0.60	0.72	–0.60	–0.60	–0.60	–0.49	–0.34	–0.20	–0.06
0.70	0.74	–0.60	-0.60	–0.60	–0.53	–0.39	–0.25	–0.13
0.80	0.76	–0.60	–0.60	–0.60	–0.57	–0.43	–0.30	–0.18
0.90	0.78	–0.60	–0.60	–0.60	–0.60	–0.47	–0.35	–0.23
1.00 or more	0.80	–0.60	–0.60	–0.60	–0.60	–0.51	–0.39	–0.28

Leeward roof : C_3 = -0.50, for all values of H/W.
Leeward wall : C_4 = -0.40, for all values of H/W.
Negative values indicate external suction on building surface.

The members should be designed in accordance with their allowable stresses as per ISI.

MATERIALS

The selection of material depends on its density, allowable stresses and economics. Initially, wood and wood piped structures were quite common. Trusses, columns, beams or purlins made of these materials have different load bearing capabilities and, in most instances, if maintained, would remain sound. The two nomographs (Figures 4 and 5) on beam and column loads are presented to get acquaintance with the strength characteristics of wood and sizes required.

There are the various materials available in the country for fabricating the structure of greenhouses. The most commonly used are wood, steel and aluminium.

Wood: This is very common where the cost of wood is comparatively very low and is available in plenty. The construction of the structure is also simple, as local artisans are able to complete the job. It is easy to construct greenhouses with wood structure. The nomographs (Figures 4 and 5) on beam and column loads are presented to help the erectors in selection of size.

TABLE [illegible]
Shape Coefficient, C, for External Wind Loads on Roof Building with Vertical Sidewalls (For Designing Principal Arch Ribs and Other Main Members)

F/w	Windward Quarter of Roof	Centre Half of Roof
0.10	0	–0.80
0.15	0	–0.85
0.20	0	–0.90
0.25	0.10	–0.95
0.30	0.19	–1.00
0.35	0.29	–1.05
0.40	0.39	–1.10
0.45	0.48	–1.15
0.50	0.58	–1.20

Leeward quarter of roof: –0.85, for all values of f/w.
Windward wall and leeward wall: Use Table.
Negative values indicate external suction on building surface.

TABLE 3
Shape Coefficient, C, for External Wind Loads on Curved Roof Building (For Designing Principal Arch Ribs and Other Main Members)

F/w	Windward Quarter of Roof
0.10	0.12
0.15	0.18
0.20	0.25
0.25	0.31
0.30	0.37
0.35	0.43
0.40	0.49
0.45	0.55
0.50	0.60

Centre half of roof: –0.70, for all values of f/w.
Leeward quarter of roof –0.58, for all values of f/w.
Endwalls +0.70, for all values of f/w.
Negative values indicate external suction on building surface.

Steel: Today, in construction of greenhouses, high tensile strength pipes and tabular steel pipes and zinc plated bolts and nuts for anchorage are widely used. There has been a considerable development in construction of greenhouse structures with prefabricated steel structures. These are commercially available. Galvanised steel or zinc coated pipes are preferred to avoid corrosion.

TABLE 4

Approximate Live Load Factor of Purlin Material Used in Greenhouse Structure (The Load is Based on a 7.3 m Span)

Cross Section	Material Size in mm	Kg/line Aluminium
L	Angle 65 × 65 × 5	
	a) Aluminium	40
	b) Steel	60
[	Channel 50 × 100 × 4	
	a) Aluminium	200
	b) Steel	300
I	I Beam 53 × 75 × 4	
	a) Aluminium	90
	b) Steel	175
○	Pipe 32 mm O.D.	
	a) STD steel	40
	b) High tensile steel	90
□	Tabular steel 50 sq mm	
	a) Aluminium	40
	b) Steel	112
▨	Wood - 50 × 100 (FIR)	40
▨	Wood - 50 × 150 (FIR)	78

Aluminium: It is light in weight, easy to handle and is not adversely affected by most greenhouse conditions. Aluminium pipes, angles, channels and T sections are commercially available. These can be used for trusses, purlins and columns. Table 4 gives the load bearing capacities of different materials. These values are helpful in designing the trusses and purlins.

REFERENCES

Chandra, P., Ojha, T. P. and Bohra, C. P. Proceedings of the Summer Institute on Greenhouse and Environmental Control. Tech. Report No. CIAE/EME/85/35, published by CIAE, Bhopal, 1985.

—oo(O)oo—

7

Greenhouse Cladding Material, Shade Nets and Ventilation

A. S. Jadhav, M. T. Patil and P. V. Patil

INTRODUCTION

GLASSHOUSE technology is perpetuating steadily in India. This technology enables climate control and favourable conditions for crop growth. The changes in climate positively affect the growth and yield of crops. During summer, they need cooling to combat the high temperatures and problems like leaf discoloration, leaf scorch and desiccation. The plants may also have stunted growth or may shed leaves and flowers. Ventilation helps reduce the effect of excess heat, but may not be enough to eliminate the total effect of excess heat and needs some shade.

Greenhouse structures have distinct effects on climatic factors, namely, temperature, light, carbon dioxide and humidity. The plant response to specific environmental factors is related to the physiological processes and to yield and quality. Controlled environment is one of the promising effects created in the greenhouse.

Light—its duration, quality and intensity—is a fundamental component for plant growth. Greenhouse covering material directly influences both light quality and intensity. Each covering/cladding material has its special qualities and specifications. Some materials allow virtually all light to pass through while others screen or filter out certain wavelengths. Most covering materials designed for greenhouses contain acceptable light quality properties. These properties can change over a period of time; however, in addition, some translucent material used in greenhouses, but not designed for them, do filter out essential wavelengths for plant growth.

Light transmission measures how much light passes through the covering and reaches the plant. Some coverings are tinted or used in conjunction with shade cloths. Although most crops prefer all light, they cannot hold up to the heat generated via the greenhouse effect. That is when we start reducing light transmission with shading compound, shade cloth, etc. Some covering materials undergo structural changes when exposed to UV light for a period of time. This can result in ageing, yellowing and other forms of reduced light transmission. For example, fibreglass is notorious for the UV breakdown that causes the fibre to swell and the resin to turn yellow. This UV breakdown can ultimately result in reducing light transmission by as much as 75 per cent. Greenhouse covering materials can last anywhere from one growing season to infinity.

TYPES OF CLADDING MATERIAL

Glass: Greenhouses that existed prior to 1950 were covered exclusively by glass. It is one of the most expensive types today. Although they have been known to last for a hundred years and longer when properly maintained, the average cost per year is still more than that of plastic film greenhouses. Historically, it is preferred due to its light transmissivity. Transmissivity of glass differs only slightly over a long period of use. Positioning of glass panes allows condensation, which usually forms inside the panes, to drain out of the greenhouse. Additionally, reduction of relative humidity is provided by the naturally high rates of air exchange. Generally, one or two air changes/hour occur in greenhouses.

Plastic: Plastic is more advantageous than glass because of its lightness and corresponding lighter structure, mobility, flexibility, impermeability, resistance to hail and simple design, resistance to wind loads, lower cost, etc.

Film Plastic

Polyethylene: Most widely used as a greenhouse film. It is available in widths up to 7 metres. The film is tough, flexible and low in cost. It can withstand temperatures as low as 30 to 50° C, but temperatures of 80 to 90° C will cause it to melt. Above 60° C, it loses much of its strength. The film is UV-stabilised treated, otherwise it breaks down within three to five months. The life of the UV-stabilised 200 micron film is about 2 to 3 years.

Polyvinyl: Similar to glass, it has a high degree of solar remittance, passing 89 to 91 pb of the incident light. It has better resistance to weathering and may last up to 5 years. Narrow widths are available. Cost of this film is two to three times higher than polyethylene.

Polyvinyl Fluoride: It has good weather standing conditions but is higher in cost. It is currently used as a surface coating to fibreglass panels.

Polyester: It is widely used in the textile industry. It has low light transmissivity compared to glass. It is very strong and can last up to 15 years. The cost is about three to four times higher than that of polyethylene.

Rigid Panels

Fibreglass Reinforced Panels: These panels consist of glass fibre reinforced polyester. Because of their greater strength, corrugated panels are preferred for roofing. Fibreglass material is guaranteed against only natural mechanical breakdowns for up to 15 years depending upon the quality of the product. Corrugations increase heat loss by 10 to 15 per cent due to the increased exposed area. Corrugated panels are, therefore, not popular in colder areas.

Acrylic Panels: It has high light transmission, excellent flexibility, strength and weatherability as compared to glass. These panels have a very high cost.

Polycarbonate Panels: It is stronger and has poor light transmissivity.

SHADE NET

Agro shade nets are light in weight but are strong and stable. They are manufactured from high density polyethylene (HDPE) plastic. These nets are treated with the best known ultraviolet

TABLE 1
Life Expectancy of Several Cladding Materials

No.	Material	Life Expectancy (Year)
1.	Glass	30+
2.	Double polyethylene UV treated	3-5
3.	Acrylic SDP	15+
4.	Polycarbonate SDP	20+
5.	Fibreglass	15-20

TABLE 2
Light Transmission Percentage of Different Cladding Material

No.	Material	Single Layer	Double Layer
1.	Glass (3 mm)	88	78
2.	Polyethylene (100 micron)	89	79
3.	Acrylic SDP (16 mm)	—	—
4.	Polycarbonate SDP (16 mm)	84	73
5.	Polycarbonate (6 mm)	84	73
6.	Polyvinyl fluoride	91	84
7.	Polyester (Mylar)	86	78
8.	Fibreglass (100 micron) corrugated	79	62

stabilisers in the highest possible concentrations that are technically known. This ensures long life of the produce. Shade nets are used to protect valuable plants against excess sunlight, heat, cold, wind, frost, insects, birds, etc. They are available in three colours, black, green and white, and in different shades, that is, 50, 75 per cent. They are available in 3 metres width and 50 metres length, that is, one bundle of 150 m^2. The cost of shade nets is Rs. 18 to Rs. 21 per square metre as per the colour and the shade percentages.

Applications of Shade Nets

Tea, coffee, cardamom, pepper, spices, vegetable nurseries; indoor plants; fruit plants; floriculture; forest nursery stock; cattle sheds, fishponds, general purpose shading, soil support, drying of various agro products; wind breaker.

VENTILATION

Greenhouses can be cooled by any of the methods commonly used in commercial, industrial or residential buildings. However, large cooling loads imposed on greenhouses by the sun and the high cost of installing and operating air-conditioning systems reduce the practical option for cooling greenhouses to ventilation, evaporative cooling and shading. Ventilation may be used with or without evaporative cooling, but evaporative cooling cannot be used without ventilation.

Ventilation can be mechanical (driven by simple propeller fans) or natural (driven by thermal buoyancy and the wind). Natural ventilation is acceptable when screening to prevent pest entry is

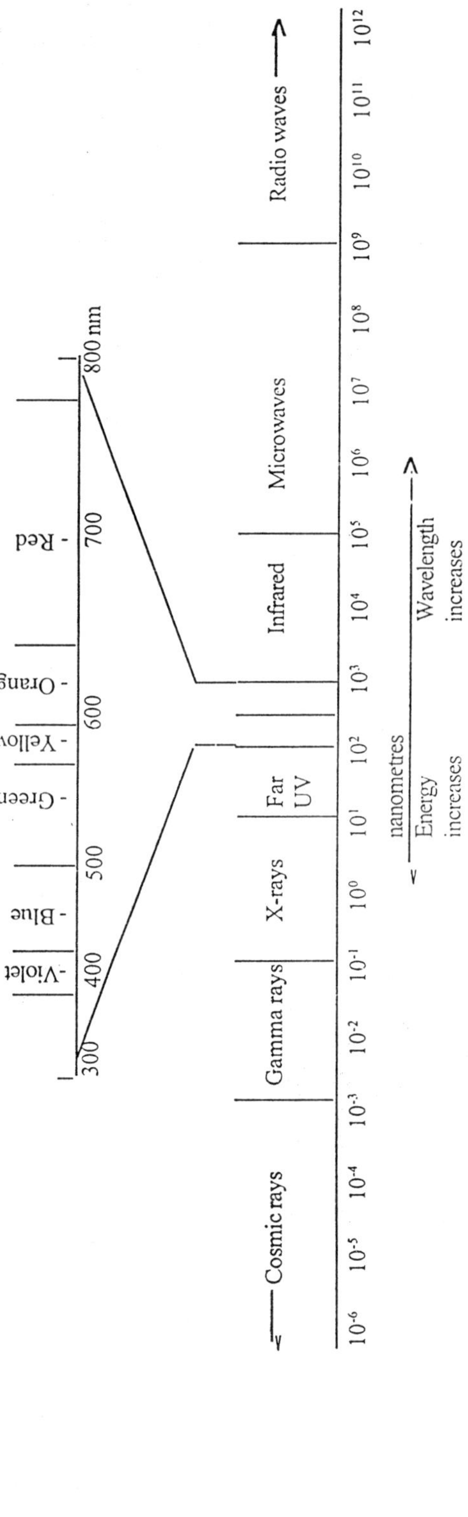

Figure 1. The electromagnetic spectrum.

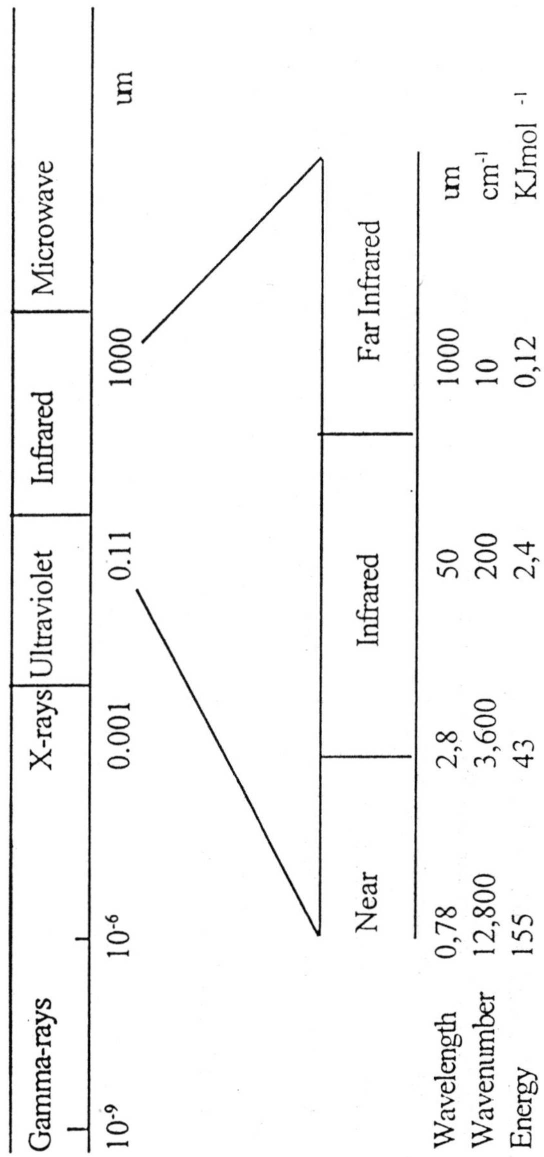
Gamma-rays
X-rays
Ultraviolet
Infrared
Microwave
10-9
10-6
0.001
0.11
1000
um
Near
Infrared
Far Infrared
Wavelength 0,78 2,8 50 1000 um
Wavenumber 12,800 3,600 200 10 cm-1
Energy 155 43 2,4 0,12 KJmol -1

Figure 2.

not a necessity and evaporative cooling is not required for temperature control. These restrictions may limit applications of natural ventilation in hot climates. However, fan ventilation uses a significant amount of electricity (perhaps 20-30 Kwh/m^2 of floor area per year in hot climates) and can be noisy, which makes it unsuitable for some locations. Natural ventilation systems if properly designed and controlled can provide rates of air exchange comparable to rates provided by mechanical ventilation systems. The primary difference from the perspective of the crop is that naturally ventilated systems generally do not provide temperature control to within the close tolerances possible when fan ventilation is employed and they are not compatible with evaporative cooling.

The conventional method for lowering temperature is accomplished by drawing cooler outside air into the greenhouse. The cooling range should constitute the requirements of different seasons. Ventilation must be considered a year-round process. During winter, a reduced capacity is needed to avoid excessive chilling. Winter ventilation is often used to reduce relative humidity.

Natural Ventilation by Wind

Slight winds provide sufficient ventilation for many greenhouse situations provided the vents are properly controlled. As one method of control, roof vents remain closed and the air crosses the greenhouse from the upward wall vent to the leeward wall vent. Air above the vents becomes quite warm and remains virtually unmixed but does not affect the plants and the result is an increased effective ventilation rate. If roof vents are opened, care is essential to avoid an open upwind roof vent and a closed downwind roof vent. Wind above the ground usually moves at a greater speed than wind near the ground and wind striking an upward vent at ridge level may reverse flow through the ridge vent (bringing the warmest air down to plant level) and suppressing the air flow through the upward sidewall inlet. The major advantage of natural ventilation is its simplicity, with no need for electricity and no noise of operation.

Forced Cooling System Using Electrical Exhaust

It is used for cooling during periods of high temperature. Ventilation is provided by electrical exhaust fans, which are set in the side.

—oo(O)oo—

8

Economics of Polyhouse Cultivation of Cut Rose for Export

M. Sudha

INTRODUCTION

FLORICULTURE in India, in the recent past has been transformed from a simple garden activity for meeting the domestic demand to an important agri-business enterprise catering to the growing export demand. A number of factors contributed to this change; important among them besides the growing international demand for modern cut flowers, are the growing urbanisation, technology innovations and the changed policy environment in the country. The liberalisation drive adopted by the Union Government during the early 90s made the import of seed and exotic plant material easy, thereby paving the way for the transfer of 'state-of-the-art' technology of polyhouse cultivation of crops, especially those of cut flowers, in view of the perceived comparative advantage in their production.

Popular as hi-tech floriculture, the protected cultivation of cut flowers predominately consists of production of roses for export purposes; though, the cultivation of other cut flowers like Lily, Gerbera and Carnation (under polyhouse structures) is becoming popular as agri-business ventures in recent times. A suitable agro-climate has been the primary production factor that encompassed the desired comparative advantage for India, in terms of reduction in cost of establishment and production of cut flowers. The locational advantage in terms of closeness to international markets helped the Indian growers to access markets like the Netherlands, Japan, Singapore and Australia between September and February, the peak winter season in traditional cut flower producing areas (Attawar, 1995).

The perceived comparative advantage in the production factors envisaged cost effective production, resulting thereby in according priority status for hi-tech floriculture as an export oriented agri-business industry in India, since the early 1990s. The hi-tech floriculture in the country today comprises a total of over 134 export-oriented production houses, specialising primarily in cultivating cut roses for export markets, since the last five years, involving an initial investment of over Rs. 250 crore. Initiated to meet the growing export demand, hi-tech floriculture production has been highly dependent on several non-production factors as well, such as the appropriate selection of markets, time, method and mode of sale and above all, the fast changing global trade order.

Keeping in view the economic significance of this industry, an attempt has been made in this chapter to provide an overall review of the status and economic perspective of this 'sunrise industry', with specific reference to the technology options, the profitability of investment, the markets and marketing strategies for exports, analysis of prices, constraints faced by the growers and the industry's future prospects.

CHANGING TREND OF WORLD DEMAND AND SUPPLY FOR FLORICULTURE PRODUCTS

Cultivation of flowers for commercial purposes is common to many countries. Japan and the U. S. A. represent the traditional and by far the largest markets for flowers (worth over NFI 10,000 million) followed by Germany (NFI 5,000 million), while both China and India dominate in terms of area under flower crops (with around 60,000 hectares each) (Anonymous, 2002). Though, most of the flower producing countries are also self sufficient in meeting their domestic demand, there has been a surge in global demand in view of the changing utility value of flowers as gift articles and as symbolic of human feelings and status. Countries like Germany, which have a sizeable domestic market, represent the main consuming lot as they fulfil over 70 per cent of the domestic demand by imports from other countries. On the other hand, countries like Colombia and Kenya, with small home markets, represent the supply side of the flower trade, with large volumes of exports. The Netherlands, which could be considered the central point for the global flower trade, is a country that has both a large home market as also a large export share.

In terms of products, roses are the main traded product, accounting for 23 per cent of global trade (Figure 1), followed by carnation and chrysanthemums, while Germany is the main market and The Netherlands the world's largest exporter. In international trade, exports from The Netherlands to Germany form the principal component of global cut flower trade, accounting for over 43 per cent of European Union exports.

INTERNATIONAL TRADE IN CUT FLOWERS

International trade, to a large extent, is organised along regional lines. Asia-Pacific countries are the main suppliers to Japan and Hong Kong. New Zealand sells 70 per cent of its exports to Japan, while Taiwan sells over 90 per cent. Hong Kong's principal suppliers are China, Taiwan (China), Malaysia, Singapore and New Zealand. African and other European countries are the principal suppliers to Europe's main markets. Kenya sends over 60 per cent of its exports to the Dutch auctions; Zimbabwe 80 per cent, and Zambia over 90 per cent. The U. S. A. is supplied mainly by other countries in the hemisphere for which the U. S. A. in turn is their main market; Colombia and Ecuador being typical with over 70 per cent of their exports going to the U.S.A.

GLOBAL TRENDS IN IMPORT OF FLORICULTURE PRODUCTS

The world trade in cut flowers and floricultural products has doubled within the last one decade, from a value of US$ 3,351 million to over US$ 7,663 million. Although Germany continues to hold the fort as the main consuming market for the global trade, there appears a sharp decline in its share from over 37 per cent in 1991 to 19 per cent by 2000. Among the traditional flower consumers, the U. S. A. has been consolidating its share of imports over the last two decades. The value of

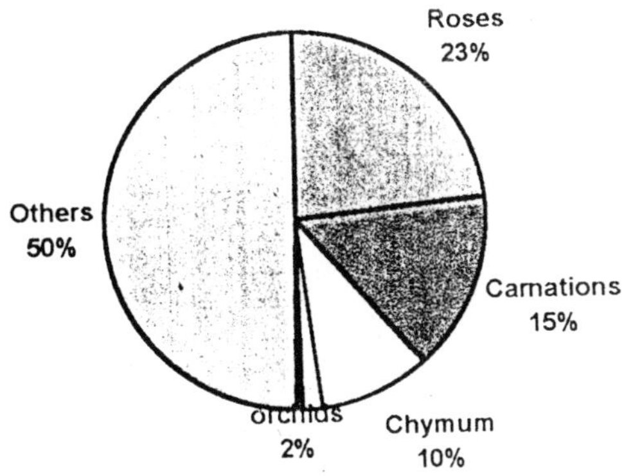

Figure 1. Per cent share of flowers in the international market (by value, 1997).

U. S. imports of flower products registered a near fourfold increase from US $ 403 million to US $ 1,361.97 million between 1991 and 2000. In terms of percentage share in the global market, the U. S. A. increased its share from 12 to 17.8 per cent. Among the other countries, the U. K. (from 9.1 to 11 per cent), the Netherlands (from 5.5 to 9.7 per cent) and Japan (from 4.3 to 5.1 per cent) have increased their share in the global import of the floriculture market. (Table 1).

GLOBAL TREND IN EXPORT OF CUT FLOWERS

In terms of exports, The Netherlands holds the prime place as the major supply centre, accounting for over 64 per cent of the annual exports in the beginning of the 1990s. Though there has been some decline in its share, the Netherlands still holds the top position in world exports accounting for nearly 50 per cent of the global exports. Colombia appears to have gained a substantial global share in exports of cut flowers. New entrants like Colombia, India and Kenya appear to be consolidating their share in the world export market. This trend indicates that the global demand for cut flowers is on the increase and there is a scope for non-traditional growers to substantiate the global flower trade in future. (Table 2).

GLOBAL TREND IN CONSUMPTION OF CUT FLOWERS

People buy flowers for a number of reasons: as a gift (for a birthday or a special day such as Mother's Day, Valentine's Day, Thanksgiving), for weddings and funerals, or for own use to brighten up the home. Among these, gift giving (to congratulate, apologise, commiserate, express love and affection) is the most frequently stated reason for buying flowers (over 70 per cent of all purchases in the United Kingdom and Austria; over 50 per cent on an average). As such, flowers compete with wine and chocolates, which are bought for the same reason.

Preferences for flowers differ markedly by country. While the rose remains the most popular item in France, accounting for 57 per cent of the total demand, the mixed bouquet is gaining popularity everywhere making up to 42 per cent of the total demand in Germany and 30 per cent of the total flower demand in the United Kingdom. Carnations are the most popular cut flower in the United Kingdom (22 per cent of the total) and in Italy (24 per cent) (Protrade, 1996).

Overall consumption levels also differ markedly by country. Germany is the largest market, followed by Italy, France and the United Kingdom. In terms of per capita consumption per day, the

TABLE 1

Main Import Markets for Cut Flowers by Value (CIF), 1991-2000 (in US$ million)

Country	1991		1995		1999		2000	
	Value	% Share	Value	% Share	Value	% Share	Value	% Share
World	3351.7	100.0	3716.8	100.0	7914.63	100.0	7662.90	100.0
Germany	1245.3	37.2	1124.9	30.3	1678.48	21.2	1457.70	19.0
U. K.	305.6	9.1	360	9.7	867.97	11.0	845.20	11.0
France	337.8	10.1	358.9	9.7	885.28	11.2	834.38	10.9
Netherlands	185.3	5.5	310.4	8.4	747.03	9.4	741.98	9.7
Italy	134	4.0	118.9	3.2	389.37	4.9	379.18	4.9
Belgium	77.3	2.3	95.8	2.6	290.40	3.7	280.39	3.7
U. S. A.	403.5	12.0	622.8	16.8	1281.10	16.2	1361.97	17.8
Japan	143.8	4.3	216.1	5.8	383.83	4.8	391.62	5.1
Switzerland	141.2	4.2	168.7	4.5	343.75	4.3	362.30	4.7
Others	413.9	12.3	340.3	9.2	1047.00	13.2	1008.12	13.2

Source: Compiled from Anonymous, 2002 and Singh, 2002.

TABLE 2
Global Export Trends of Cut Flowers by Country (in US$ million)

Country	1991		1995		1999		2000	
	Value	% Share	Value	% Share	Value	% Share	Value	% Share
World	3351.7	100.0	3716.8	100.0	7914.63	100.0	7662.90	100.0
Netherlands	2147.4	64.1	2102.2	56.6	4078.07	51.5	3810.62	49.7
Israel	145.9	4.4	158.7	4.3	166.99	2.1	154.60	2.0
Kenya	53.6	1.6	103.5	2.8	156.47	2.0	164.20	2.1
Ecuador	22.1	0.7	102.2	2.7	211.34	2.7	215.98	2.8
Italy	113.2	3.4	92.3	2.5	295.94	3.7	268.93	3.5
U. S.A.	23.0	0.7	28.6	0.8	217.61	2.7	220.87	2.9
Costa Rica	15.0	0.4	23.3	0.6	160.72	2.0	161.06	2.1
Belgium	6.6	0.2	21.6	0.6	273.80	3.5	267.61	3.5
Germany	16.9	0.5	20.0	0.5	199.97	2.5	197.97	2.6
India	2.0	0.1	11.4	0.3	23.49	0.3	26.97	0.4
Colombia	NA		NA		550.15	7.0	570.34	7.4

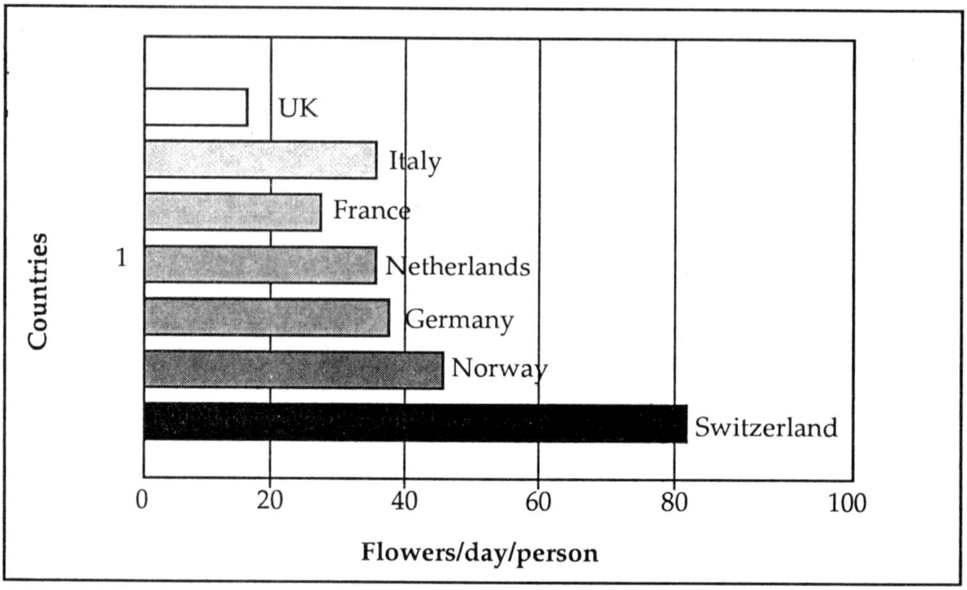

Figure 2. Average per capita consumption of flowers.

Swiss spend by far the largest amount per person; at 87 ECUs, 50 per cent more in 1995 than the Norwegians, who are second at 54 ECUs. The Dutch and the Germans buy 40 ECUs worth on an average, Italians 29 ECUs and the French 27 ECUs. In the United Kingdom, per capita consumption at 16 ECUs lags behind the European countries (Figure 2).

There are indications that consumers are becoming more demanding, particularly in Germany and Switzerland. This is manifested in two ways. There is more attention for the conditions under which flowers are being grown and there is a search for new varieties and novelties. The retail market is becoming more sophisticated. Thus, there appears ample scope for new entrants with new types of flowers to capture the growing global market for cut flowers. It is in this context that there arises a need to assess the status and changing profile of the competitive edge of India to meet the new global order with regard to the floriculture trade.

Status of Polyhouse Cultivation of Cut Flowers for Export in India

Flowers have a special cultural significance in India, both as components of religious importance and also of decorative value. Used mainly in the form of loose flowers, which are made into garlands or used for adorning the hair of Indian women. Flowers are said to have two distinct components of demand, one being the seasonal one, while the other is the more steady daily one (Kolavalli, 1997). The seasonal demand depicts the demand, which reaches a peak in specific days during festivals, while the steady demand largely encompasses the daily requirements for offerings at worship or adorning the hair of women.

In view of the steady and a rather inelastic demand, growing flowers as a part of crop combinations has been very popular with growers especially in the peri-urban areas. Cultivation of flowers fetch higher returns and also provide gainful employment opportunities for farm women, as compared to any other group of crops (Nageswara Rao, 1997; Varadarajan, 1995; Subrahmanyam,

1983). However, despite the fact that the cultivation of flowers is known since time immemorial, definite or systematic statistical information of the area under this group of crops was not available in India till the early 1980s. Use of flowers for vase purposes, popular as cut flowers, is of recent origin and hence the documented evidence of acreage under this group is not maintained systematically. As per the latest available statistics, a total of around 88,607 hectares (Singh, 2002) is under flower cultivation and this includes the area under both traditional and vase purposes. The area under floriculture depicts an annual increase of 13 per cent over the 1989 figures.

AREA AND PRODUCTION TREND OF FLOWERS IN INDIA

Documented evidence of the cultivation of cut flowers for vase purposes can be traced back to around 1985. The statistical information on the spread and production of flowers indicates wide spatial and temporal variations across different time periods. The area under flower crops was only around 26,000 hectares till the early 1980s with Tamil Nadu, Karnataka and West Bengal representing the main growing States. However, there has been a sudden increase in flower area, with an all time high of around 87,000 hectares reported in the year 1995, which could be attributed to substantial increase in the flower area in Tamil Nadu. The area under flowers in 1998-99 is around 73,972 hectares, with a total production of 418,848 tonnes. Out of the total area of 88,607 hectares reported under flowers during 1999-2000 (Singh, 2002), around 500 hectares, which is hardly around one per cent of the total area, is said to be under protected cultivation of flowers. This is a recent phenomenon, with the introduction and expansion of area under hi-tech floriculture.

EMERGENCE AND GROWTH OF HI-TECH FLORICULTURE IN INDIA

Initiated to meet the growing international demand, hi-tech floriculture was given a priority status with preferential credit, delineation of Export Promotion Zones (EPZs), and incentives in the form of special concessions and policy support. Although more than 177 hi-tech floriculture projects were approved, there are around 134 projects which have been established and over a 100 which are actually functional. Karnataka with more than 50 projects, Pune with around 33 projects, Hyderabad with around 15 projects and Haryana with 28 projects represent the main hi-tech floriculture (rose) production centres of the country.

COMPOSITION AND TREND IN EXPORT OF FLORICULTURAL PRODUCTS

A significant feature of the flower trade in India is the growing exports, which indicate a multifold increase within the last one decade. Floricultural exports constituted an important component of horticultural crop exports and account for around 1.32 per cent of the total horticultural exports. Within the group of floricultural exports, it can be seen from the Table 3 that live plants followed by dried plants accounted for the major chunk up to the early 1990s.

However, the trend changed from 1992-93 onwards, with the increase in cut flower exports. The export of cut flowers constituted hardly 0.88 per cent of the total exports during 1988-89, but increased to 21.6 per cent by 1997-98 and to 27.89 per cent by 1999-2000. This increase is essentially a result of the favourable policy support and the increased export performance by the hi-tech floricultural units set up during this period. In view of the growing significance of floriculture, the Government of India had allotted over Rs. 100 crore as budgetary support towards the promotion of this sector in the Eighth Plan period.

Despite the increased support and an increasing trend in exports, the hi-tech floriculture sector as an agri-business activity is yet to be stabilised under Indian conditions. Dwindling year-to-year exports and increasing number of growers defaulting credit repayment in recent times

TABLE 3
Trend in Export of Floricultural Products (Rs. lakhs)

Description	Cut Flowers	Dried Flowers	Live Plants	Dried Plants	Bulbs	Total
89-90	4.90	25.54	315.44	167.64	43.30	556.82
Per Cent	0.88	4.59	56.65	30.11	7.78	100.00
91-92	39.90	643.51	401.53	239.87	120.58	1445.39
92-93	108.99	735.38	305.66	262.32	78.34	1490.69
93-94	99.64	1092.13	244.20	243.98	149.53	1829.48
94-95	299.82	1957.81	579.74	128.08	102.34	3067.79
95-96	1003.30	3645.56	795.32	358.28	192.16	5994.62
96-97	1808.65	3233.91	912.12	256.22	128.43	6339.33
97-98	1894.12	5237.30	671.20	483.54	298.08	8584.24
Per Cent	21.60	61.01	7.82	5.63	3.47	100.00
98-99	2511.65	5104.01	—	1816.50	271.49	9703.65
99-00	2912.74	4435.31	—	2925.45	169.96	10443.46
Per Cent	**27.89**	**42.47**	—	**28.01**	**1.63**	**100.00**

Source: *APEDA Statistics*, 1999-2000.

has left doubts of success of the industry. This could be attributed to a number of production and non-production factors common to any teething nascent industry. An attempt has been made in the following section to bring forth the specific features of the industry, the peculiarities of technological features and associated economic factors that decide the viability and profitability of the industry so as to draw some inferences for the future prospects of this industry.

SPECIFIC FEATURES OF THE INDUSTRY/PRODUCTION CENTRES/TECHNOLOGY OPTIONS

In view of the high initial investment, the initiation of export oriented production centres was identified in such a way that they fulfilled the objective of cost effective production. Thus, Karnataka, especially the Hosur region and Hyderabad in the South, Pune in Maharashtra in the West, Punjab and Haryana in the North, and Bengal in the East were identified as the main growing centres of hi-tech floriculture.

Being a new technology, hitherto not common to Indian conditions, the establishment of hi-tech floriculture production units involved several specific features, such as the selection of an appropriate polyhouse technology, identification and selection of the right variety and the right growing method.

(i). The Polyhouse Structure

A polyhouse is a framed structure, covered with transparent or translucent material like UV stabiliser film and is large enough to grow crops economically. Such cultivation facilitates artificial control of sunlight, humidity and temperature, provision to include drip irrigation, misters and fertigation.

The initial investment in floriculture projects was directly correlated to the type of polyhouses and their attendant features. The main source of polyhouse design were the traditional production centres like Holland, Israel or France thereby adding up to the initial high cost of the projects. The technology options available for the growers today range from the least expensive indigenous designs followed by imported designs with special growing media. The most popular type of polyhouse designs are (1) the saw-toothed, top vent type, (2) the open top type, and (3) the tunnel type; each with a specific advantage. The saw-toothed top vent type is by far the most ideal for Indian conditions, while the open top and tunnel types help in increasing flower quality. Though more than 60 per cent of the growers opted for the saw-toothed type and mostly opted for indigenous fabrication, some growers imported the polyhouses on a turnkey basis from France or Israel involving a very high cost. The initial investment in a polyhouse, which accounts for the major proportion of the total project cost thus shows wide variations from as low as Rs. 40 lakh a hectare to over Rs. 221 lakh/ha.

(ii). The Exotic Variety

The second important feature of export-oriented production of cut flowers is the selection and sourcing of the planting material. Since the production is for export markets, it was mandatory that the exotic varieties of flowers in demand in the international market be obtained from the right source. The ruling rose varieties are Grand Gala and First Red, among the red coloured ones and Nobless and Ravel among the pink fleshed ones.

Being exotic, these varieties are protected under Intellectual Property Rights/Plant Breeders' Rights and include a component of royalty in their price. The price per plant varies from 3.5 to 4 DFL, of which nearly 50 per cent is charged as royalty depending on the stage of product cycle the variety is in. The cost of planting material itself accounts for over Rs. 45 to 50 lakh per hectare, as a total of over 65,000 plants are required per hectare. The Indian industry can be grouped so far as a red based industry, which primarily concentrated on Holland as the main market centre.

(iii). Growing Technology

Growing technology was not familiar to the Indian growers and hence the majority opted for high priced consultants, thereby adding substantially to the initial costs. Use of expensive fertilisers and pesticides, inappropriate methods of cultivation, lack of standard methods of cultivation and non-suitability of some of the varieties for Indian growing conditions were the major causes for recurring failures of the cut rose producers.

ECONOMICS OF CUT ROSE CULTIVATION FOR EXPORT

The economic evaluation of cut flower production under protected conditions would include the evaluation of the investment options available, the costs, returns and profits in cultivation of cut flowers, the market and marketing process and the constraints thereof.

Economic Evaluation of Investment Options

The establishment and cultivation cost of hi-tech floriculture for exports is a highly capital intensive option and has a direct correlation with the type and extent of technology selected. The investment requirements for polyhouse-grown cut flowers is nearly 10 times higher in comparison to that of open-cultivated flowers due to the technological requirements (Subrahmanyam, 1997; Sivaraman, 2000). Very few studies portray the economics of investment in polyhouse-grown cut flowers, and those which are available, are restricted to roses alone, as the rose forms the major component of the cut flower sector so far. Though the rose is the most important among the cut flowers grown, only around 200 hectares is the area under cut rose, as compared to over 5,567 hectares under open-cultivated rose.

Establishment Cost of Hi-Tech Rose Production Unit

The cost of establishment for a hectare of polyhouse-grown cut rose production unit is Rs. 150 lakh (Table 4), with most of the technological components of indigenous origin. Studies in the past have placed the establishment cost at around Rs. 225 lakh/ha, with a range of Rs. 176 lakh to Rs. 446 lakh, depending on the technological choices made (Sudha, 2000; Chengappa, 1999) (Table 5). The technology options revolved around the type of polyhouse selected and the scale of production, ranging from an all-indigenous low cost polyhouse for small-scale production up to 2 hectares [Indigenous Least Cost and small scale, (ILS)], to an imported design of more than 4 hectares [Imported High Cost on medium scale, (MHM)] and for imported designs with special growing medium such as rock wool blocks grown on an average size of around 3 hectares [Imported with Very High Cost (MVM)]. As could be seen from Tables 5 and 6, the cost of erecting a polyhouse is the major component of the total cost, and accounted for over 33 per cent of the total project cost (Anonymous, 1997). The cost of the polyhouse varied from Rs. 30 lakh/ha for an Indian design to Rs. 66 lakh/ha for a French design and to Rs. 183 lakh/ha for an imported Holland design. Following closely, is the cost of the exotic planting material, accounting for 21.33 per cent of the establishment cost.

A high degree of variation in costs due to variations in adopted technology, absence of

TABLE 4
Establishment Costs per Hectare of Hi-Tech Rose Farm

Particulars		Value (Rs. Lakhs)	% to Total
I.	**Land & Buildings**	**14.0**	**(9.33)**
	Land and site development	7.0	(4.66)
	Buildings for storage, etc.	7.0	(4.66)
II.	**Greenhouse and Accessories**	**50.0**	**(33.33)**
	Poly house 10,000 sq.m.	40.0	(26.66)
Poly film, shade nets, side covers, etc.		4.0	(2.66)
	Fertigation equipment and installation	3.0	(2.00)
	Drip irrigation system	1.5	(1.00)
	Fumigation structures, irrigation pipes, etc.	1.5	(1.00)
III.	**Cold Storage, Refrigerated Van, etc.**	**24.0**	**(16.00)**
IV.	**Cost of Planting Material**	**32.0**	**(21.33)**
V.	**Other Fixed Costs**	**22.0**	**(14.32)**
	Preliminary, pre-operative, contingency, etc.	8.0	(5.33)
	Margin money, public issues	10.0	(6.66)
	Miscellaneous	4.0	(2.66)
VI.	**Cost of Planting and Maintenance up to First Harvest**	**8.0**	**(5.33)**
	Total Establishment Costs.(I+II+III+IV+V+VI)	**150.0**	**(100.00)**

TABLE 5
Establishment Cost of Cut Rose Producing Units (Rs./ha)

Features	(ILS)*	(MHM)	(MVM)	(ILL)	Pooled	2000-2001
Total project cost	189.87 (100)**	248.08 (100)	438 (100)	178.9 (100)	225.01 (100)	150 (100)
Cost of polyhouse	50.86 (26.8)	73.47 (29.6)	221 (50.5)	42.8 (24)	69.44 (30.9)	50 (33.3)
Equipment and machinery	27 (14.22)	26 (10.5)	32 (7.3)	10 (6)	25.45 (11.3)	24 (16)
Cost of plants	43.9 (23)	56 (22.6)	54 (12.3)	32.5 (18.2)	47.24 (21)	32 (21.33)
Technology transfer fee	6.93 (3.6)	6.36 (2.6)	19.6 (4.5)	2.8 (1.6)	7.26 (3.2)	–
Other capital costs	61.18 (32.2)	86.25 (34.8)	111.4 (25.4)	90.8 (51)	75.62 (33.6)	44.0 (29.33)

* Technology classifications.
** Percentage to total; 2000-2001 figures are based on personal communication with producers and NABARD estimates.

standard practice with regard to the selection of polyhouse structures have been the major factors for high variability in costs across farms. However, the comparative advantage for India is visible in terms of lower initial costs for erecting a protected greenhouse as also in the use of plastic in place of the traditionally used glass. The cost of establishing a greenhouse (with glass or polythene) was in the range of Rs. 2,000 per square metre in the U. S. A. as against Rs. 350 per square metre for structures using wooden poles as in Kenya (Sudha, 2000; Wen Fei, 2000; Anonymous, 1997). The high cost of establishment of over Rs. 200 lakh could be attributed to the high dependence on imported technologies (Anonymous, 1997); growers' lack of knowledge about the technology options, high costs incurred on technology transfer and exploitation at various levels, which lead to insurmountable technology transfer costs.

The selection of varieties for cultivation also had an impact on the high establishment costs. Growers often relied upon export consultants for the selection of specific varieties suitable for foreign markets and thereby opted for imported exotic varieties paying high royalties for the planting material. The cost of a rose plant varied from DFL 1.5 to over DFL 3, amounting to around Rs. 56 to 78 per plant. With a population requirement of around 65,000 plants per hectare, the total investment in planting material accounted for over Rs. 32 lakh. There were other non-production costs as well, such as, raising of public issues and allotment for the margin money, etc., which contributed towards the overall increase in the cost of establishment.

Cost of Annual Maintenance of Hi-Tech Rose Farms

The average annual cost of maintenance for an export oriented rose farm works out to Rs. 67 lakh, with a range of Rs. 59.49 to Rs. 99.75 lakh/ha for different technology options, as has been presented in Tables 6 and 7.

TABLE 6
Costs and Returns from a Hectare of Polyhouse-Grown Cut Roses

		(Rs./Ha)		
No.	**Particulars**	**Value (Rs. Lakhs)**		**Per Cent to Total**
A.	**Production costs**	**21.26**		**31.62**
a.	Cultivation Expenses	7.51		11.17
b.	Utilities	2.87		4.27
c.	Grading and packing	3.68		5.47
d.	Administrative overheads	2.27		3.38
B.	**Selling costs**	**22.46**		**33.42**
a.	Freight	13.38		19.91
b.	Commission and other marketing costs	9.08		13.51
I.	**Variable costs (A+B)**	**43.72**		**65.04**
II.	**Fixed costs**	**23.49**		**34.95**
	Total cultivation costs (I+II)	**67.21**		**100.00**
	Yield and Sale Details			
			Export Sale	**Domestic Sale**
III.	**Yield (Lakh stems)**	**15.57**	**9.73**	**5.84**
IV.	Price (Rs./spike)	6.01	8.4	2.1
V.	Gross Returns (Rs. lakhs)	94.00	81.73	12.26
VI.	**Net Returns**	**26.79**		
VII.	**BCR**	**1.40**		

The annual maintenance costs include the cost on the manure, fertiliser and pesticides. High doses of imported chemicals and pesticides also added to the high cost of maintenance. A majority of the projects were operated by well established corporate houses, thereby incurring high expenses on administrative expenses, such as maintenance of a corporate office in urban areas or cities and on communicating between city and farm. It could be seen that the administrative expenses accounted for over 5 per cent of the annual cultivation costs. The expenditure on utilities was also higher for the hi-tech farms as the expenditure on maintaining the temperature and humidity controls in the computer-operated polyhouses was substantially higher than the regular crop cultivation.

The component of labour cost was only around 4 per cent of the total cultivation costs under Indian conditions indicating the cost effective production under Indian conditions compared to the high percentage of over 24 on labour in developed countries (Bhattacharya, 1997).

The annual maintenance costs also include the cost of selling, which is inclusive of packaging, freight, commission and other export costs. On an average, the hi-tech farms incurred around

TABLE 7
Costs, Returns and Profits per Hectare of Hi-Tech Rose Units Under Different Groups (1997-98)

No.	Parameters (Rs.lakhs)	Group 1		Group 2		Group 3		Group 4		Pooled	
		ILS	% to Total	MHM	% to Total	MVM	% to Total	ILL	% to Total		% to Total
1.	Production expenses	6.54	16.45	9.30	21.19	8.90	14.51	6.10	12.45	7.51	17.18
2.	Utilities	2.38	5.99	2.77	6.31	7.33	11.94	2.52	5.15	2.87	6.56
A.	**Sub-total (1+2)**	**8.92**	**22.44**	**12.07**	**27.50**	**16.23**	**26.45**	**8.62**	**17.60**	**10.38**	**23.74**
3.	Grading & packing	3.19	8.02	3.56	8.11	7.28	11.86	3.89	7.94	3.68	8.41
4.	Administration	7.65	19.24	6.14	14.01	7.87	12.82	7.36	15.02	7.20	16.47
B.	**Sub-total (A+3+4)**	**19.76**	**49.70**	**21.77**	**49.62**	**31.38**	**51.13**	**19.87**	**39.91**	**21.26**	**48.62**
C.	Export cost	20.00	50.30	22.10	50.38	30.00	48.88	29.47	60.09	22.46	51.38
D.	Total production cost(B+C)	39.76	100.00	43.87	100.00	61.38	100.00	49.34	100.00	43.73	100.00
E.	Depreciation	7.76		12.79		20.0		7.44		10.14	
F.	Interest	11.97		13.0		18.37		17.21		13.35	
G.	**Total cost (D+E+F)**	**59.49**		**69.66**		**99.75**		**73.99**		**67.21**	
H.	Yield (No. of stems in lakhs)		14.53		16.05		18.00		17.15		15.57
I.	Gross Returns (Rs. lakhs)		80.34		95.46		144.0		120.10		94.01
	Average Price (I/H) (Rs./spike)		**5.52**		**5.95**		**8.0**		**7.00**		**6.01**

ILS = Indigenous low cost and small scale; MHM = Imported high cost on medium scale; MVM = Imported very high cost on medium scale; ILL = Indigenous low cost on large scale. Administrative costs include the interest on working capital also.

Rs. 22 lakh per hectare as selling costs, with a range of Rs. 20 lakh to Rs. 30 lakh a hectare, depending on the yield obtained and the percentage exports undertaken by the hi-tech farm. On an average, the hi-tech rose farms were producing a total of over 157 lakh stems/ha, out of which at least 62 per cent were of export quality. The production technology adopted influences the total yield and also the percentage of export quality flowers produced by a farm (Chengappa, *et al*, 1998; Tilekar and Nimbalkar, 1999). Efficient production systems depend on the capacity of the farm to achieve a higher number of export quality stems per square metre. The average stems per square metre vary with the variety as well, but the average under Indian conditions is around 167 stems (Anonymous, 1997; Sudha, 2000).

Profitability of Hi-Tech Rose

The profitability of cut roses for export purposes depends largely on the proportion of export quality flowers produced and the price realised. The cost per spike of cut rose produced and marketed was in the range of Rs. 4.98 to Rs. 5.80 per spike when produced for the export market and the realisation was Rs. 8.20/spike to Rs. 9/spike in export sale. The overall profitability of the hi-tech rose farms could be in the range of over Rs. 27 lakh, net of all overheads, if the farm is able to export at least 62 per cent of the total production (Table 6). On an average, the hi-tech floriculture fetched a net realisation of up to Rs. 27 lakh a hectare and a BCR of over 1.40. However, this was not a standard figure, as a number of projects were not able to achieve such high export quantities.

From the point of the quantum of investment made, this may be a low rate of dividend, categorising this agri-business venture into a low profit, high-risk enterprise.

Markets, Marketing Channels and Practices

Marketing Channels, Costs and Options Followed by Hi-Tech Rose Units: The export -oriented rose units operate in two distinct market networks, namely, one for the domestic market and the other for the export market. Their sale to the international markets is mainly through market intermediaries, such as export traders, who act as the forwarding agents to the international markets. Direct export of rose spikes is also common but involves additional costs at every assembly centre. In addition to the packing, transport and clearing and forwarding costs, the growers need to incur additional expenses such as transport charges at the international port and commission charges at the market. However, identifying a market intermediary at international markets helps realise higher prices in direct sale and also avoids market commission (Figure 3). The preference for a specific market during a specific time was largely dependent on the colour of the flower and the season. Since marketing costs were very high, the growers' preference for a specific market was also dependent on the quantity of the produce being exported.

The traditional markets were Holland and Europe, while Singapore, Australia and Japan are the new markets for the hi-tech floriculture units. However, the Holland market is considered a safe one, irrespective of the colour of the flower, stem length and season. Growers' first preference for a market is based on price stability and the capacity of the market to absorb large volumes. Nearly 43 per cent of the Indian growers opted for the Holland market as the best for all colours, while 86 per cent opted for Holland especially for the reds (Sudha, 2000).

Singapore, an upcoming market, is identified as the second best for reds, while Europe is the third best market for cut roses. The market preference for pastoral colours like Confette and Prophyta gets divided between Holland and Japan. September and February (Valentine's Day) are the two major peaks for flower sales. Over 71 per cent of the growers prefer Holland for reds during Valentine's

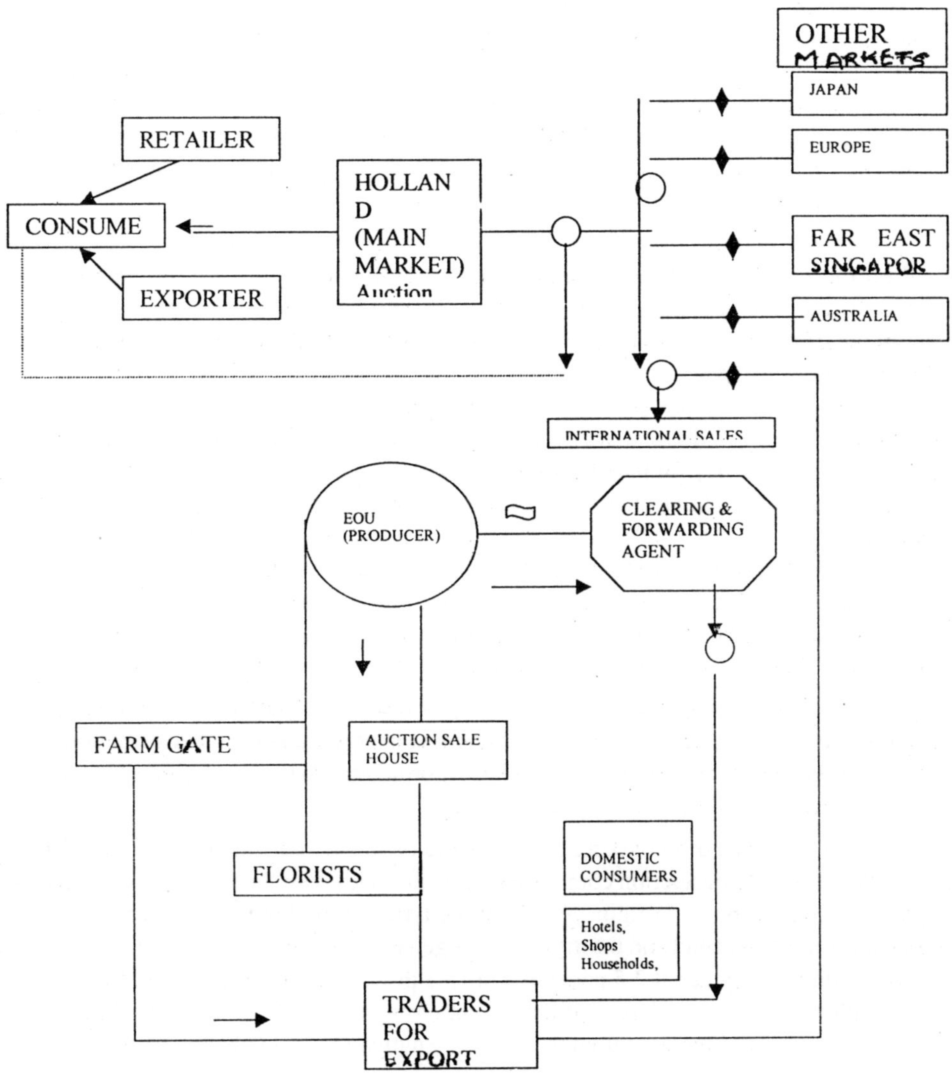

Figure 3. Marketing channels followed by hi-tech rose units in export and domestic sale of roses.

Day. At least 29 per cent were of the opinion that the Japanese market is specially preferred during Valentine's Day in February, even for pastoral colours.

Marketing Costs and Margins

Marketing is a specialised activity in cut roses. Being highly perishable, the flowers have to be harvested, graded and packed in special ways. The harvested flowers are graded based on the stem length as long (60 cm and above), medium (50 cm) and small (35 to 50 cm), and wrapped in cardboard wrappers in bundles of 20 numbers. These bundles are kept in buckets with water and cold stored at 0 to 2° C. These are then packed into specified size cardboard boxes based on the ultimate destination. For markets of Holland and Europe the packing is in long boxes, accommodating up to 400 spikes. The packaging for the Japanese market is distinctly different, with a large number of shorter length cut flowers per box. Marketing costs and practices are complex and significantly different in different markets.

The specificities of export marketing costs in cut rose marketing is presented in Table 8, based on the costs prevalent at Alsmeer market, Holland. Three types of marketing costs are charged for cut rose at Holland markets. The first type includes 'handling charges', charged per single spike of rose, 'lot charges' charged for a specific number of flowers in a pre specified stem length and 'bucket charges'. A 'lot' indicates a whole bunch of flowers of a single colour of a uniform stem length. Bucket charges are for a single bucket, which accommodates 160 spikes of a single uniform length, and a total of 18 buckets make a trolley for the spike lengths above 60 cm, while 27 buckets make a trolley for lengths lower than 60 cm. Trolley charges are for a single trolley, that is, the growers pay for a full trolley, for 18 buckets of 160 stems each (2,880 stems) in the spike length of more than 60 cm or for a total of 4,320 spikes or 27 buckets each with 160 spikes. The growers need to pay for the full trolley even if the number of spikes is lower than 2,880 in the >60 cm category or less than 4,320 spikes in the <than 60 cm category. Similarly, even for a single additional bucket over and above the prescribed numbers, one additional trolley charge needs to be paid. Hence, it is very important for the growers to plan their harvests and export quantities well in advance.

The second type of costs is charged per week, irrespective of the number of times the market is accessed. These include the (1) document charges, and (2) the distribution costs. Hence, it could be inferred that the quantity of flowers exported or sold play a significant role in contributing to the market charges while accessing the Holland market.

The third type of marketing costs is based on the actual volume of flowers sold. These are the (1) PVS levy, (2) the auction commission, which varies with the season, and (3) the import duty varying with the country of origin. All the three types of marketing costs are charged for auction sales, while only the second and the third are charged in case of direct sales. The auction commission and import duty are higher at 7.5 per cent and 15.3 per cent of the gross realisation in case of direct sales. Table 9 presents the details of marketing costs in the case of auction sales and direct sales. Two volumes, the minimum required for >60 cm length and <60 cm length have been worked out.

Marketing costs, like handling charges, lot charges and bucket charges work out to around 1.5 to 2 guilders per spike when growers participate in auction sales. The major chunk of the marketing costs is accounted for by the auction commission and import duty, both of which change, based on the season. The commission is higher during the season when Indian growers normally access the international market. The total market costs in auction sales worked out to 17 per cent of the gross price realised during the lean period and 20 per cent of the gross price during the peak

TABLE 8
Marketing Costs and Their Variations in Alsmeer Market

No.	Item	Rate	Auction Sale		Direct Sale	
			Lean Season	Peak Season	Lean Season	Peak Season
Type I Costs (Lot-based charges)						
1.	Handling	ƒ0.027/spike	Remains same for all seasons		Not applicable	
2.	Bucket charges		160 stems of a single colour in single length		Not applicable	
3.	Lot charges	ƒ6.00/lot	A lot is all spikes in single colour and single length		Not applicable	
4.	Trolley charges	ƒ12.00/trolley	A trolley is 18 buckets for more than 60 cm length; 27 buckets with less than 60 cm length		Not applicable	
Type II Costs (Costs/week)						
1.	Documentation costs	ƒ85/week	Remains same for all seasons and all modes of sale			
2.	Distribution costs	ƒ35/week	Remains same for all seasons and all modes of sale			
Type III Costs (as a % of gross returns)						
1.	PVS levy	0.625% of GR	Remains same for all seasons and modes of sale			
2.	Commission	4.7% (lean) 7.5% (peak)	Changes with lean and peak season but remains same for different modes of sale			
3.	Import duty	12% (AS) 15.3% (DS)	Changes for auction and direct mode of sale			

period, for various volumes sold. An additional tariff up to 15 per cent is charged on the flowers from India, thereby making the marketing costs account for 33 to 35 per cent of the gross realisation.

In direct sales, the handling charges, lot charges and trolley charges could be avoided; however, the commission and the import duty were higher. Assuming the same gross realisation for the lean and peak seasons indicates that the marketing costs would account for nearly 59 per cent of the gross realisation in the case of the lean season, thereby making it uneconomical to take up direct marketing. On the contrary, when the price realised is higher (180 guilders was obtained based on the peak season price realised by the best varieties like First Red and Grand Gala), the direct sale could work out to be highly profitable. In order to maximise profits during the peak seasons, growers often plan their harvests by adopting specificmarketing strategies.

TABLE 9
Marketing Costs for Imported Cut Roses In Alsmeer Market, (Holland) (Cost/Spike)*

No.	Item	Rate	Auction Sale		Direct Sale	
			Lean Season	Peak Season	Lean Season	Peak Season
Type I Costs (Lot-based charges)						
1.	Handling	ʃ 0.027/spike	0.027	0.027		
2.	Lot charges	ʃ 6.00/lot	0.002	0.002		
3.	Trolley charges	ʃ 12.00/trolley	0.004	0.004		
	Sub total		0.033	0.033		
Type II Costs (Costs/Week)						
1.	Documentation costs	ʃ 85/week	0.03	0.03	0.03	0.03
2.	Distribution costs	ʃ 35/week	0.012	0.012	0.012	0.012
	Sub total	-	**0.042**	**0.042**	**0.042**	**0.042**
Type III Costs (as a % of Gross Returns)						
1.	PVS Levy	0.625% of GR	0.5	0.85	0.5	0.85
2.	Commission	4.7% (lean) 7.5% (peak)	3.76	6.35	3.76	6.34
3.	Import Duty	12% (AS) 15.3(DS)	9.6	16.2	20.65	20.65
	Sub Total	ʃ	**13.92**	**23.39**	**25.33**	**28.26**
Total marketing costs		ʃ	**13.92**	**23.42**	**25.33**	**28.26**
Marketing costs in Rupees/spike**		@Rs21=1DFL Dollar	**2.92**	**4.92**	**5.31**	**5.93**
Gross realisation		Rs/spike	**16.80**	**28.35**	**16.80**	**28.35**
Net realisation		Rs/spike	**13.85**	**23.43**	**11.49**	**22.43**
Market Charges as a % of Gross realisation			**17.35**	**17.35**	**31**	**20**

*Cost/spike is given in Dfl cents; ** The marketing costs/spike is given in rupee equivalent on suitable conversion.

MARKETING STRATEGIES

The hi-tech rose producers not only need to organise efficient production, but also select an efficient marketing strategy to operate in the international trade. This could be attributed to the frequent fluctuations in the international prices. At least two strategies are popularly adopted by the growers in rose marketing. These are, (1) flushing, and (2) direct marketing.

Flushing: This refers to a production practice where a large number of rose stems are made ready for harvest at a time. Pruning and bending back the requisite number of stems at a time ensures production planning as the rose stems would be ready for harvest exactly 40 days after the bending. Thus, rose growers can integrate production with marketing through the practice of flushing. By cutting the required number of shoots on a given day, growers can plan to maximise their harvests to meet a specific occasion. This practice is popularly known as flushing.

Rose being a fashion driven flower, the market demand is typically high during specific dates. These are:

- Mother's Day
- Thanksgiving Day
- Christmas Day (25 December)
- New Year's Day (1 January)
- Valentine's Day (14 February)

Both arrivals as well as prices peak on Valentine's Day. The monthly average price of different varieties at the Holland market presented in Figure 4, indicates that irrespective of the colour or variety, the market price has two distinct peaks, one during February, coinciding with Valentine's Day while the other is during September. Hence, by following specific strategies like flushing, growers try to maximise the harvest during these special occasions and reap high profits.

Direct Sale: Since the marketing costs are high and the marketing practices so complicated, growers avoid auction sales. They identify specific agencies and market the flowers directly to these parties. In such strategies, growers can sell lower volumes and avoid higher costs like trolley charges and lot charges, which remain the same even for small volumes. Direct sale is a common method in sales to the Singapore and Australian markets.

Ideal Methods of Production and Marketing: The growers often prefer to have direct sales either in Holland auction houses or operate in the Singapore market, as it would fetch higher prices. The Singapore market was preferred as it is accessible throughout the year. Preference for the Japanese market was the least as the marketing practices are very cumbersome and expensive. During lean seasons, the preference was for the commission sales to other markets. Growers also expressed the feeling that the domestic market sale is a very important component of their marketing strategy. Though the existing auction centre was inadequate to absorb all the flowers, it was inevitable to access the same (Table 10).

International Price Analysis for Better Marketing Strategy for Hi-Tech Roses

Marketing strategy also involves price analysis, as any forecast about the probable price on a specific occasion can help the growers plan an effective marketing strategy.

Behaviour of World Prices: Prices realised in the international market have a direct bearing on the profitability of hi-tech rose farms. Though the exports as well as the export earnings from cut flower roses from India have been increasing over the years at an annual compound growth of 2.3 per

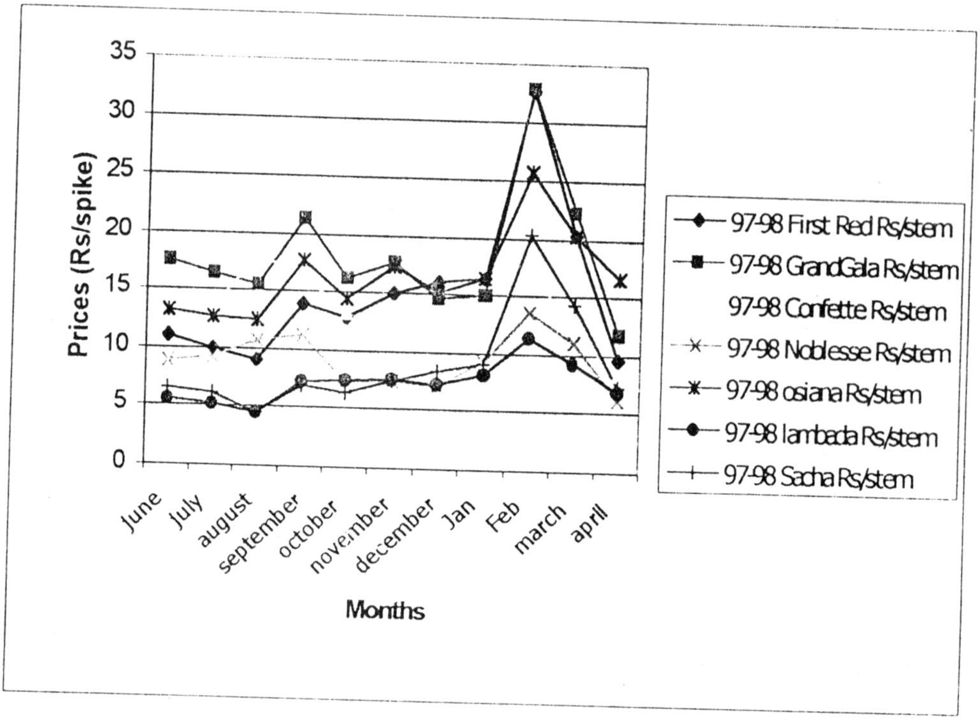

Figure 4. Pattern of monthly price hike for different varieties of cut rose at Alsmeer Market, Holland.

cent, this forms only 0.07 to 0.1 per cent of the total sales in the global trade. Despite the temporal variation, the prices of cut roses show some typical seasonal variations. It could be seen that in terms of price realisation per spike of cut rose, red varieties like First Red and Grand Gala ranked first especially during Valentine's Day. On the contrary, during September, the highest was for Grand Gala, followed by Osiana and Lambada. This trend indicates that during Valentine's Day, the consumer preference is for red varieties while during September, the consumer preference is for colours other than red.

PRICE ANALYSIS AND FORECAST

Analysis of the arrivals and prices at the international market centres can be of value for cut-rose producers and exporters for planning their production and harvests. Using econometric methods such as ARIMA models, it is possible to predict the future prices and thereby plan the marketing strategy well in advance. The analysis of arrival and price of specific varieties such as the First Red indicates an increasing trend in the coming years. Imports of variety First Red at the Alsmeer market centre at Holland showed a compound growth of over 101 per cent per annum, indicating a future scope for this variety. The arrivals in the year 1990-91 was 0.291 million stems, while it was 35.78 million spikes in the year 1994-95, and 78.23 million stems in 1997-98, indicating a near ten-fold increase within one year from 1990-91 to 1991-92 and a four old increase from 1991-92 to 1992-93. The arrivals of First Red show a steady but slow rate of increase after 1993-94. This trend

TABLE 10
Ideal Methods of Marketing Cut Rose for Exports

No.	Marketing Channel	Remarks
1.	Producer-(C&F)-Holland Auction-Wholesaler-Retailer-Consumer	Very high and complex marketing costs
	Producer- (C&F)-Wholesaler(CA)-Retailer-Consumer	Needs to identify specific period when the market is high, but some marketing costs can be avoided in international sales.
2.	Producer- (C&F)-Retailer-Consumer	Direct sale, highly preferred by producers as the marketing costs can be avoided, popular at Singapore and Australian markets.
3.	Producer- (C&F)-Consumer	Similar as the earlier one, avoids marketing costs.
4.	Producer-Domestic Auction-Florists-Consumer	Low volumes are transacted, often a distress sale, preferred to keep the cash inflows flowing.
5.	Producer-Procurer (specialist market intermediary)- (C&F)-International Sales	Most preferred, specialist player could be another producer, a merchant exporter, a trader or a florist. This is equivalent to on farm sales.

in the arrival of imported First Red could be attributed to the increased area and popularity of the variety in countries like India, Kenya and other African nations. The net importer's price, after deducting the import costs, was 0.76 DFLs during February 1991, and the same has been decreasing steadily over the years (Table 11).

The Auto Regressive Integrated Moving Average Method (ARIMA), an econometric model, has been used to forecast the price of First Red variety to integrate the production process and maximise profits.

The model indicated a seasonal behaviour, following a multiplicative model of (2,1,2) (0,1,1). The identical model could be specified as:

First Red arrivals

$$(1-0.33103\,B-0.269634\,B^2)(1-B)(1-B^{12})\,Yt =$$
$$(0.401790)(0.9186921)$$
$$9.949+(1-0.7182\,B^1-0.2796\,B^2-0.3009\,B^{12})a_1$$
$$(0.36204)(0.335106)(1.61462)$$

First Red prices

$$(1+0.31672\,B-0.0763\,B^{12})(1-B^{12})\,Yt =$$
$$(-1.6978)(0.52241)$$
$$-0.00692+(1-0.12451\,B^1-0.745716^2-0.9797\,B^{12})a_1$$
$$(0.7391)(4.5208)(0.6231)$$

TABLE 11
Arrivals and Prices of cv. First Red in Alsmeer Market

Year	First Red	
	Arrival (million stems)	Price (Rs./stem)
1988-89	na	na
1989-90	na	na
1990-91	0.291	16.24
1991-92	2.424	21.46
1992-93	8.87	20.10
1993-94	24.84	18.06
1994-95	35.78	16.83
1995-96	49.36	16.34
1996-97	56.16	14.34
1997-98	78.23	14.70
CGR	101.00	

The price predictions based on this model were used for planning a production process. Hi-tech rose farms identify an export plan and follow the method of 'flushing' an appropriate proportion of the farm area and the associated costs well in advance.

Production can be planned by flushing the desired number of plants or plant population for arriving at the optimum stems for export on a specific occasion. The costs and return for a specified decision, such as 25 per cent flushing during September, followed by 50 per cent flushing during Valentine's Day or New Year's Day, can be estimated based on the number of stems bent and the forecasted price (Table 12). In view of the high investment and high costs involved in export, an integration of production with international price is essential to minimise the risks (Sudha, 2002).

CONSTRAINTS AND PROBLEMS OF HI-TECH FLORICULTURE

Production Oriented

Based on the discussion so far, the production based constraints faced by the hi-tech farms can be summed up as

High initial investment and production expenditures, due to absence of standard technology options;

- Absence of specificity of varieties or type of flowers to be grown;
- Inconclusive impact of technology on quality of flowers;
- Inconsistent and low domestic demand;
- Absence of standard methods of marketing or market destinations;
- Cumbersome methods and absence of clear insurance cover against natural calamities;

TABLE 12
Net Realisation Per Spike Due to Integrated Production Management

Mode of Sale	Cost/Price	September	October	December	January	February
	Price	14.53	13.85	16.14	17.04	24.52
	Import Duty	2.18	2.08	2.42	2.56	3.68
	Net Price	12.35	11.77	13.72	14.48	20.84
(A) Costs and Returns in Auction Sale						
	Cost/Spike	**Net Realisation per Spike in Different Months**				
(1) 25% Flush	6.87	5.48	4.90	6.85	7.62	13.97
(2) 50 % Flush	6.87	5.48	4.90	6.85	7.61	13.97
(3) 75 % Flush	9.93	2.42	1.84	3.79	4.55	10.91
(4) 100 %	11.09	1.26	0.68	2.63	3.39	9.75
(B) Costs and Returns in Direct Sale						
	Cost/Spike	**Net Realisation per Spike in Different Months**				
(1)	7.31	5.04	4.46	6.41	7.17	13.53
(2)	7.32	5.03	4.46	6.40	7.17	13.53
(3)	10.72	1.63	1.06	3.00	3.77	10.13
(4)	12.01	0.35	-0.23	1.71	2.48	8.84
(C) Costs and Returns in 50% Auction and 50% Direct Sale						
	Cost/Spike	**Net Realisation per Spike in Different Months**				
(1)	7.09	5.26	4.68	6.63	7.39	13.75
(2)	7.09	5.26	4.68	6.63	7.39	13.75
(3)	10.32	2.03	1.45	3.40	4.16	10.52
(4)	11.55	0.80	0.22	2.17	2.94	9.29

making it essential to minimise the costs at every stage. Though a number of projects opted for very high cost polyhouses and growing methods, the definite impact of technology on quality of produce cannot be proved conclusively. Cost minimisation strategies through adoption of low cost technology options for polyhouse structures, use of biofertilisers and pesticides should be the objective of the industry for long run sustenance. The Kenyan experience with use of wooden poles for polyhouse construction helps in reducing the cost of establishment. Adoption of three-variety combinations on an average size of around 3 hectares, with a project cost in the range of Rs. 150 lakh and an annual production cost up to Rs. 45 lakh/ha fetching a yield of around 145 stems is an economically viable option of hi-tech floriculture (Anonymous, 2002; Tilakar and Nimbalkar, 2000).

Market Oriented

- Uncertainty of markets/destinations;
- Lack of sufficient cargo space and regularity of flights;
- High marketing costs and cumbersome procedures;

- Complex marketing methods and costing procedures;
- Absence of uniform procedures and assistance at international markets;

thereby forcing the growers to seek alternate marketing channels and methods. Adopting an appropriate market strategy, therefore, becomes an essential feature of hi-tech floriculture. Emergence of 'specialist retailers' has been a recent phenomenon, which has helped by decentralising the production from marketing functionaries.

Price Related

- Frequent fluctuations both in the exchange rate as also the market price;
- Absence of method of price predictions and few peak points.

Policy and Developmental Support for Hi-Tech Roses

Floriculture did not receive the desired focus for developmental inputs till the last decade. Though some State Governments initiated efforts in this direction through the introduction of special schemes, a major thrust started only since the Eighth Plan (1992-1997) period. A scheme on commercial floriculture was introduced on the recommendation of the Trivedi Committee in 1987-88, aiming to initiate plans for strengthening floriculture. However, subsequent to the Eighth Plan budgetary allocations, organisations like the National Horticulture Board (NHB) introduced several schemes for building infrastructure and post-harvest handling. Other organisations such as the Agricultural and Food Products Export Development Authority (APEDA) also provide several incentives to the farmers to encourage them to export floricultural products. Modernising auction centres and creating cold storage facilities at airports are some of the initiatives taken up by APEDA. Starting up of Infrastructure parks by the State Governments in Karnataka and Tamil Nadu for hi-tech floriculture is the latest in the developmental activities for this sector.

Concept and Measure of Competitive Advantage

One of the main reasons for the initiation and expansion of hi-tech floriculture is the perceived competitive advantage under Indian conditions. While climatic suitability, abundance of primary factors of production and closeness to international markets encompassed a relative position of advantage for India for cultivation of export-oriented cut flowers, the competitive advantage was envisaged through cost effective production.

Export competitiveness indicates the ability of a firm to produce commodities with a certain level of efficiency, so that they can compete in the international market through competitive price. Conventional economic measures of competitive advantage are by the Nominal Protection Coefficient (NPC).

NPC is given as a ratio of domestic price of the commodity to its world reference price, adjusted to transfer costs. Commodities with NPC of more than one are said to be protected by the policy regime relative to the situation that would prevail under free trade.

$$NPC = \frac{p^d}{p^r}$$

where P^d = is the domestic price, and

P^r = the international reference price adjusted to transfer costs.

Domestic market price is the wholesale market price of the commodity in a selected market. The reference price is the international price adjusted for transfer costs, marketing and trading margins including the processing charges if any, to make the commodity equivalent to the internationally traded commodity. The NPC for cut rose varieties grown under Indian conditions was less than one, thereby indicating the competitive advantage for cut flower production in India (Chengappa, 1999; Sudha, 2000) (Table 13).

Comparing the competitive advantage of several nations in the field of hi-tech floriculture, Mishra (1997) observes that, besides the agroclimatic suitability to grow a number of horticultural crops, India also possesses the advantage in terms of cheaper land and labour, research capability and the required policy support for growing as a leading nation in hi-tech floriculture. Though the industry suffered initial setbacks due to faulty choices of technology, the turnaround has already been initiated for exploiting the untapped potential.

ISSUES DUE TO WTO AND NEW GLOBAL ORDER

In addition to the technology and cost-related constraints, the Indian cut flower industry also has to face other constraints due to the prescriptions of the WTO and GATT agreements. Some of the important issues and their implications for the industry are:

1. Need for Intellectual Property Rights or Associated Legislation

First and foremost among the prescriptions of the WTO, which have a direct bearing on the Indian industry, is the urgent need for adoption of appropriate intellectual property regimes. Absence of the same could hinder the easy access for exotic plant varieties required for the industry's success. Indian legislation protecting the plant varieties and breeders' rights is in the process of being implemented.

The absence of the same has not hampered the Indian growers till now, as the industry so far has been based on the well-accepted First Red and Grand Gala varieties, and the growers pay the requisite royalties for acquiring the same. However, adopting such legislation could help Indian growers and or researchers to develop new varieties for the international market.

2. Removal of Most-Favoured Nation Status

The prescriptions of WTO also imposed the removal of according any special status, such as the Most Favoured Nation status, to any country under the international market access clause. This clause specifies that if a nation provides any special incentive for any of its allies, a similar facility be extended to all other trading partners as well. Thus, if such a clause is implemented and adhered to, the Indian flower industry stands to gain, as the differential tariff at the international market would become null and void and thereby the competitive edge of Indian flowers would increase. (The African countries get a special priority under the rome Convention between the EU and Africa).

3. Restrictions on Export Subsidies

One of the clauses under TRIPs prescriptions requires the removal of any form of subsidies for promoting export of agricultural commodities. This clause is perceived to have adverse implications on the export incentives provided for the cut-flower industry. However, there are no such incentives or subsidies given for the Indian flower industry, which need to be modified. Some of the incentives provided can be covered under the 'green box' policies or designated as 'schemes

TABLE 13
Nominal Protection Coefficient (NPC) for Hi-Tech Rose Varieties Grown in Bangalore (1998-99)

No.	Item	Variety								
		First Red	Grand Gala	Konfette	Noblesse	Osiana	Prophyta	Lambada	Sacha	Baronesse
	Alsmeer									
1.	Cents/stem net price (DFL)	76.05	83.36	61.88	42.14	75.45	32.35	33.48	40.10	24.26
2.	Alsmeer Price Rs./stem (Reference Price)	15.97	17.51	12.99	8.85	15.85	6.77	7.03	8.42	5.09
3.	Expected wholesale price (at farm gate)	3	3	3	3	3	3	3	3	3
4.	Handling and transport	2	2	2	2	2	2	2	2	2
5.	FOB price (3 + 4)	5	5	5	5	5	5	5	5	5
6.	Freight, port clearance	2.3	2.3	2.3	2.3	2.3	2.3	2.3	2.3	2.3
7.	Domestic reference price (5 + 6)	7.3	7.3	7.3	7.3	7.3	7.3	7.3	7.3	7.3
8.	NPC (7/2)	0.46	0.42	0.56	0.82	0.46	1.08	1.04	0.87	1.43

for reducing export marketing costs' approved for a specific period of time. The export freight subsidies also get categorised as 'non-actionable subsidies' under the 'green box' policies and can continue without any adverse implications.

STRATEGY FOR FUTURE EXPANSION

Diversification

Diversification into alternate flower crops which have export demand is the urgent need of the hour for the industry. Kenya, an important competitor for the Indian cut flower industry, caters to over 60 per cent of the global cut flower demand and produces over 40 different types of cut flowers. The organisational pattern followed by Kenya in integrating small and marginal farmers into export-oriented production needs to be emulated for future prospects of the Indian cut flower industry. Diversification into flowers other than rose is an immediate requirement for success of the Indian industry. Though there are sporadic efforts at growing carnations, Gerberas and other new ornamental crops, there have not been sufficient volumes produced forexport purposes.

Need to Develop Indian Varieties

Though the Indian cut flower industry is on a progressive growth path in recent years and has been able to achieve a turnaround, the long term goal of the industry needs to be on the development of Indian cut rose varieties which will hold promise at the international markets. This requires the establishment of appropriate intellectual property rights laws and associated safeguards to protect our germplasm and also help in developing the same to suit the international markets.

Thrust on Alternate Processes

There is a need for the industry to concentrate on developing other suitable products, which can be patented and marketed internationally. These include products like 'coconut husk blocks' serving as growing media, 'biological fertilisers and growth regulators' which could be patented. Attempts also need to be made in converting the excessive production of rubber into suitable propagation media or containers that can support rose plants.

CONCLUSIONS AND RECOMMENDATIONS

Hi-tech floriculture, an agri-business venture, has been initiated in India since the last one decade, primarily due to a perceived comparative and competitive advantage, to meet the growing international demand. Hi-tech floriculture in India is synonymous with the protected cultivation of rose for exports, though hardly around one per cent of the total area under rose is currently under protected cultivation. Initiated with a very high investment of over Rs. 250 crore, this export-oriented production is based mainly in the four major metropolis. The capital-intensive investment is also technology intensive and requires high maintenance costs and is a high-risk venture.

The global trade in floriculture, on one hand, is expanding with increasing demand and scope for new entrants like Kenya and India to capture the market; however, on the other hand, Indian growers so far have not been successful in doing so because of the high cost of marketing and high technology transfer costs and failure of technology to adapt to local situations, leading to low capacity utilisation.

Despite the competitive advantage at the international markets, Indian floriculture is yet to be stabilised due to the absence of standard cost minimisation efforts, increased capacity utilisation and appropriate marketing strategies. Understanding the global price trends and forecasting prices could be of value for integrating production with marketing.

The changing global trade order in the form of WTO and TRIPs prescriptions have immediate impact on the industry, through the need for legislative and institutional support. Anticipated withdrawal of trade barriers that created differential tariffs could be beneficial to the progress of the industry.

Although hi-tech floriculture at the moment depicts a high cost, high risk, low profit venture, adoption of suitable measures such as cost minimising the technology option, integrating production and marketing through proper price predictions and efforts at increasing the volume of exports hold a promise for its revival. Diversification into cut flowers other than rose is the most important strategy for tapping the potential.

REFERENCES

Anonymous. 'ITC, based on AIPIH-Union Fleurs, *Yearbook of the International Horticultural Statistics* and Produktschap voor Siergewassen (The Hague), 1997a.

Anonymous. *Market News Service.* Various issues of the publication of the International Trade Centre, UNCTAD, WTO, Geneva, 1998.

Anonymous. *The World Cut Flower Industry: Trends and Prospects.* ILO, Sectoral Activities Programme, Industrial Activities Branch working paper, SAP 2.8/wp. 139, 2002.

Anonymous. *Viability of Indian Cut Flower Exports Industry.* TATA Consultancy Services report prepared for APEDA, 1997b.

Attawar, Manmohan. 'Export of cut flowers and ornamental plants from India'. *Advances in Horticulture.* (Eds.). Chadha, K. L. and Bhattacharjee, S. K. 12: 1082-1107, 1995.

Bhattacharya. 'Floriculture management'. *Floriculture Today* 2(7): 15-23, 1997.

Chengappa, P. G., Govinda Reddy, D. M. and Nethra, N. N. 'Cut flower Industry in Karnataka. Analysis of trade and constraints'. Project report submitted to the Directorate of Marketing and Inspection (DMI), Ministry of Rural Areas and Employment, Government of India, New Delhi, 1998.

Iyer, Arun. 'Blooming for Valentine's Day'. *Economic Times,* January 28, 2002.

Kolavalli Shashi; Atheeq, L. K. and Xavier, Jacob. *Floriculture Industry in India.* New Delhi, India: Oxford and IBH Publishing Co. Pvt. Ltd., 1991.

Mishra, J. P. 'High-tech agriculture in India'. *Agricultural Economics Research Review* 10(2): 230-243, 1997.

Nageswara Rao, A. 'Higher employment and income potential of horticulture'. *IJAE* 52(3): 584-591, 1997.

Singh, H. P. 'Floriculture industry in India: its perspective'. *Indian Horticulture* 46(4): 51-56, 2002.

Subrahmanyam, K. V. 'Production and marketing research on ornamental crops'. In: *Advances in Horticulture.* (Eds.). Chadha, K. L. and Bhattacharjee, S. K. New Delhi, India: Malhotra Publishing House, 1995.

Sudha, M. 'Economics of protected cultivation of cut-rose for international markets: Role of price prediction models in integral production with marketing'. *Agricultural Economics Research Review* 14(1): 1-15, 2001.

Sudha, M. Competitive advantage of hi-tech floriculture around Bangalore—An economic analysis. Ph.D. Thesis. Department of Agricultural Economics, University of Agricultural Sciences, Bangalore, 2000.

Tilekar, S. N. and Nimbadkar, C. A. 'Self generated marketing system for rose cut flowers produced under hi-tech agriculture'. In: *Agricultural Inputs and Output Market Reforms in India* (Conference Proceedings). (Eds.). Ramesh Chand and Arora, V. P. S. New Delhi, India: Agricultural Economics Research Association, pp. 169-179, 1999.

Varadarajan, S. and Ravindran, N. *Economics of Flower Marketing in Coimbatore City.* Department of Agricultural Economics, TNAU, Coimbatore (Mimeo), 1976.

—oo(O)oo—

9

Insect Pests of Ornamental Plants and Their Management

P. Karuppuchamy and S. Palaniswamy

INTRODUCTION

ORNAMENTAL plants are grown widely to enhance the beauty of house and office premises, community parks and gardens. Moreover, growing of ornamental plants is highly remunerative and has got great export potential. At present, the estimated area under flower crops is 65,000 hectares with a production of 3 lakh tonnes of loose flowers and over 500 million cut flowers with stems. The growing interest in ornamentals is not only for meeting domestic needs but also for export-oriented floriculture. This has resulted in the increase in the area of cultivation besides intensification of floriculture both in the open and in greenhouses. Numerous insect and non-insect pests attack different parts of ornamental plants, thereby, reducing the quality and quantity of production. This chapter deals with the information on insect pests of major ornamental crops and their management.

PESTS OF ORNAMENTAL SHRUBS

ROSE (*ROSA* SP.)

Roses are grown throughout the world for cut flowers and for making essential oils, rose water and *gulkand.* Rose plants are prone to attack of many pests at almost every stage of plant growth and development. The important pests and their management are furnished below.

Aphids

Aphis gossypii Glover
Macrosiphum rosaeformis Das.
Macrosiphum rosae L.
Macrosiphum euphorbiae (Thom.)
Myzus persicae (Sulzer)

Chaetosiphon tetrapodes Walker.
Cinera sp.
Rhodobium porosum Sand.
(Aphididae: Hemiptera)

of which, *Macrosiphum rosaeformis* and *Macrosiphum rosae* are the common species. They are pear-shaped, soft bodied, light green to dark blackish green aphids and attack rose plants in plains and hills, respectively. *Macrosiphum rosae* infests tender shoots (Agarwala, 1983) and *Macrosiphum euphorbiae* infests shoots and buds of rose in India (Agarwala, 1983; Agarwala and Raichauduri, 1981; Nair, 1975). In North India, *Macrosiphum rosaeformis* occurs on roses from November to April. The nymphal development of its wingless form is completed in 11-14 days whereas winged forms complete their nymphal development in 14-19 days. *Aphis gossypii* attack cultivated varieties of roses from September to December with the peak period of activity in October. This aphid multiples well under temperatures of 20-25° C. The aphid, *Chaetasiphon tetrapodes,* sucks sap from tender shoots (David, 1957) and *Cinera* sp. infests shoots of roses in Himachal Pradesh (Sood and Kakar, 1990; Sohi and Singh, 1995).

Aphids of different species are found in clusters on the tender portions of shoots, buds, flowers and leaves. Both nymphs and adults feed by sucking the cell sap. As a result, tender shoots wither, buds fall prematurely and the flowers show malformation and fading. Severe damage to the top of the plant may reduce the number of flowers produced. Aphids also excrete honeydew on which sooty mould grows and some species transmit viruses. Aphids feeding on flowers make them unmarketable and their presence is a nuisance on the plants because they leave cast skins stuck to the plant when they moult, which affect the value of the plant/flower.

Management

- The parasitic wasps, *Aphidius rosae* Hal. and *Aphelinus* sp. parasitise the aphid and coccinellid beetle *Coccinella septumpunctata* Linn. and predate upon it (Atwal and Dhingra, 1971; Das, 1918; Dhingra, 1968).
- The larvae of many species of syrphids can control aphid population rapidly. The adult flies are often seen on or hovering near flowers. The slug-like, pale green to yellow maggots can consume as many as 400 aphids during their development.
- The larvae of green lacewing, *Chrysoperla carnea,* are voracious feeders of aphids and can consume up to 425 aphids per week.
- All species of aphids can be controlled with spraying of Malathion 50 EC at the rate of 875 ml or Dimethoate 30 EC at the rate of 750 ml or Phosphamidon 85 WSC at the rate of 187.5 ml in 250 litres of water per hectare (Atwal, 1986; Bhattacharjee and De, 2003).
- Spraying of Methyl Demeton 25 EC at 2 ml/l or neem oil at 3 per cent or Phosalone 35 EC at 2 ml/l or application of Carbofuran 3 G at 5 g/plant also controls the pest (Regupathy *et al*, 1994).

Thrips

Rhipiphorothrips cruentatus Hood
Lefroyothrips lefroyi (Bagn.)
Megalurothrips usitatus (Bagn.)

Physothrips andrewsi (Bagn.)
Retithrips syriacus (Mayet)
Taeniothrips fulvus Ananthakrishnan and Jagdish
Taeniothrips lefroyi (Bagn.)
Taeniothrips rhopalantennalis Schum.
Thrips tabaci Lindamen
Thrips hawaiiensis Morgan
Thrips florum Schm.
Thrips melanurus Bagn.
(Thripidae: Thysanoptera)

Lefroyothrips lefroyi was recorded infesting rose flowers in Himachal Pradesh (Sood, 1990). *Megalurothrips usitatus* feeds on flowers in many parts of India. *Physothrips andrewsi* also lacerates petals of rose and destroy its beauty (Bagnall, 1921). *Retithrips syriacus* attacked leaves show silvery appearance, turn brown, curl up and then fall away (Patel, 1988; Raghava Rao *et al*, 1964). Their feeding also destroys the flowers.

Rhipiphorothrips cruentatus is the most common species attacking roses and other host plants like *Lagerstroemia* sp. and *Punica granatum* Linn. The nymphs are reddish in colour whereas the adults are dark brown or black in colour. The pest breeds throughout the year except in winter when it pupates in the soil. The adult appears in March and lays eggs on the undersurface of leaves in plant tissues. *Taeniothrips fulvus* feeds on leaves and flowers of rose in Himachal Pradesh (Sood, 1990). *Taeniothrips lefroyi, Taeniothrips rhopalantennalis, Thrips tabaci* and *Thrips hawaiiensis* infest leaves and flowers from March to November and hibernate in the pupal stage from December to March (Ananthakrishnan, 1973; Butani and Dadlani, 1983; Lakshminarayan, 1968; Nair, 1975).

Both nymphs and adults of aphids feed by rasping the tissues and sucking the sap which oozes out from the wounds. Affected leaves are deformed with brown or silvery patches or burnt margins. The affected leaves get distorted, wither and drop down. In case of severe infestation, flower buds shed prematurely and show brown patches on the petals. The females lay 60-100 eggs during their life stage. The nymphs hatch in 2-7 days, feed during the first two instars and then pass through two non-feeding pseudo-pupal stages in protected parts of the plant such as leaves or flowers, in leaf litter or in soil. The nymphal and pupal periods are completed in 9-20 and 2-5 days, respectively. The females can reproduce with or without fertilisation.

Management

- Yellow or blue sticky traps are most useful for detecting the first infestation of thrips especially when rose is grown in greenhouse conditions.
- The predatory phytoseiid mites, *Euseius* sp., *Neoseiulus* sp. and minute pirate bug *Orius* sp. feed on several species of thrips.
- Spray Malathion 50 EC at the rate of 500 ml in 500 litres of water per hectare (Atwal, 1986).
- Spray neem oil 3 per cent or fish oil rosin soap at 2.5 per cent or Endosulfan 35 EC or Phosalone 35 EC at 2 ml/l of water.
- Soil application of Aldicarb 10 g at 8-10 kg or Thimet 10 g at 4-10 kg or Carbofuran 3 g at 12-30 kg/ha will be effective and economical only when they are used simultaneously to control thrips and nematode infestation (Randhawa and Mukhopadhyay, 1998).

Scales and Mealybugs

Scales

Lindingaspis rossi (Mask.)
Aonidiella aurantii Mskll.
Aspidiotus dictyospermi var. *arecae* Newst.
Aspidiotus transparens Gr.
Aspidiotus orientalis Newst.
Chionaspis sp.
(Diaspididae: Hemiptera)

Saissetia hemisphaericum Targ.
(Coccidae: Hemiptera)

Iceria purchasi Mask.
Iceria aegyptiaca Mask.
Iceria formicarum Newst.
(Margarodiae: Hemiptera)

Mealybug

Ferrisia virgata Cockll.
(Pseudococcidae: Hemiptera)

Lindingaspis rossi appears as reddish brown, waxy, scale like pox marks on the lower portion of the old stems and at times on younger shoots also. The nymphs and adults suck the sap from shoots and stem. The affected portion of the plant dries up. In severe infestation the whole plant may die. *Aonidiella aurantii* is flat, circular and reddish in colour and measures about 2 mm in diameter. This scale is commonly seen on the young shoots and leaves causing weakening of the plants (Ramesh, 1994). The infested plant bears a few small flowers (Ali, 1968; Butani, 1974; Nandakumar *et al*, 1988). Greenhouse as well as garden roses are prone to attack by this pest. They appear more severely before and after the monsoon and are often choosy in their preference to certain cultivars of roses. Some insects preferred rose cultivars are Versailles, Incense, Kordes Perfecta, First Prize and Camelot are among the Hybrid Teas, Zorina, Zambra, Ice White, Red Pinocchio in Floribundas. No rose plant is immune to it (Ramakrishnan, 1984).

The scales *Aspidiotus dictyospermi, Aspidiotus orientalis* and *Aspidiotus transparens* infest shoots and leaves (Nair, 1975) whereas *Chionaspis* sp. infests tender shoots (Ayyar, 1919). *Iceria aegyptiaca, Iceria formicarum* and *Iceria purchasi* infest twigs of roses in different parts of India (Glover, 1934; Nair, 1975; Prabhakar Rao, 1950; Ramachandra Rao and Cherian, 1944; Srinivasan *et al*, 1974; Sohi and Singh, 1995). The mealybug *Ferrisia virgata* infests shoots of roses in South India (Nair, 1975).

Management

- Early detection of scales by visual inspection of the plants will help to follow pest management practices. The presence of ants may be an indicator for the presence of scales.
- Cutting of twigs and leaves infested by scales and burying them will minimise scale incidence.
- Rub off the scales with a used toothbrush or cotton soaked in kerosene or diesel or methylated spirit.

- The grubs of green lacewing *Chrysoperla carnea* feed on the immature stages of scales. Several species of predatory ladybird beetles such as *Chilocorus* sp., *Rhizobius* sp., *Cryptolaemus montrouzieri* feed on soft scales in green houses (Rice Mahr *et al*, 2001).
- Two sprays with Malathion 50 EC or Dimethoate 30 EC at the rate of 2 ml/l (Pirone *et al*, 1960) or fish oil rosin soap at the rate of 25 g/l or Endosulfan 35 EC at the rate of 2 ml/l of water, first at the time of pruning and again during March-April, effectively control scales.
- Application of granular insecticides like Carbofuran 3G at 1.5 kg a.i./ha and Disulfoton and phorate 10G at 3.0 kg a.i./ha gave good control of scales (Srinivasan *et al*, 1974). Application of Carbofuran 3G at the rate of 5 g/plant is also recommended (Regupathy *et al*, 1994).

The above chemicals and predatory ladybird beetles will also control mealybugs.

Black Fly

Aleurocanthus spiniferus Q.
Aleurocanthus rosae Singh
(Aleurodidae: Hemiptera)

Both nymphs and adults suck the sap from leaves resulting in crinkling of leaves. The adults lay eggs scattered on the undersurface of leaves. The nymphs are brownish yellow and the puparia jet-black with a cottony fringe all around and globular secretions sticking to tips of the spines. This pest has a countrywide distribution and occurs as a serious pest in some places.

Management

- Spraying of Malathion 50 EC at the rate of 875 ml or Dimethoate 30 EC at the rate of 750 ml or Phosphamidon 85 WSC at the rate of 187.5 ml in 250 litres of water per hectare controls whitefly (Atwal, 1986).

Leafhopper

Amrasca devastans Dist.
Edwardsiana rosae (L.).
Motschulskyia serrata (Mats.).
Zygina (= *Hypericiella*) *bicornia* Dwor.
(Cicadellidae: Hemiptera)

Amrasca devastans, Motschulskyia serrata and *Zygina bicornia* suck the sap from leaves and cause stippling on leaves (Sohi *et al*, 1988; Sohi and Mann, 1986). *Edwardsiana rosae* is whitish to light yellow, with yellow abdomen and sucks the sap from leaves causing yellowing and withering of the attacked surface giving the plants a sickly appearance.

Management

- Spray methyl demeton 25 EC or Phosalone 35 EC at the rate of 2 ml/l or neem oil at 3 per cent or apply Carbofuran 3G at 5g/plant.

Red Cotton Bug

Dysdercus sp.
(Pyrrhocoridae: Hemiptera)

The red cotton bug also causes moderate to serious damage to flowers of roses in India (Verghese and Prasad, 1983).

Management

- Spray Monocrotophos 36 WSC at the rate of 1 ml/l or Quinalphos 25 EC at the rate of 1.5 ml/ l of water.

Red Spider Mite

Tetranychus cinnabarinus Boisd.
Tetranychus urticae Koch
(Tetranychidae: Acari)

Spider mites are found in large colonies on the undersurface of leaves covered with fine, silky webs. They develop through egg, larva, protonymph, deutonymph and adult stages. Spider mites can produce a generation in 7-14 days in warm temperatures. In cooler temperatures, generations take more than 40 days and below 54° F development stops. In green houses, adults may go into diapause for several months hibernating in cracks and crevices from September through February-March in temperate climates. These mites feed by penetrating the plant tissue with sharp stylets and removal of cell contents. The chloroplasts disappear and the small amount of cellular material coagulates to form an amber mass. In the palisade layers, only the penetrated cells are damaged. Adjacent cells show no evidence of injury. The spider mites are 0.5-0.6 mm long. Feeding of spider mites results in white specks on leaves which later coalesce and produce white patches resulting in reduced photosynthetic activity (Karuppuchamy, 2001). Ultimately, affected leaves become mottled, turn yellow and fall. Infested flower petals lose their brightness resulting in a direct loss to the grower. Cultivars like Summer Snow, First Prize, Oregold and Maria Cellas are more susceptible than certain others (Ramakrishnan, 1984). Dhooria (1999) recorded lowest mite damage on Golden Times and Pavarroti. Some polyantha and floribunda roses completely lose their leaves prematurely due to the attack of this pest. The damage is more pronounced during warm and dry seasons. In the polyhouse, the prevailing high humidity and temperature is very favourable for the population build up of spider mites.

Management

- Frequent visual inspection of plant parts is the best method for detecting infestation of spider mites. Because of their small size and cryptic behaviour, they are often overlooked until their feeding damage or webbing becomes apparent.
- The exotic phytoseiid mite, *Phytoseiulus persimilis* was found to be the most promising predator for controlling *Tetranychus urticae* in rose under both polyhouse and field conditions (Ghosh, 2001). Efforts have been made to establish *Phytoseiulus persimilis* in India by importing it in 1994 under the All India Co-ordinated Research Project on Biological Control of Crop Pests and Weeds (Krishnamoorthy, 1988). Although the potentiality of this predator has been demonstrated under Indian climatic conditions, it has failed to establish under field conditions.
- Release of laboratory reared colonies of the indigenous obligatory predatory mite, *Amblyseius longispinosus* has been found to be effective in controlling *Tetranychus urticae* in roses (Mallik *et al*, 1998). The predatory potential of

Amblyseius tetranychivorus against *Tetranychus urticae* under both laboratory and polyhouse conditions was demonstrated (Ghosh *et al,* 2000). Field and Hoy (1984) tried the biological control method and reported that strains of *Typhlodromus* (= *Metaseiulus*) *occidentalis* could be integrated into the pest control programme without being seriously affected by chemicals used against other pests or against *Tetranychus urticae* itself during the establishment of the predator. This predatory mite *Typhlodromus occidentalis* is smaller than *Phytoseiulus persimilis* and develops best at cooler temperatures, but it tolerates a wide range of humidity (40-80 per cent). It is able to regulate spider mite population at low densities and for longer periods of time than *Phytoseiulus persimilis*, although it cannot control spider mite population quickly. It can also survive for longer periods without prey (Rice Mahr *et al*, 2001).

- Frequent spray with sulphur helps to control this mite. Several acaricides such as Dicofol, Morestan and Tetradifon have been reported to be very effective in controlling rose mites (Pirone *et al*, 1960; Ramakrishnan, 1984). Eight ml of Dicofol is thoroughly mixed with 10 litres of water with 10-15 drops of any sticker and application of this chemical every five or six days may be sufficient to control the mites. Spraying the affected parts with methyl demeton at 0.1 per cent is also suggested. Spraying of Monocrotophos at 1 ml/l, Dicofol at 2 ml/l or wettable sulphur at 3 g/l are very effective in controlling rose mites (Dhua, 1999). Dhooria (1999) reported that spraying of Abamectin 1.8 EC or Ethion 50 EC or oxydemeton methyl 25 EC at 2 ml/l of water was found to be effective in reducing the population of *Tetranychus urticae* under polyhouse conditions. Ethion and oxydemeton methyl caused more than 50 per cent mortality of eggs whereas Abamectin was ineffective as an ovicide but was very effective in killing emerging larvae.

Leaf Cutter Bees

Megachile anthracina Smith.
Megachile disjuncta Fab.
(Megachilidae: Hymenoptera)

The adult bees are black in colour. These insects cut circular or semicircular cuttings or oval shaped patches from the leaf margins which result in defoliation occasionally. Female bees carry pieces of leaf to their nest built in the soil, in rotten wood or in crevices. Each female forms its own nest provided with a paste of honey and pollen in which eggs are laid (Ramesh, 1994).

Hairy Caterpillars

Orgyia (= *Notolopus*) *postica* Walker.
Euproctis fraterna Moore.
Porthesia scintillans Wlk.
(Lymantriidae: Lepidoptera)

The larva of *Orgyia postica* is a hairy caterpillar with brown head, a pair of long pencils of hairs pointing forwards from the prothorax, tuft of yellowish hair laterally on the first two abdominal segments, yellowish tuft of hairs dorsally on the first four abdominal segments and long brown hair dorsally from the eight abdominal segments. The adult is a brown coloured moth with stout abdomen.

The larva of *Euproctis fraterna* is reddish brown with a red head surrounded by white hairs and long tufts of hair all over the body and a long pre-anal tuft. The adult is a yellowish moth with a pale transverse line on the forewings. The larva of *Porthesia scintillans* is yellow coloured with a brown head, a yellow stripe with central red line on the body and tufts of black hair dorsally on the first three abdominal segments. The adult is yellowish with spots on the edges of fore wings (Regupathy *et al*, 1994). The larvae completely defoliate the foliage.

Castor Semilooper

Achaea janata L.
(Noctuidae: Lepidoptera)

The moth is stout, greyish brown with wavy lines on the forewings and white blotches on the hind wings. The larva is a semilooper with varying shades of colour, black head, a red spot on the third abdominal segment and a red tubercle in the anal region. The larva feeds voraciously on the foliage.

Looper

Agathia sp.
(Geometridae: Lepidoptera)

The moth is green in colour. The larva completely defoliates the foliage.

Slug Caterpillar

Parasa lepida Cr.
(Cochlidiidae: Lepidoptera)

The adult is a green moth with a brown band at the base of the forewings. The larva is slug like, ventrally flat, greenish body with white lines and four rows of spiny scoli tipped red or black (Regupathy *et al*, 1994). The larvae defoliate the foliage.

Management of Defoliators

- Spray Methyl Parathion 50 EC at the rate of 1 ml/l or Endosulfan 35 EC at the rate of 2 ml/l of water.

Gram Pod Borer

Helicoverpa armigera (Hubner).
(Noctuidae: Lepidoptera)

The female moth lays eggs on younger buds. The young larva bores into a bud and eats the internal contents. Perforated petals are seen when the flower opens (Ramesh, 1994).

Management

- Handpicking of the larvae in the early stages of infestation.
- Spray Endosulfan 35 EC at the rate of 2 ml/l or dust Carbaryl 10 per cent at 20-25 kg/ha or NPV at the rate of 250 LE/ha or *Bacillus thuringiensis* or Methomyl at the rate of 2 ml/l or Cartap hydrochloride at the rate of 1 g/l of water.

Flower Chafer Beetle

Oxycetonia versicolor F.
Oxycetonia albopuncta (Fab.).
Chiloloba acuta Wied.
(Cetonidae: Coleoptera)

Oxycetonia versicolor is a red coloured beetle with black marking which feeds upon the growing points making irregular holes and punches on the leaves. The grubs feed on the roots. Flowering is affected if not controlled in time. *Oxycetonia albopuncta* has been recorded from Himachal Pradesh (Sood, 1990). *Chiloloba acuta* feeds on leaves and flowers.

Management

- Hand picking and destruction of chafer beetles during daytime (Anonymous, 1999).
- Certain indigenous plant materials were tested for their antifeedant and insecticidal properties against the beetles. Spraying of petroleum ether extract of dried rhizomes of *Acorus calamus, Centretheum anthelminticum* and dried and immature seeds of *Linum usitatissimum* at concentrations of 0.1 to 0.5 per cent, 0.2 per cent and 2 per cent, respectively, were found very useful (Verma and Singh, 1980).
- Treat the soil around the trunk with Carbaryl 10 per cent at the rate of 30 g/plant (Ramesh, 1994).
- Spray methyl parathion 50 EC or Endosulfan 35 EC or Malathion 50 EC at 2 ml/l of water.

Digger Wasp

Crabro sp.
(Crabronidae: Hymenoptera)

Digger wasps damage the plants after pruning. They dig a hole into the pith of the stem through the cut end and make a nest there. The pruned stem cuttings show the signs of drying up from the tip.

Management

- Add a few drops of Dimethoate 30 EC or Phosphamidon 85 WSC in the hole and plug with a small piece of cotton for killing the insect (Bose and Yadav, 1993).
- Kill the insect by passing a flexible wire through the hole.
- Add a drop or two of ether into the hole and plug with a piece of cotton (Ramesh, 1994).

Ash Weevils

Myllocerus dentifer Fab.
Myllocerus discolor Boh.
Myllocerus richardi Nathan.
Myllocerus setulifer Desb.
Myllocerus undecimpustulatus maculosus Desb.
Myllocerus viridanus Fab.
(Curculionidae: Coleoptera)

Adult weevils measuring about 3-4 mm long, feed on the petals and skeletonise them completely causing moderate to heavy damage to rose in India (Nair, 1975; Tewari, 1983; Verghese and Prasad, 1983). The population of these species increases from April to November and declines thereafter.

Management

- Spray Endosulfan 35 EC or Methyl Parathion 50 EC at the rate of 2 ml/l of water.

Bud Borer

Argyroploce aprobola Meyr.
Eucosma zelota Meyr.
(Eucosmidae: Lepidoptera)

Larva of *Argyroploce aprobola* bores into the bud and feeds on petals of flowers and *Eucosma zelota* spins up the leaves.

Management

- Spray Dichlorvos 76 EC at the rate of 1 ml/l of water.

Leaf Feeder

Acleris extensana
(Tortricidae: Lepidoptera)

Yellowish green larva with black head and brown prothorax ties up tender shoots and feeds by scrapping, bores into buds and flowers (Regupathy *et al*, 1994).

White Ants or Termites

Microtermes obesi Holmgren.
Odontotermes obesus (Ramb.).
(Termitidae: Isoptera)

They feed on roots and rootlets of rose plants. The damage is more severe in sandy and sandy loam soils. The attacked plant withers and dies. The pest is active throughout the year (Butani and Dadlani, 1983).

Management

- Use of undecomposed farmyard manure should be avoided.
- Use of neem cake acts as a repellent as well as manure.
- Drenching the rose beds with Chlorpyriphos 20 EC or TC at the rate of 0.1 per cent controls termites in established gardens or before planting (Bose *et al*, 1999).

Other Insects

The adults of *Adoretus versutus* Harold and *Adoretus lasiopygus* Bur. (Scarabaeidae) feed on leaves of roses in many parts of India (Deol, 1974; Fletcher, 1917). *Anthaxia baconis* Thomas (Buprestidae) prefers to feed on leaves of ornamental hedging rose. Its infestation is confined to Himachal Pradesh (Hameed, 1978). The grubs of *Apogonia rauca* Fletcher (Scarabaeidae) feed on roots and adults feed on leaves of rose in South India (Patil and Veeresh, 1984). The beetle *Calaspis hypochlora* Lef. (Chrysomelidae) feeds on leaves (Nair, 1975). The grubs of *Celosterna scabrator* var. *spinator* Fab. (Cerambycidae) bore into the stem and twigs whereas adults gnaw the bark on twigs (Beeson, 1931). The beetle *Holotrichia consanguinea* (Blanch) (Melolonthidae) feeds on leaves from June to August. The beetles emerge from soil with the first shower of the monsoon in June. After emergence they feed on leaf lamina leaving behind veins and veinlets (Bindra and Singh, 1971). The weevil, *Hyperstylus* sp. feeds on petals in Himachal Pradesh (Sood, 1990). The adults of *Sthenias grisator* Fab. (Cerambycidae) gnaw at the stem and twigs (Fletcher, 1917; Nair, 1975).

Halcystogramma hibisci St. (Gelechiidae), *Soritia leptalina* Koll. (Zygaenidae), *Taragamma siva* Lef. (Lasiocampidae), *Thalassodes veraria* Guen. and *Hyposidra successaria* Wlk. (Geometridae), *Staropus alternus* W. (Notodontidae), *Spodoptera litura* F. and *Parallelia algira* L. (Noctuidae) feed on leaves. *Peronia schallerianana* Linn. (Tortricidae) roll the leaves. *Cacoccia pomivora* Meyr. is a leaf roller in Assam. *Ulodermis trigrapha* Meyr. bores into the bud in Assam. *Platypepla aprobola* rolls the leaves in Gujarat (Nair, 1975). *Serica calcuttae* Br. (Melolonthidae) is a defoliator and *Mylabris pustulata* Thumb.(Meloidae) feeds on flowers. The chrysomelid beetles *Cryptocephalus dodecospilus* Suff. and *Calaspis hypochlora* Lef. feed on the leaves.

JASMINE (*JASMINUM* SP.)

Jasmine is attacked by several insect and non-insect pests. Some of them are important which cause considerable damage to jasmine and have been reported by several authors (David and Kumaraswamy, 1975; Easwaramoorthy *et al*, 1980; Tewari and Mohan, 1981; Vyas, 1998). The work done on jasmine pests and their management is reviewed hereunder.

Jasmine Bud Worm

Hendecasis duplifascialis Hmpsn.
(Pyraustidae: Lepidoptera)

This is a major pest of jasmine, first reported by Hampson in West Africa, India and Srilanka in 1896. Later, David (1958) and Prasad *et al* (1983) recorded it in South India and Delhi, respectively. Greenish larva with a black head bores into immature jasmine buds and feeds on floral structures and in severe cases webbing of buds is also noticed. Kalshoven (1950) stated that *Hendecasis duplifascialis* bored flower buds. Dried flowers with frass and characteristic webbing of the adjacent flower buds are symptoms of *Hendecasis duplifascialis* infestation. The larva bores into a number of flower buds (David, 1958; Prasad *et al*, 1983). Amutha (1994) studied the population fluctuation of this pest on five cultivars of jasmine and noticed the prevalence of this insect throughout the year. The adult is a small white moth with black palpi and three lines on the abdomen and the hind legs (Regupathy *et al*, 1994).

Management

- Prasad *et al* (1983) observed *Cantheconidia furcellata* Wollf. (Pentatomidae) preying on the early instar larva of *Hendecasis duplifascialis*. The other natural enemies include the parasitoids *Tetrastichus* sp., *Bracon brevicornis*, *Elasmus* sp. and *Telenomus* sp. (Amutha, 1994).
- Neem cake extract 5 per cent was found to be effective followed by neem seed kernel extract (NSKE) 5 per cent and neem oil 1 per cent (Nelson *et al*, 1993).
- Spraying of palmarosa oil at 0.1 per cent and jatropha oil at 0.5 per cent were effective in recording the lowest bud damage by bud worm (Isabel, 1996).
- When one-day pupae were dipped in Diflubenzuron, it inhibited 83.3 per cent adult emergence (Winstone and David, 1986a). Diflubenzuron (Dimilin 25 WP) at 100 ppm and above caused cent per cent mortality of *Hendecasis duplifascialis* larvae under laboratory conditions by ingestion (Winstone and David, 1986b).
- Quinalphos 25 EC at 0.1 per cent was significantly effective in controlling the bud worm (Reddy *et al*, 1978).

- Spray Dimethoate 30 EC at the rate of 2 ml/l of water (Ramesh, 1994).
- Spray Monocrotophos 36 WSC at the rate of 2 ml/l of water at fortnightly intervals during flowering (Anon., 1999).
- *Hendecasis duplifascialis* can also be effectively controlled by the application of Deltamethrin at 25 g a.i./ha or Cypermethrin at 150 g a.i./ha (Dhandapani *et al*, 1989).
- Amutha and David (1995) reported that Cypermethrin applied at 75 g a.i./ha was the most effective treatment followed by Deltamethrin at 125 g a.i./ha and Diflubenzuron at 100 g a.i./ha. Spraying of Monocrotophos 36 WSC at 0.1 per cent and Fenvalerate at 0.02 per cent was also effective against bud worm (Isabel, 1996).

Gallery Worm

Elasmopalpus jasminophagus Hmpsn.
(Phycitidae: Lepidoptera)

It is also a serious pest causing webbing of terminal leaves, shoots and flowers. The larva is greenish with red head and thorax and lateral brown streaks on the body. The larva grows to a length of one inch and makes tunnels of silk and excreta outside the buds and feeds on them and pupates within the tunnel (David, 1958). The moth is narrow, elongate and dark grey with pale hind wings. This pest prefers *Jasminum auriculatum* (Amutha, 1994). *Elasmopalpus jasminophagus* infestation occurs during the early stages, particularly in the vegetative buds and the larva causes numerous holes in the terminal leaves in folded condition (David and Venugopal, 1962).

Management

- Spray Dimethoate 30 EC or Malathion 50 EC at 2 ml/l of water (Bose *et al*, 1999).

Blossom Midge

Contarinia maculipennis Felt.
(Cecidomyiidae: Diptera)

This pest was first reported on *Jasminum sambac* L. in Andhra Pradesh (Rao *et al*, 1954, 1955) and also recorded on *Jasminum auriculatum* and *Jasminum sambac* in Tamil Nadu (David, 1958). The maggot is seen within the buds at the base of the corolla. The attack by the maggot causes swelling at the base of the bud, stunting and final drying.

Management

- Among the plant products tested, palmarosa oil at 0.1 per cent and jatropha oil 0.5 per cent were effective in recording lowest per cent bud damage by *Contarinia maculipennis* (Isabel, 1996).
- Spraying of Monocrotophos 36 WSC at 0.05 per cent and 0.1 per cent are effective for the control of blossom midge (Reddy *et al*, 1978). David and Natarajan (1988) reported that soil application of Carbofuran 3G at 40 g per bush and foliar application of Cypermethrin at 25 g a.i./ha, Fentrothion 0.1 per cent and Monocrotophos 0.08 per cent at fortnightly intervals effectively reduced the purple discoloration and drying of flower buds on *Jasminum*

sambac. Carbofuran at 40 g/bush also reduced blossom midge incidence (Gunasekaran, 1989). Studies on chemical control also revealed that Monocrotophos 0.72 kg a.i./ha followed by Cypermethrin 75 g a.i./ha effectively controlled this pest (Amutha, 1994). Isabel (1996) reported Monocrotophos 0.1 per cent to be highly effective followed by Fenvalerate 0.02 per cent.

Leaf Web Worm

Nausinoe geometralis (Guen.)
Nausinoe neptis Gram.
(Pyralidae: Lepidoptera)

Among the leaf feeders, *Nausinoe geometralis* is a major pest in jasmine (Tewari and Mohan, 1981). The moth is brown with hyaline patches on both wings. The female lays 15 to 20 eggs singly on the leaves and these hatch in 3 to 4 days. The larvae attack the lower bushy and shaded portions of the plants in the rainy season and terminal shoots during dry and sunny weather. They skeletonise the leaves by scraping away the parenchymatous tissues, become full grown in 12 to 15 days and pupate beneath webbings spun over the leaves and the branches for 6 to 7 days. The pest is common in South India where it completes its life cycle in 22 to 24 days in July and August (Nair, 1975). The peak infestation is from July to December.

Management

- The green lace wing, *Chrysoperla carnea* prefers to feed on the first instar larvae of *Nausinoe geometralis* (Isabel *et al*, 2001b). Parasitoids like *Elasmus brevicornis* Gah. and *Apanteles* sp. and the predator *Chlaenius hamifer* Chadoir were found to attack the larvae of webworm. Spiders, namely, *Marapissa* sp., *Phidippus pujabensis* Tikadar and *Salticus manjitus* Tikadas were reported to prey on *Nausinoe geometralis* (Shukla and Sandhu, 1983).
- *Bacillus thuringiensis* sub sp. *kurstaki* (Delfin®) was effective in bringing 50 per cent mortality even at low concentration of 0.08 mg/100 ml when compared to another formulation Agree® at the rate of 0.2 mg/100 ml in laboratory conditions.
- Among the plant products, palmarosa oil at 0.1 per cent and jatropha oil at 0.5 per cent were effective in recording lower per cent leaf damage by leaf web worm (Isabel, 1996; Isabel *et al*, 2001a).
- Plants should be sprayed with Malathion 50 EC at 2 ml/l of water to control the pest (Pirone *et al*, 1960).
- Monocrotophos proved to be the most effective insecticide causing 90 per cent mortality of the pest and the effectiveness persisted up to 14 days after application (Sandhu and Shukla, 1984).
- Endosulfan 35 EC at 350 g a.i./ha or Quinalphos 25 EC at 250 g a.i./ha was found to be effective against this pest (Gunasekaran, 1989).

Leaf Roller

Glyphodes unionalis Fb.
Glyphodes celsalis Wlk.
(Pyraustidae: Lepidoptera)

The greenish larva of *Glyphodes unionalis* knits together the tender leaves with silk and scrapes their green matter in its early stages and eats bulk leaf tissues in the later stages. The faecal matter is collected in the webbing (David, 1958). The adult moth is one inch long, white with a thin brown line along the coastal margin of the forewings. Webbings of terminal portions of the twigs and the presence of excreta below the webbings are only due to the attack by *Glyphodes unionalis.* Symptoms of *Glyphodes celsalis* infestations on leaves are folding along the margins and leaf tip in a typical horizontal manner ventrally.

Management

- Spray Methyl Parathion 50 EC at the rate of 2 ml/litre of water (Ramesh, 1994).

Jasmine Eriophyid Mite

Aceria jasmini Channa.
Calacarus jasmini
Phyllocoptes salmaliae
(Eriophyidae: Acari)

Among these eriophyid mites, *Aceria jasmini* is the most serious recorded on *Jasminum pubescens* by David (1958) and on *Jasminum auriculatum* by Sundararaj *et al* (1967) and Mohan and Udayagiri (1985). It produces felt-like hairy outgrowths on the leaf surface, tender stems and buds. The growth of the plant is stunted and production of flowers is suppressed due to the attack by *Aceria jasmini* (David, 1958). The incidence of *Aceria jasmini* was reported to increase from March and reached a peak during July (Jose *et al*, 1999b). *Calacarus jasmini* and *Phyllocoptes salmaliae* were reported on *Jasminum sambac* in West Bengal.

Management

- The cultivar Parimullai of *Jasminum auriculatum* released by Tamil Nadu Agricultural University, Coimbatore, is resistant to gall mite or erineum mite.
- Soil application of neem cake at the rate of 100 g/plant and spraying of neem oil 2 per cent or water extract of pongamia kernel 4 per cent significantly checked the damage by gall mite, *Aceria jasmini* on *Jasminum auriculatum* (Rani and Mohan, 1994). Soil application of neem cake at 250 kg/ha combined with foliar application of NSKE 10 per cent was effective in controlling *Aceria jasmini* (Umapathy and Rajendran, 1999).
- Fortnightly spray of Monocrotophos 36 WSC at 0.5 kg a.i./ha, Phosalone 35 EC at 0.7 kg a.i./ha or wettable sulphur at 1 kg a.i./ha was found to be effective against *Aceria jasmini* (Mohan and Udayagiri, 1984). Mallikarjunappa *et al* (1999) reported that a mixture of Ethion 50 EC at 0.05 per cent and Dimethoate 30 EC at 0.03 per cent effectively reduced this mite in Bangalore. Jose *et al* (1999a) reported that Diafenthiuron at 0.1 per cent was the most effective chemical followed by bromoprophylate at 0.1 per cent, Dicofol 18.5 EC at 0.1 per cent and Monocrotophos 36 WSC at 0.1 per cent in controlling *Aceria jasmini.*
- Dust sulphur at the rate of 20 kg/ha or spray Tetradifon at the rate of 2 ml/litre of water (Ramesh, 1994).

Red Spider Mite

Tetranychus cinnabarinus Boisduval

Tetranychus cucurbitae Rah.

(Tetranychidae: Acari)

Tetranychus cinnabarinus feeds on the lower surface of leaves of jasmine and produce white specks which later coalesce to produce white patches resulting in reduced photosynthetic activity leading to yield reduction (Karuppuchamy *et al*, 1986). *Tetranychus cucurbitae* often attack the leaves of *Jasminum sambac* and *Jasminum auriculatum* (David, 1958).

Management

Spraying of neem seed kernel extract caused high mortality and reduction in the fecundity of *Tetranychus cinnabarinus* (Mansour and Ascher, 1983).

Spraying with either Dicofol 18.5 EC at 2 ml/l of water or wettable sulphur 35 SP at 3 g/l or Dinocap 48 EC at 2 ml/l of water effectively controlled *Tetranychus cinnabarinus* (Karuppuchamy *et al*, 1986). Isabel (1996) reported that among various acaricides tested, Acarol and Diafenthiuron showed significant control.

Tingid Bug

Corythauma ayyari Dist.

(Tingidae: Hemiptera)

The tingid bug and its nymphs feed on the sap of the leaves from their ventral sides especially near the midribs. The infested regions of the leaves turn yellowish which ultimately dry and drop.

Management

- Spray Endosulfan 35 EC at the rate of 2 ml/l of water.

Whitefly

Dialeurodes kikaldyi (Kotinisky)

Dialeurodes citri R & H.

Dialeurodes vulgaris K.

Bemisia giffardi Kot.

Aleurotrachelus sp.

(Aleurodidae: Hemiptera)

The nymphs and adults of these species infest the lower surface of leaves causing slight discoloration to them. Govindarajan *et al* (1974) recorded *Aleurotrachelus* sp. causing severe damage to *Jasminum auriculatum* and, to a lesser extent, on *Jasminum grandiflorum* and *Jasminum multiflorum* (Gaud *et al*, 1991). *Bemisia giffardi* occurrence on *Jasminum sambac* and its distribution in Bihar and Tamil Nadu was reported by Jesudasan and David (1991).

Management

- Spraying of Fenitrothion 0.1 per cent, Fenthion 0.1 per cent or Monocrotophos 0.04 per cent controlled *Aleurotrachelus* sp. on jasmine (Govindarajan *et al*, 1974).

Jasmine Bug

Antestiopsis cruciata Fb.
(Pentatomidae: Hemiptera)

This bug at times is found in large numbers feeding on the tender shoots and buds and prevents flower formation. Nymphs and adults suck the sap from leaves causing yellowing and drying of leaves. The nymphs are dark brownish-black and round. The adults are dark brown, shield-shaped with orange and white markings on the wings.

Management

- Spray Endosulfan 35 EC at the rate of 2 ml/l of water (Ramesh, 1994).

Flower Thrips

Thrips orientalis Bagn.
Dendrothrips jasminum Ram and Marg.
Frankliniella sulphurea Sch.
(Thripidae: Thysanoptera)
Haplothrips ceylonicus Schm.
Haplothrips ganglbaneri Schm.
(Phloeothripidae: Thysanoptera)

The adults of *Thrips orientalis* are black and nymphs are yellowish in colour. They feed and breed in the buds and flowers and rarely on the tender leaves. They occur throughout the year and cause severe leaf damage in Kerala. *Dendrothrips jasminum* and *Frankliniella sulphurea* are also recorded on jasmine in South India. The affected flowers give a decayed appearance and fall off prematurely. *Haplothrips ceylonicus* and *Haplothrips ganglbaneri* infest the flowers.

Management

- Spray Malathion 50 EC at the rate of 2 ml/l or Methyl Parathion 50 EC at 1 ml/l of water.

Other Pests

Helicoverpa armigera (Hubner)(Noctuidae) attack *Jasminum sambac* during peak flowering period (Amutha, 1994). The roots of jasmine were found to be affected by the larvae of *Lachnosterna* (*Holotrichia* sp.) (Melolonthidae) (Perumal *et al*, 1971). Nair and Nair (1974) recorded mealy bug *Planococcus lilacinus* Ckll (Pseudococcidae) and David and Kumaraswami (1975) recorded shoot fly, *Sycophila* sp., infesting *Jasminum grandiflorum*. The sphingids *Deilephila nerii* Linn. and *Acherontia styx* W. feed on leaves occasionally. California red scale *Aonidiella aurantii* Mask.(Diaspididae) infest tender shoots and buds causing them to fade. The black hairy caterpillar *Pericallia ricini* Fb. (Arctiidae) cause damage by feeding on leaves. The cyclamen mite, *Steneotarsonemus pallidus* Banks (Tarsonemidae), was found to severely damage *Jasminum sambac*. The pale green plant hopper, *Flata ocellata* Fb.(Flatidae), feeds on the terminal shoots (David, 1958). The oviposition punctures made by the adults of *Ricania fenestrata* Fb. (Fulgoridae) cause withering of shoots (David, 1958). The wingless grasshopper, *Orthacris simulans* B. (Acrididae) eats tender leaves.

HIBISCUS (*HIBISCUS ROSASINENSIS*)

Mealybug

Maconellicoccus hirsutus (Green).
(Pseudococcidae: Hemiptera)

Mealybug is a serious pest of hibiscus. Nymphs and adult congregate in the terminal growing points and suck the sap resulting in severe crinkling and curling of leaves. In severe infestations, the plants become stunted. The nymphs and adults are pinkish in colour.

Management

1. Australian ladybird beetle, *Cryptolaemus montrouzieri* (Mulsant), is an efficient predator of *Maconellicoccus hirsutus* which is able to contain the pest (McComie *et al*, 1997).
2. Spray of fish oil rosin soap at the rate of 25 g/l or Dichlorvos 76 EC at the rate of 1 ml/l of water.

Whitefly

Bemisia tabaci Guen.
Trialeurodes vaporariorum (Westw.)
(Aleurodidae: Hemiptera)

Nymphs and adults suck the sap from the lower surface of leaves causing yellowing and drying of leaves.

Management

- The entomogenous fungus *Paecilomyces fumosoroseus* is used for the management of *Bemisia tabaci* on *Hibiscus* in Florida (Osborne and Landa, 1994).
- Spray Methyl Demeton 25 EC or Methyl Parathion 50 EC at the rate of 1 ml/l of water.
- One or two releases of *Encarsia formosa* commencing at the time of infestation control *Trialeurodes vaporariorum* under glasshouse conditions (Sell and Kuo-Sell, 1989).

POINSETTIAS (*EUPHORBIA PULCHERRIMA*)

Whitefly

Trialeurodes vaporariorum (Westw).
Bemisia argentifolii Bellows and Perring
Bemisia tabaci Guen.
(Aleurodidae: Hemiptera)

Both nymphs and adults suck the sap from the underside of the leaves and cause yellowing and drying of leaves besides development of sooty mould on honeydew secreted by whiteflies.

Management

- Release of coccinellid predator, *Seriangium parcesetosum* Sicand may be used in multiple biological control of silver leaf whitefly (Ellis *et al,* 2001).
- Both commercially produced and locally produced *Encarsia formosa* reduced *Bemisia tabaci* populations considerably with maximum nymph parasitism of 60-80 per cent in greenhouses in Ohio (McMahon *et al,* 1994).
- Myco-insecticide containing *Beauveria bassiana* (Balsamo) effectively controls *Bemisia argentifolii* in the greenhouse (Olsen and Oetting, 1999).

- The entomogenous fungus *Paecilomyces fumosoroseus* is used as a main component in the integrated pest management of *Bemisia tabaci* on poinsettia in greenhouses in Florida (Osborne and Landa, 1994).
- Spray Methyl Parathion 50 EC or Methyl Demeton 25 EC at the rate of 1 ml/l or Phosalone 35 EC or Malathion 50 EC at the rate of 2 ml/l of water.

CROTONS (*CODIAEUM VARIEGATUM*)

Mealybugs

Planococcus (*Pseudococcus*) *citri* Risso.
Planococcus longispinosus
Planococcus lilacinus (Ckll).
Ferrisia virgata (Ckll).
(Pseudococcidae: Hemiptera)

Mealybugs occur on the lower surface of leaves, base of petioles and at nodes of tender shoots. The body of the insect is covered with a white, wax-like mealy substance. As a result of infestation, plants give withering appearance. *Ferrisia virgata* is abundant in drought conditions.

Scales

Saissetia coffeae Walker.
Parasaissetia nigra (Neit.).
Coccus depressum (Ckll).
Parlatoria pergandii C.
Parlatoria proteus var. *gloveri* Pack.
L. auriculatus Gr.
Pinnaspis minor Mask.
(Coccidae: Hemiptera)

Iceria aegyptiaca D.
(Margarodidae: Hemiptera)

Scale insects feed from the lower surface of leaves along the midribs. The affected plants lose their vigour and the leaves dry off.

Management of Mealybugs and Scales

- Spray Quinalphos 25 EC at the rate of 2 ml/l or Fenitrothion or Fenthion at the rate of 1 ml/l of water.

Other Insects

Heliothrips haemorrhoidalis B. and *Ayyaria chaetophora* K. (Thripidae) attack crotons by sucking the sap from leaves. The beetle *Sthenias grisator* Fb. (Cerambycidae) girdles the stems of crotons (Nair, 1975).

OLEANDER (*NERIUM* SP.)

Sphingid Caterpillar

Deilephila nerii L.
(Sphingidae: Lepidoptera)

The larva is stout, elongated with a spine at the anal region. The large larva feeds voraciously on the leaves resulting in defoliation of the plant (Nair, 1975; Regupathy *et al*, 1994). The adult is a robust, green moth with yellow markings.

Leaf Butterfly

Euploea core C.
(Danaidae: Lepidoptera)

The larva is blackish brown and striped with six tentacles, two each near the anterior and posterior ends and a pair at the middle. The adult is a black butterfly with a series of white spots on the margin of the wings. The larvae cause damage by feeding on the leaves.

Management

1. Hand picking and destruction of the larvae.
2. Spray Endosulfan 35 EC or Phosalone 35 EC at the rate of 1 ml/l or Monocrotophos 36 WSC at the rate of 2 ml/l of water (Anonymous, 1999).

Other Pests

The beetles of the stem girdler *Sthenias grisator* Fb. (Cerambycidae) bore into the stem (Nair, 1975). The loopers *Agathia hemithearia* Guen., *Agathia lactana* F. (Geometridae) and woolly bear *Pericallia ricini* F. feed on leaves. The scales *Aspidiotus transparens* Gr., *Parlatoria calianthiana* Ber. & Leon. (Diaspididae) and mealy bug *Nipaecoccus vestator* (Mask.) (Pseudococcidae) also attack oleander (Nayar *et al*, 1976).

BARLERIA (*BARLERIA CRISTATA* L.)

Bud Worm

Trichoptilus sp.
(Pterophoridae: Lepidoptera)

The larva bores into the bud and feeds on the internal contents. The adults are small with plume-like wings.

Management

- Spray Endosulfan 35 EC at the rate of 1 ml/l of water or dust Carbaryl 10 D at the rate of 15-20 kg/ha (Ramesh, 1934).

Leaf Folder

Syngamia latimarginalis W.
(Pyraustidae: Lepidoptera)

The larvae roll the leaves and feed on them.

Management

- Collection and destruction of leaf rolls along with larvae.
- Spray Endosulfan 35 EC at the rate of 2 ml/l of water or dust Carbaryl 10 D or Fenitrothion 5 D at the rate of 20 kg/ha.

Leaf Miner

Trachys sp.
(Burprestidae: Coleoptera)

Pseudoanapomyza alternanthera (Segur).
(Agromyzidae: Diptera)

Adults of *Trachys* sp. damage by cutting notches in leaves and sepals. The larva mines into the leaves and gives the appearance of scorched blotches. In case of severe infestation, the plant gives a dried appearance. The maggots of *Psuedoanapomyza alternanthera* mine the leaves.

Management

- Spray Phosalone 35 EC at the rate of 2 ml/l of water.

Tingid

Habrochila laeta D.
(Tingidae: Hemiptera)

The nymphs and adults feed on leaves, as a result yellowish spots appear. In case of severe infestation, leaves become brown, shrivel and dry off.

Management

- Spray Dimethoate 30 EC or Malathion 50 EC at the rate of 1 ml/l of water.

Other Pests

The larvae of *Precis hierta* F., *Precis almana, Precis atlites, Precis lemonias* Linn., *Precis orithya* Linn. (Nymphalidae), *Eustrotia mianoides* Hmpson (Noctuidae), *Euproctis fraterna* M. (Lymantriidae) and *Craspedia* sp. (Geometridae) feed on leaves (Nayar *et al,* 1976). The scale insects *Icerya* sp. (Margarodidae), *Orthezia insignis* D (Ortheziididae), *Cerococcus hibisci* Gr. (Astrolecaniidae), *Saissetia hemisphaericum* (Targ.) (Coccidae) and the aphid *Aphis gossypii* G. (Aphididae) also attack barleria. The wingless grasshopper *Orthacris simulans* B. (Acrididae) feeds on leaves. The bud worm *Trichoptilus* sp. (Pterophoridae) and leaf folder *Syngamia latimarginalis* Walk. (Pyraustidae) also attack barleria (Nayar *et al,* 1976).

PESTS OF ORNAMENTAL HERBS

BEGONIA (*BEGONIA* SP.)

Aphids

Aphis gossypii Glover.
Myzus persicae (Sulzer).
(Aphididae: Hemiptera)

Nymphs and adults cause damage by sucking the sap from the under-surface of leaves.

Thrips

Thrips tabaci Lind.
Heliothrips haemorrhoidalis (Bouche).
(Thripidae: Thysanoptera)

Thrips cause damage by sucking the sap from the leaves. The infested leaves show irregular, brownish red streaks on the upper surface of leaves and roughened brownish spots on the under-surface.

Management of Aphids and Thrips

- Spray Dimethoate 30 EC at the rate of 1 ml/litre or Phosphamidon 85 WSC at the rate of 0.5 ml/l of water.

Tarsonemid Mites

Polyphagotarsonemus latus Banks.

Steneotarsonemus pallidus Banks.
(Tarsonemidae: Acari)

For details see under Cyclamen.

Mealybug

Planococcus citri Risso.
(Pseudococcidae: Hemiptea)

Both nymphs and adults suck the sap and devitalise the plant. The mealybug is elliptical in shape, with short bristles or filaments all around the margin. The body is covered with a waxy mealy substance.

Management

- Spray Malathion 50 EC or Methyl Parathion 50 EC at the rate of 2 ml/l of water.

Whitefly

Trialeurodes vaporariorum (Westw.)
(Aleurodidae: Hemiptera)

Nymphs and adults suck the sap from the under-side of leaves and cause yellowing and drying of leaves. They secrete honeydew, on which sooty mould fungus grows later.

Management

- Spray Malathion 50 EC at the rate of 2 ml/l or Phosalone 35 EC or Methyl Parathion 50 EC at the rate of 1 ml/l of water.

Black Vine Weevil

Otiorhynchus sulcatus Fab.
(Curculionidae: Coleoptera)

Grubs feed on the smaller roots and tunnel through the longer roots and feed on the internal contents. Sometimes they also feed at the base of the stem. The affected plants ultimately wilt and die.

Management

- Laboratory, greenhouse and field studies showed that the use of entomogenous fungus *Metarhizium anisopliae* gave significant control of black vine weevil (Moorhouse *et al*, 1990).

LOTUS (*NELUMBO NUCIFERA*)

Aphid

Siphocoryne nymphaease Linn.
(Aphididae: Hemiptera)

The aphids suck the sap from shoots and floral buds.

Management

- The infested leaves and flower stalks may be kept immersed in water for at least one or two days, which will help to drown the aphids.
- Spraying of jet of water may knock the aphids into the water where they will be eaten by fish.

Thrips

Scirtothrips dorsalis Hood.
(Thripidae: Thysanoptera)

Scirtothrips dorsalis is a serious pest of lotus. They suck the sap from leaves.

Management

- Spraying of lotus leaves containing larvae of thrips with Acetamiprid, Fipronil, Imidacloprid, Methiocarb and Fenitrothion + Fenpropathrin controlled all the larvae and adults within 16 hours of treatment (Chinling *et al*, 1999).

Case Worm

Nymphula sp.
(Pyralidae: Lepidoptera)

The caterpillar has tracheal gills and feeds on leaves of the lotus.

Other Pests

The hairy caterpillar *Semrya conspersa* M. causes appreciable harm at times, especially in the ceded districts and N. Circars (Ayyar, 1963). The tobacco caterpillar *Spodoptera* sp. also feeds on leaves.

PESTS OF ANNUALS

CHRYSANTHEMUM (*CHRYSANTHEMUM MORIFOLIUM*)

Chrysanthemum has been a crop with many pest problems. Well over 35 pests are known to be the cause of unsaleable blossoms and low quality pot plants. The important pests of chrysanthemum and their management are described hereunder.

Aphids

Myzus persicae (Sulz).
Microsiphoniella sanborni Gillette.
Brachycaudus helichrysi Kalt.
Aphis gossypii Glover.
Aphis fabae Scopoli.
Aulocorthum circumflexum
(Aphididae: Hemiptera)

Apterous forms of *Myzus persicae* are yellowish green or rarely reddish. Alates are usually dull reddish, rarely green, measures about 1.4-2.6 mm. *Microsiphoniella sanborni* is large and maroon or dark brown in colour. Newly hatched nymphs of *Brachycaudus helichrysi* are dark green and the adults feeding on bark are chocolate coloured. Adults of *Aphis gossypii* are yellowish green to black in colour, 3 mm long with red eyes. Nymphs are globular, yellowish green or brown in colour. *Aphis fabae* is a coal black or very dark olive green aphid measuring about 1.6-2.6 mm.

Aphids on the terminal shoots and on the under side of the leaves feed by sucking the sap. Feeding results in stunting, leaf curling, withering of flowers and severe infestations resulting in death of the entire plant. Severe damage to the top of the plant may reduce the number of flowers produced. Aphids also excrete honeydew on which sooty mould grows. Aphids feeding on flowers make the plants unmarketable. Some species of aphids may also transmit viruses.

Management

- Aphids show varietal preference in feeding and reproduce at different rates on different cultivars of chrysanthemum. For example, 40 times more *Myzus persicae* were found on cultivar Tuneful than on Portrait. *Myzus persicae* were found to reproduce much faster on Tuneful than on Golden Princess Anne (Rice Mahr *et al*, 2001).
- Providing screens on vents and doors will help to prevent aphids from getting into the greenhouse, but often the most effective screens reduce air flow. The maximum hole size to exclude *Aphis gossypii* is 341 μ.
- Weeds may harbour aphids both inside and outside the greenhouse. Pull weeds or use commercially available herbicides or use barriers to prevent weed growth.
- Natural enemies such as green lacewing, *Chrysoperla carnea*, syrphids and ladybird beetles keep aphid populations in check.
- A single aqueous spray of Vertalec, a commercial formulation of *Verticillium lecanii* at the rate of 15 g/litre (10^8 CFU/g) consistently and effectively controlled *Myzus persicae* (Gardner *et al*, 1984). Frequent ultra low volume applications of the entomogenous fungus *Verticillium lecanii* are used against *Myzus persicae* and *Aphis gossypii* on glass house chrysanthemums. Application of *Verticillium lecanii* once in two weeks as a mist (600-800 l/ha) gave good control of both species of aphids and could be integrated with potentially harmful fungicides (Helyer and Wardlow, 1987).
- Neem formulations can be used in rotation with regular insecticides for the control of aphids.
- Spray Permethrin 50 EC or Malathion 50 EC at the rate of 1 ml/l of water or Endosulfan 35 EC or Methyl Parathion 50 EC (Ramesh, 1994) or Monocrotophos 36 WSC at the rate of 2 ml/l of water (Regupathy *et al*, 1994).
- Do not use Dimethoate for aphid control in chrysanthemum (Ramesh, 1994).

Thrips

Thrips tabaci Lind.
Thrips nigrophilosus
Thrips hawaiiensis Morgan
Frankliniella tritici (Fitch)
Frankliniella intonsa
Frankliniella occidentalis Pergande
Hercirothrips sp.
Microcephalothrips abdominalis (Crawford)
(Thripidae: Thysanoptera)

Haplothrips ramakrishnae K.
(Phloeothripidae: Thysanoptera)

Thrips are very small insects and commonly hide in flowers, buds and leaf axils and often go unnoticed until damage appears. They are slender, white to black in colour and feed on growing points causing mottling and distortion of leaves and also leaf silvering due to separation of the upper epidermal tissue from the rest of the leaf. The insect also damages flowers of summer blooming cultivars. The affected flowers look as if dried due to scorching heat. *Frankliniella occidentalis* is of special importance because it transmits the tomato spotted wilt virus and Impatiens necrotic spot virus to many floricultural crops. It is replacing many of the other thrips as the predominant thrip pest in greenhouses throughout the world (Rice Mahr *et al*, 2001). *Microcephalothrips abdominalis* infests flowers by causing abrasions on petals, due to which the attacked part turns brown. Heavy infestation results in damage to corolla, stamens and developing seed, leading to premature flower drop (Ananthakrishnan, 1973).

Management

- Among the germplasms screened, Chandrika, Bangalore Local Yellow Double, M-7, Pankaj and Yellow Star are least susceptible to *Microcephalothrips abdominalis* (Rami Reddy *et al*, 2001).
- Sanitation: This is an important cultural practice for preventing thrips problems. Remove weeds that can act as a refuge for thrips or harbour virus from inside and outside the greenhouse (Sharma and Bhattacharjee, 2002).
- Yellow or blue sticky traps can be used for mass trapping of thrips. For western flower thrips, *Frankliniella occidentalis*, which pupates mainly in the soil, place the traps near the soil to catch large numbers of newly emerged thrips before they move into the crop. Blue traps may be more effective for mass trapping than yellow traps.
- Release of nymphs and adults of the predatory minute pirate bug, *Orius insidiosus* (Say), reduced the western flower thrips, *Frankliniella occidentalis*, on chrysanthemum in the Netherlands (Fransen *et al*, 1993). Woets (1973) observed *Thrips tabaci* in greenhouses and subsequently Ramakers (1978) observed *Amblyseius* sp., *Orius minutus* L. and *Entomophora* sp. as effective biocontrol agents against thrips. Adults of the predator *A. cucumeris* feed on 3-6 immature thrips per day whereas *Orius* sp. consume 5-20 thrips per day, living up to 25 days as an adult. *Entomophthera* sp. causes significant thrips mortality outdoors in Europe, but in greenhouses it usually changes quickly to the resting sphore stage (Rice Mahr, 2001).
- Spray Dimethoate 30 EC at the rate of 1 ml/litre of Fenitrothion 50 EC at the rate of 3.5 ml/litre of Malathion 50 EC at the rate of 2 ml/litre or apply Aldicarb 10G at the rate of 5 g/sq. m. area.
- Spinosad, derived from *Saccharopolyspora spinosa* is very effective against thrips and has low to moderate impact on beneficials (Rice Mahr, 2001).

Leaf Miner

Phytomyza (= *Chromatomyia*) *sygenesiae*(Hardy)
Liriomyza trifolii (Burgess).
Liriomyza huidobrensis Blanchard.
Liriomyza pusilla Meigen.
(Agromyzidae: Diptera)

The larva of *Phytomyza sygenesiae* mines into the inner tissues of the leaf just under the epidermis, making tunnels and disfiguring the leaf. The mines made by them are irregular. *Liriomyza trifolii* is a major pest of commercially grown chrysanthemums throughout the world (Lindquist, 1983). The maggots of *Liriomyza trifolii* tunnel within the leaves between the upper and lower surfaces and pupate at the end leaving characteristic trails behind (Parrella and Jones, 1984). Mature larvae of *Liriomyza trifolii* usually vacate the leaf to pupate on the ground, whereas *Liriomyza huidobrensis* may either remain stuck to the leaf or pupate on the ground.

Management

- *Liriomyza trifolii* is attracted more to short day (SD) grown chrysanthemums than long day (LD) grown ones (Knodel *et al*, 1985).
- The cultivars Albatross, Bright Yellow, Tuneful, Colonel Comfort, Divinity, Nob Hill, Pink Marble and Sea Foam recorded less leaf miner damage than did Iceberg (Rice Mahr, 2001).
- Avoid over-fertilising chrysanthemums, since high levels of nitrogen are associated with leaf miner problems.
- Avoid introducing any new plant materials that may contain leaf miner eggs or larvae into the greenhouses.
- Female *Liriomyza trifolii* were found to fly further than males and although both sexes were trapped on yellow sticky cards, predominantly males were caught. Traps placed up to 26 metres apart gave adequate information to determine population levels (Jones and Parella, 1986).
- Herbert *et al* (1984) described three non-insecticidal methods of controlling *Liriomyza trifolii*. The first is trapping of adults of *Liriomyza trifolii* on yellow sticky boards giving 50 per cent control, second by using fungicide Pyrazophos at 0.5 ml/l and third by inter-planting field bean (*Vicia faba*) as a trap crop.
- The braconid wasp *Opius dimidiatus* is one of the most abundant larval pupal parasitoids of *Liriomyza trifolii* (Rice Mahr, 2001).
- Soil drenches of crude neem extract at 0.4 per cent caused significant mortality of late instars and pupae of *Liriomyza trifolii* both in research and commercial greenhouses in chrysanthemum for a period of 3 weeks (Larew *et al*, 1985).
- Foliar treatment with Methamidophos, Cypermethrin and Permethrin gave 90 per cent control of *Liriomyza* but Methamidophos was found to be phytotoxic (Alverson and Gorcuch, 1982). Ladieu and Helyer (1983) found Pyrazophos effective in reducing the damage by *Phytomyza* leaf miner by 27 per cent to that of control. Recently, Abamectin and Spinosad have been found effective against leaf miners.
- Pyrazophos at 0.0225 per cent and Triazophos at 0.04 per cent gave good control of larvae of *Phytomyza syngenesiae* and *Liriomyza trifolii* on chrysanthemum under glasshouse conditions (Lindquist *et al*, 1984).
- Storage of chrysanthemum cuttings at 1.18º C for 15 days or more resulted in 100 per cent mortality of larvae of *Liriomyza trifolii* and prevented egg hatching and mine formation. Mine formation was also prevented by dipping cuttings

in Pyrazophos 30 EC for 1-5 minutes, while longer periods resulted in phytotoxicity. Efficient control of first instar larvae was achieved with Pyrazophos and Chlorpyriphos 4 EC. Chlorpyriphos also controlled adults, as did Permethrin and Cyfluthrin. Pyrazophos was the only insecticide which gave good control of second and third instar larvae (Medina and Paris, 1985).

- Providing a 40 mesh insect proof net prevents the entrance of *Liriomyza trifolii* and *Liriomyza huidobrensis* into the greenhouses (Gokkes, 1999).

The American Bollworm

Helicoverpa armigera Hb.

(Noctuidae: Lepidoptera)

The larva cause damage by feeding on the leaves, young buds and flowers.

Chrysanthemum Caterpillar

Hydylepta (= Lamprosema) *indicata* F.

(Pyraustidae: Lepidoptera)

The larvae damage the tender leaves and flower heads of chrysanthemum in South India.

Management of American Boll Worm and Chrysanthemum Caterpillar

- Handpicking of larvae in early stage of infestation.
- Spray the crop with Endosulfan 35 EC at the rate of 2 ml/l or dust with Carbaryl 10 D at the rate of 20-25 kg/ha or spray *Bacillus thuringiensis* or Methomyl at the rate of 2 ml/l or cartap hydrochloride 1 g/l. In case of *Helicoverpa* spray HaNPV at the rate of 250 LE/ha.

Red Spider Mite

Tetranychus urticae Koch.

(Tetranychidae: Acari)

The spider mites occur on the lower surface of leaves causing white specks in the early stage. They occur in the hot season and damage leaves and buds which give a pale appearance.

Management

- Spray Dicofol 18.5 EC at the rate of 2 ml/l or wettable sulphur at the rate of 3 g/l of water.
- Release predatory phytoseiid mite *Phytoseiulus persimilis*.

Broad Mite or Yellow Mite

Polyphagotarsonemus latus Banks

(Tarsonemidae: Acari)

The yellow mites are very small, measuring about 150 μm long. Damage is normally noticeable only when large numbers of mites are present, and takes the form of distorted or brittle, down-curved leaves with a shiny undersurface. Adult females live for about 10 days and produce about five eggs daily. The mite life cycle is completed in less than one week at 20° C allowing a rapid population increase (Sunderland *et al*, 1992).

Management

- The predatory phytoseiid mite, *Amblyseius* sp. is used for the control of *Polyphagotarsonemus latus* (Wardlow, 1989; Karuppuchamy *et al*, 1994).
- The acaricides Fenbutatin Oxide and Tetradifon applied as high volume sprays controlled *Polyphagotarsonemus latus* and was safer to the predatory mite, *Phytoseiulus persimilis* (Hassan *et al*, 1987).
- Spray Dicofol 18.5 EC at the rate of 200 g a.i./ha or wettable sulphur 35 SP at the rate of 500 g a.i./ha (Karuppuchamy and Mohanasundaram, 1987).

Bihar Hairy Caterpillar

Diacrisia obliqua Walker.
(Arctiidae: Lepidoptera)

The larvae skeletonise the leaves from the upper portion.

Management

- Spray Endosulfan 35 EC or Quinalphos 25 EC at the rate of 2 ml/l of water.

Gall Midge

Diarthronomyia chrysanthemi Ahlberg.
(Cecidomyiidae: Diptera)

This pest damages young plant parts mostly. The development on chrysanthemum in the glasshouse continues throughout the year with eight overlapping generations. Soon after hatching, the young larva enters into the leaf and stimulates the formation of small pimple-like galls. The infested foliage is distorted and becomes unsaleable in the market.

Management

- Application of Diazinon 25 EC at 0.15 per cent, Monocrotophos 40 at 0.1 per cent, Fenitrothion 50 at 0.15 per cent or Methidathion 40 EC at 0.1 per cent three times at 10-day intervals, the first application coinciding with the appearance of the pest (Baranowski, 1985).

Other Pests

The tobacco caterpillar, *Spodoptera litura* F.(Noctuidae) causes damage by feeding on leaves. The white grub, *Holotrichia consanguinea* (Blanch)(Melolonthidae) cuts the underground portion of the stem and roots causing sudden wilting of healthy plants during dry hot months. The larvae of the stalk borer, *Papaipema nebris* Guen. (Noctuidae) and earwig, *Forficula auricularia* L. (Forficulidae) cause damage by feeding on flowers and leaves.

Besides, mirid bugs *Creontiades pallidifer* Wlk. feed by sucking sap from tender portions of the plant and leaves. The affected parts show irregular spots. In case of severe infestation, the affected leaves drop and dry up (Ramesh, 1994). An eriophyid mite, *Paraphytoptus chrysanthemi* Keifer, is associated with phyllody in the U. S. A. and France (Breaky and Batchelor, 1950; Favard, 1955). The beet armyworm and the corn earworm are also recorded in Florida grown chrysanthemums. Two species of tingids, namely, *Cadmilos retiarius* Dist. and *Galeatus scrophilus* Saunders (Nayar *et al*, 1976) and the greenhouse whitefly, *Trialeurodes vaporariorum* (Westw.) (Larew and Locke, 1990), also attack chrysanthemum.

CARNATION (*DIANTHUS CARYOPHYLLUS*)

Red Spider Mite

Tetranychus urticae Koch.
Tetranychus cinnabarinus Boisduval.
(Tetranychidae: Acari)

The spider mite is the most serious pest on carnations which feeds on the under-sides of leaves, sucks the sap, and eventually the leaves turn pale and have a dusty coating and fine webs. In severe attacks, plants become stunted and flowers are also invaded. Plant growth, crop quality, yield and vase life of carnation flowers decreases with increasing mite populations.

Management

- Initiate management measures when the density of mite reaches 1/leaf (ETL).
- Use varieties having straight and flat leaves, which are proved to be resistant.
- Cleanliness and elimination of weeds will reduce the incidence of the mite.
- Proper ventilation and water are the most important factors to prevent pest attack in greenhouses.
- Spray Dicofol 18.5 EC at the rate of 2 ml/l or wettable sulphur at the rate of 3 g/l (Ramesh, 1994). Propargite was the most effective compound against adults, eggs and larvae (Pandolfo, 1984).
- Chlorfenapyr 10 EC at 500 ml/ha sprayed at monthly intervals effectively suppressed the carmine spider mite *Tetranychus cinnabarinus* for a period of 3 weeks in carnations under polyhouse conditions. The mortality of the mite in the treated plots ranged from 90-95 per cent. Fluvalinate 25 EC at 50 ml/ha also suppressed *Tetranychus cinnabarinus* but the effectiveness lasted for only one week (Valunj *et al*, 1999).

Aphids

Aphis gossypii Glover.
Myzus persicae (Sulz).
(Aphididae: Hemiptera)

Aphis gossypii feeds on flower buds, leaves and terminal shoots by sucking the sap and also transmits mottle virus. The aphid transmits 10 per cent virus after one hour of feeding and 100 per cent transmission takes place if aphids feed for 12 hours (Singh and Singh, 1989). *Myzus persicae* sucks the sap from the leaves and disfigures the young growth. In severe infestations, they leave sticky deposits or honeydew on the leaves and flower buds.

Management

- Spray Diazinon 20 EC or Dimethoate 30 EC at the rate of 2 ml/l of water before the opening of flower.

Thrips

Thrips flavus Shrank.
Thrips hawaiiensis (Morgan).
Thrips tabaci Lind.
Haplothrips coloratus Trybom.

Haplothrips nigricornis Trybom.
(Thripidae: Thysanoptera)

Thrips tabaci sucks the sap from leaves, causing them to turn yellow and patchy, often with black specks and slight crinkling. A severe attack adversely affects the growth. They also cause streaks in the flowers, making them unmarketable. The dark and pink cultivars suffer most by this species (Bhatt, 1993). The other species of thrips destroy the beauty of flowers by scraping the flower petals. The attacked flower shows brown abrasions (Sood and Kakar, 1990; Wang, 1982; Sohi and Singh, 1995).

Management

- Spray Dimethoate 30 EC at the rate of 1 ml/l or Fenitrothion 50 EC at the rate of 3.5 ml/l or Malathion 50 EC at the rate of 2 ml/l of water or apply Aldicarb 10G at the rate of 5 g/sq. m. area.

Leaf Beetle

Aulacophora nigripennis
(Chrysomelidaes: Coleoptera)

The beetles feed on the leaves.

Management

- The beetles can be controlled by planting *Dianthus pungens* as a trap crop for carnation. When this trap plant was used, the mean number of resident *Aulacophora nigripennis* per 100 carnation plants per day decreased from 1.5 to 0.1 and the rate of damaged carnations decreased from 55.8 to 4.2 per cent under greenhouse conditions (Uda *et al*, 2001).

Cotton Boll Worm

Helicoverpa armigera Hubner.
(Noctuidae: Lepidoptera)

The larvae feed on leaves, flower buds and flowers. The larvae bore into buds by making round holes and feed inside. The attacked buds fail to open. The attack by this pest is more on carnation when it is grown near fields of Egyptian clover (Singh, 1985).

Management

- The moth of *Helicoverpa armigera* in carnation greenhouse is controlled by using yellow fluorescent lamps (Yase, 2001).
- Spray Endosulfan 35 EC at 2.5 l or Fenitrothion 50 EC at 2.125 l/ha using 312.5-375 litres of water (Anonymous, 1991).

Carnation Tortrix Moth

Tortrix pronubata Meyr.
(Tortricidae: Lepidoptera)

The larvae after hatching find a suitable site among the leaves and then fasten some of the leaves together with silken thread and feed inside the shelter. Injury is usually found around the growing points of the shoots, but the larvae will feed on any of the leaves and may also bore into the flower buds.

Management

- A mixture of Carbaryl and Dicofol is recommended for the control of larvae of tortrix moth in nurseries and beds (Arroyo, 1963).

Carnation Fly

Hylemya briennescens
(Anthomyiidae: Diptera)

The tiny, white maggots bore into the leaves and growing tips of young plants, more particularly those plants growing outdoors. They may be responsible for the death of young plants by tunnelling down the apical shoot.

Management

- Apply Carbaryl 4G or Lindane 6G or Phorate 10G at the rate of 10 kg/ha to the soil followed by light irrigation.
- Spray Malathion 50 EC at the rate of 2 ml/l of water.

Other Insects

The maggots of *Liriomyza trifolii* (Burgess) (Agromyzidae) make serpentine tunnels within the leaves between the upper and lower surface. Variegated cutworm, *Peridroma saucia* (Noctuidae) often causes damage to carnations (Rice Mahr *et al*, 2001).

ASTER (*CALLISTEPHUS CHINENSIS*)

Leaf Hopper

Macrosteles fascifrons Stahl.
Orosius albicinctus Dist.
(Cicadellidae: Hemiptera)

They suck the sap from foliage. *Macrosteles fascifrons* is a vector of virus disease caused by *Chlorogenus callistephi*. The development of pale yellowish tinge on the leaves and a large number of adventitious shoots are the main symptoms. Plants become dwarf and blooms are off colour, the rays being usually yellowish green.

Aphid

Macrosiphum artemisiae
Macrosiphum solanifolii.
(Aphididae: Hemiptera)

Aphids attack leaves, stems and flowers of asters. They suck the sap from the lower surface of leaves and from flower stalks.

Management of Leaf Hoppers and Aphids

- Spraying of Dimethoate 30 EC or Methyl Demeton 25 EC at the rate of 2 ml/l (Anonymous, 1999) or Phosphamidon 85 WSC at the rate of 1 ml/l of water.

Cyclamen Mite

Steneotarsonemus pallidus (Banks)
(Tarsonemidae: Acari)

Mites feed inside the folds of young leaves. Infested plants show stunted growth and brown scarrings along the stems. In case of heavy infestation, small, green rosettes are produced in place of flowers.

Management

- Spraying of Dicofol 18.5 EC at the rate of 2 ml/l of water keeps the pest under control.

Black Blister Beetle

Epicauta pennsylvanica
(Meloidae: Coleoptera)

This pest is also known as aster beetle. It feeds on foliage and flowers, often completely destroying the plants.

Management

- Spraying with Endosulfan 35 EC at the rate of 1 ml/l or Malathion 50 EC at the rate of 2 ml/l of water during the peak period of infestation.

Other Insects

The lacewing bug *Corythauma marmorata* (Tingidae) and cutworm, *Agrotis segetum* (Noctuidae)(Geissler and Schliephake, 1991), also cause damage to asters. Tarnished plant bug *Lygus lineolaris* (P. deB) punctures the terminal shoots below the flower buds which causes the flowers to droop and die. Root aphid *Anuraphis maidiradicis* feeds in large numbers on the roots causing yellowing, dwarfing and wilting of the plant.

MARIGOLD (*TAGETES ERECTA* L., *TAGETES PATULA* L.)

Two Spotted Spider Mite

Tetranychus urticae Koch.
(Tetranychidae: Acari)

The spider mite can become a pest late in a crop cycle primarily because it spins webs over the flowers which interfere with pollination.

Management

- Spray Dicofol 18.5 EC at the rate of 2 ml/l of water.
- Spot application of Dienochlor mixed with insecticidal soap also controls the mite (Heinz and Parella, 1990).

Leaf Hoppers

Empoasca fabae (Harris).
Macrosteles fascifrons (Stahl.).
(Cicadellidae: Hemiptera)

Both nymphs and adults suck the sap and cause cupping or rolling of leaves, wilting of shoot tips and leaflets. French marigold is more damaged than the African.

Management

Spray Dimethoate 30 EC (or) Malathion 50 EC at the rate of 2 ml/l of water.

Hairy Caterpillar

Diacrisia obliqua Walker.
(Arctiidae: Lepidoptera)

The larvae of this polyphagous pest eats away foliage.

Management

- Spray Dichlorvos 76 EC or Endosulfan 35 EC at the rate of 1 ml/l of water.

Leaf Miner

Liriomyza trifolii (Burgess)
(Agromyzidae: Diptera)

Larval feeding in the leaf mesophyll causes serpentine mines. Heavy leaf miner damage may reduce seed production because of reduced photosynthesis (Parella *et al*, 1985) or severe stunting. The young plants may be killed.

Management

- Inundative release of the parasitic wasp, *Diglyphus begini* (Ashmead) (Eulophidae: Hymenoptera), reduced *Liriomyza trifolii* populations in greenhouse marigolds grown for seed within 8 weeks of first release (Heinz and Parella, 1990).

Aphid

Aphis craccivora Koch.
Aphis gossypii Glover.
Myzus persicae (Sulzer)
(Aphididae: Hemiptera)

At high densities, feeding by these species can severely damage the apical portion of the plant and reduce the number of flowers and seed heads produced. They also excrete honeydew.

Management

- Spray Dimethoate 30 EC or Methyl Demeton 25 EC or Fenitrothion 50 EC at the rate of 1 ml/ l of water.

Greenhouse Whitefly

Trialeurodes vaporariorum (Westw.)
(Aleurodidae: Hemiptera)

At high population levels, sooty mould fungus grows on the honeydew excreted by whiteflies. Fungal growth on leaves can interfere with photosynthesis or hinder pollination when it covers the flowers' reproductive parts.

Management

- Release of *Encarsia formosa* Gahan and *Chrysoperla carnea* Stephens in greenhouses will check the population build up of whitefly (Heinz and Parella, 1990).
- Insecticidal soap can be used to reduce whitefly populations with no residual effects on natural enemies. Soap sprays are most effective against nymphs and adults but pupae are killed as well (Rice Mahr, 2001).

Tingid Bug

Cadmilos retiaris Dist.
(Tingidae: Hemiptera)

It attacks leaves of marigold. The pest appears in July and remains active up to September. The females lay eggs in a slanting position mainly on the upper surface of the leaves. The eggs hatch in 5-7 days and young nymphs moult five times during the nymphal period of 15-21 days. The nymphs and adults suck the sap and the infested leaves turn yellowish brown and finally dry up.

Management

- The eggs are parasitised by *Trichogramma* sp. The mite *Leptus* sp. (Erythraeidae) is parasitic on both nymphs and adults (Atwal, 1986).
- Spray Endosulfan 35 EC at the rate of 875 ml in 250 litres of water per hectare (Anonymous, 1991).

Other Pests

Larval feeding by various lepidopterous species such as *Trichoplusia ni* (Hubner), *Spodoptera exigua* (Hubner) and *Helicoverpa armigera* Hb. (Noctuidae) can directly damage the foliage and flowers. The adults of the blister beetle, *Epicauta mannerheim* nr. *maklin* (Meloidae: Coleoptera) feed on pollen and floral parts. Western flower thrips, *Frankliniella occidentalis* (Pergande) (Thripidae), is seldom a major pest on seed crops. However, as a pollen feeder, *Frankliniella occidentalis* at high densities depletes the pollen available for fertilisation. *Microcephalothrips abdominalis* (Thripidae) feeds on flowers during the winter throughout India (Nair, 1975). The nymphs and adults of the grasshopper, *Attractomorpha crenulata* F. (Acrididae), feed on leaves and petals in Himachal Pradesh (Sood, 1990). The maggots of *Chromatomyia horticola* Goureau make small, irregular white galleries on the leaves (Singh and Ipe, 1973). The trypetid, *Trypanea amoena* Ferfed maggots attack flowers and interfere with the opening of flower buds. The maggots and puparia are found in the thalamus at the base of disc florets (Trehan, 1947).

ZINNIA (*ZINNIA ELEGANS*)

Whitefly

Bemisia tabaci Guen.
(Aleurodidae: Hemiptera)

The nymphs and adults suck the sap from leaves. Due to feeding, leaves curl, turn yellow to brown, which ultimately shrivel and fall.

Management

- Spraying of Methyl Demeton 25 EC at the rate of 3 ml/l of water controls whitefly.

White Grub

Holotrichia consanguinea Blanchard.
(Melolonthidae: Coleoptera)

The adult beetles feed on leaves during the night. As a result of feeding, holes of various shapes are formed on leaves which give an ugly appearance. The grubs feed on rootlets. This pest is serious during June-July.

Management

- Digging flower beds during May-June controls the pest in the soil.
- Adult beetles may be controlled by spraying Carbaryl 50 WP at the rate of 2 g/l of water (Anonymous, 1990).

Gram Pod Borer

Helicoverpa armigera Hub.
(Noctuidae: Lepidoptera)

Larvae cause damage by feeding on petals and foliage.

Management

- Spray Endosulfan 35 EC or Methonyl or B.T. at the rate of 2 ml/l of water or HaNPV at the rate of 250 LE/ha.
- Hand picking of larvae in the early stages of infestation.

Zinnia Flea Beetles

Monolepta khasiensis

Monolepta signata Oliv.

Longitarsus sp.

Nisotra sp.

(Chrysomelidae: Coleoptera)

Beetles cause damage by feeding on petals and leaves.

Management

- Spray Malathion 50 EC at the rate of 2 ml/l of water.

Other Pests

The leaf miner *Liriomyza compositella* have been found making zigzag galleries on leaves and rarely on stems (Udayagiri, 1987). The blossom thrips *Microcephalothrips abdominalis* Crawford damages the flowers (Sood and Kakar, 1990). The other pests include black bean aphid, *Aphis fabae* Scopoli, the leafhoppers, *Graphocephala coccinea* and *Macrosteles fascifrons* (Stahl.), the long tailed mealybug, *Pseudococcus adonidum*, the broad mite, *Polyphagotarsonemus latus* Banks, the cyclamen mite, *Steneotarsonemus pallidus* Banks and two spotted spider mites *Tetranychus urticae* Koch.

HOLLY HOCK *(ALTHEA ROSEA* CAV.*)*

Whitefly

Bemisia tabaci Guen.

(Aleurodidae: Hemiptera)

Both nymphs and adults suck the sap from leaves. In case of severe infestation, vitality of the plant is reduced. While feeding, they excrete honeydew on which sooty mould develops.

Management

1. Spray Dimethoate 30 EC or Monocrotophos 36 WSC or Methyl Demeton 25 EC at the rate of 1 ml/l of water.

Leaf Hopper

Amrasca devastans Dist.
(Cicadellidae: Hemiptera)

Nymphs and adults feed on leaves and cause stippling.

Management

- Spray Methyl Demeton 25 EC at the rate of 3 ml/l or Phosphamidon 85 WSC at the rate of 0.75 ml/l of water (Anonymous, 1990).

Tingid

Urentis euonymus Fab.
(Tingidae: Hemiptera)

Nymphs and adults suck the sap from the lower surface of leaves. The infested leaves become pale yellow, turn brown, shrivel and fall.

Management

- Spray Malathion 50 EC at the rate of 1.25 l/ha or Endosulfan 35 EC at the rate of 875 ml/ha (Anonymous, 1991).

Gram Pod Borer

Helicoverpa armigera Hub.
(Noctuidae: Lepidoptera)

Neonate larvae feed on young leaves. The older larvae feed on flower buds and floral parts.

Management

- Spray Endosulfan 35 EC at the rate of 2.5 l/ha or Quinalphos 25 EC at the rate of 2 l/ha (Anonymous, 1990).

Spotted Bollworm

Earias vitella (Fb).
Earias insulana (Boisd).
(Noctuidae: Lepidoptera)

The larvae cause damage by boring into the growing shoots, buds, flowers and capsules. The attacked shoots wither, drop and die ultimately; the infested buds and flowers drop. The damaged capsules show holes plugged with excreta of the larva. The internal contents of the capsules are destroyed by the larva feeding on it.

Management

- Collect and destroy the early attacked buds and capsules to prevent further multiplication.
- The egg parasitoid *Trichogramma chilonis* Ishii parasitises eggs up to 15 per cent in North India. The pupal parasite *Brachymeria nephatidis* Gahan parasitises from 13-57 per cent (Atwal, 1986).
- Spray Endosulfan 35 EC at the rate of 2 to 2.5 ml/l or Malathion 50 EC at the rate of 2 ml/l of water.

Other Pests

The larva of *Sylepta derogata* Fb. (Pyraustidae) rolls the leaves and feeds on them. The maggot of *Chromatomyia horticola* Goureau (Agromyzidae) makes zigzag lines by feeding on leaves. The mirid bug, *Creontiades pallidifer* Wlk., sucks the sap from tender portions of the plant and leaves. The red cotton bug, *Dysdercus cingulatus* F. (Pyrrhocoridae), sucks the sap from leaves and tender shoots. The larva of *Crocidosema plebeiana* Zell. (Eucosmidae) also injures hollyhock (Nair, 1975).

PESTS OF BULBOUS CROPS

GLADIOLUS (*GLADIOLUS* SP.)

Gladiolus Thrips

Thrips simplex (Morison).
(Thripidae: Thysanoptera)

The gladiolus thrips were first described by Morison (1930) as *Physothrips simplex* collected on carnation flowers. It causes deformities and discoloration of gladiolus flowers and corms (bulbs) become soft and prone to decay (Denmark and Price, 2001). Moulten and Steinweden (1931) described the thrips under the name *Taeniothrips gladioli* and it was synonymised by Steel (1935). Batti (1969) places *gladioli* in the genus *Thrips*. Due to the attack, the buds and flowers remain closed or open irregularly. The infested surfaces are of unnatural glistening whitish grey in colour, because the cells in drying out become filled with air and so reflect the light. Thrips also thrive on corms in storage.

Adults emerged are milky-white, but soon turn brown except for the third segment which is light brown. The wings have a light transverse band near the base. A single female lays about 200 eggs. The egg is about 0.3 mm long, opaque white, smooth and bean shaped. Eggs are deposited in the leaf tissue and corms. The incubation period is 5-10 days. The two larval stages are light yellow and are usually found beneath the leaves or bracts. The fully developed second instar larva is about the size of the adult. The first pupal stage is distinguished from the second by having forward projecting antennae and short wing pads. The second pupal stage, which is a quiescent period, has the antennae folded over the back and has much longer wing pads.

Honey Suckle Thrips

Thrips flavus Schrank.
(Thripidae: Thysanoptera)

Adults are yellow to orange in colour. Legs and forewings are yellow to pale yellow. The hind margin of the pronotum is provided with four long setae. Antennae 7, segmented. Nymphs are white to pale yellow. Both nymphs and adults feed on the blossoms, as a result they get shrivelled and become dark.

Flower Thrips

Frankliniella sp.
(Thripidae: Thysanoptera)

Flower thrips often invade gladiolus and damage the florets.

Other Thrips

Thrips hawaiiensis Morgan.
(Thripidae: Thysanoptera)

Liothrips sp.
(Phlaeothripidae: Thysanoptera)

They feed on flowers (Sood and Kakar, 1990).

Management of Thrips

- Spraying of Malathion 50 EC at the rate of 1.25 ml/l or Diazinon 20 EC at the rate of 1 ml/l or Carbaryl 50 WP at the rate of 2 g/l of water before the bud initiation.
- At the time of flower emergence, spikes may be sprayed with Methyl Parathion 50 EC at the rate of 1 ml/l or Endosulfan 35 EC at the rate of 2 ml/l of water.
- Spray Methyl Demeton 25 EC or Dimethoate 30 EC at the rate of 2 ml/l. of water (Anonymous, 1999).

Aphids

Aphis gossypii Glover.
Myzus persicae (Sulz.).
Macrosiphum euphorbiae Thomas.
(Aphididae: Hemiptera)

Aphids suck the sap from plants and transmit virus diseases.

Management

- Spray Methyl Demeton 25 EC or Dimethoate 30 EC or Malathion 50 EC or Endosulfan 50 EC or Endosulfan 35 EC at the rate of 2 ml/l of water.

Mirid Bug

Lygus lineolaris (P. deB).
(Miridae: Hemiptera)

The adult bugs are mottled yellowish or reddish brown with flat, oval bodies, variously spotted, about 50-60 mm long. The nymphs are greenish. The bugs puncture the terminal shoot below the flower bud and inject toxic saliva, which usually causes the flower to drop and die.

Management

- Destroy the weeds in the vicinity of the crop.
- Spray Endosulfan 35 EC or Dimethoate 30 EC at the rate of 2 ml/l or Carbaryl 50 WP at the rate of 2 g/l of water before flowering. Repeat the spray if necessary.

Seed Corn Maggot

Hylemya cilicura
(Anthomyidae: Diptera)

The adults are slender, greyish with large (6 mm long) winged flies. The maggots are small and white in colour. Maggots feed on the seeds of gladioli by entering the capsules in the field or webbing the seeds in storage.

Management

- Apply Carbaryl 4G or Lindane 6G or Phorate 10G at the rate of 10 kg/ha to the soil followed by light irrigation.
- Spray Malathion 50 EC at the rate of 2 ml/l of water.
- Soak the seed material for one hour in Thiram + Malathion solution (Ramesh, 1994).

Bulb Mite

Rhizoglyphus echinopus Fum and Rob.
(Acarididae: Acarina)

Bulb mite feeds on the underground portions of bulbs, weakening the plant tissue. Feeding damage also promotes the secondary invasion of plant pathogens such as *Fusarium, Pythium* and *Rhizoctonia* (Rice Mahr *et al*, 2001). Apart from this species, *Rhizoglyphus solani* (Jefferson *et al*, 1956) and *Rhizoglypus rhizophagus* Banks may be considered as pests of gladiolus (Bald and Jefferson, 1952; Jefferson *et al*, 1956).

Other Pests

In India, the larvae of *Agrotis segetum* and *Feltia subterranea* and *Spodoptera dolichos* (Noctuidae) in the U.S.A. feed on different parts of plants and may sever plants at the base particularly when young (Magie and Poe, 1972). The greasy cut worm, *Agrotis ipsilon* Rott., tobacco caterpillar, *Spodoptera litura* damage young plants by cutting them from the base at ground level. Besides, cabbage green looper, *Trichoplusia ni* Hb.(Noctuidae) and cabbage white butterfly, *Pieris brassicae* Linn.(Pieridae) primarily feed on leaves. *Pieris brassicae* occasionally feeds on florets and spikes. The gram pod borer, *Helicoverpa armigera* Hubner (Noctuidae), feeds on spikes and florets. They also bore into seed capsules by making clear-cut holes and feed on the internal contents. In case of severe infestations, most of the capsules are found empty. The spider mites, *Tetranychus urticae* Koch and *Tetranychus bimaculatus* W. M.(Tetranychidae) feed on gladiolus leaves and make white specks on leaves, resulting in bronzing and silvering of the affected plants. The attack is severe in warm and shady conditions.

LILY (*LILIUM* SP.)

Leaf Caterpillars

Polytela gloriosa F.
Brythys crini Fb.
(Noctuidae: Lepidoptera)

The larvae of *Polytela gloriosa* feed on the green matter of leaves and cause defoliation. The moth is stout with a mosaic pattern of yellow, red and black on the forewings and black hind wings. The smooth larva is pretty looking with black, white and red spots. Round, yellowish eggs are laid in clusters of 13 to 42 on the apical portion of the under surface of leaves. They hatch in 3 to 4 days. The larval period lasts from 16 to 20 days undergoing 5 moults. Pupation takes place in soil for 15 to 20 days. There are two generations per year in North India; the pupae of the second generation hibernate in the soil (Sachan and Srivastava, 1965; Sohi and Singh, 1995).

The larvae of *Brythys crini* feed on leaves. In severe infestation, the whole plant is defoliated. The moth completes its life cycle in about 6 weeks (Khan and Abedi, 1957).

Management

- Spray Malathion 50 EC or Endosulfan 35 EC at the rate of 2 ml/l of water.

Aphids

Rhopalosiphum nymphaease Linn.
Myzus persicae (Sulzer).
Dysaphis tulipae
Neomyzus circumflexus
(Aphididae: Hemiptera)

Rhopalosiphum nymphaease sucks the sap from shoots and floral buds which results in withering of tender shoots and premature falling of buds. While feeding, aphids excrete honeydew on which sooty mould develops. The affected plants lose their elegance (Butani and Dadlani, 1983). *Myzus persicae* feeds on leaves and *Dysaphis tulipae* feeds on corms and infested corms, developing stunted and deformed plants. *Neomyzus circumflexus* is the crescent marked lily aphid which sucks the sap from leaves and tender portions of the plant and devitalises them.

Management

- Spray Malathion 50 EC or Dimethoate 30 EC at the rate of 2 ml/l of water.
- For the control of *Dysaphis tulipae* in storage, dust corms with Carbaryl dust at the rate 15-20 kg/ha before they are stored.

Lily Bulb Mites

Rhizoglyphus echinopus (Fum and Rob)
Rhizoglyphus robini
(Acarididae: Acarina)

The bulb mites feed on the bulbs and roots and destroy them. Later, they infest the scales and burrow into stems. The stem and leaves of infested plants become yellow. On the Easter lily, *Lilium longiflorum,* typical symptoms include rosetting, cessation of growth at a height of 3-6 inches, chlorosis and occasional death. Feeding damage by bulb mite also promotes secondary invasion of plant pathogens such as *Fusarium*, *Pythium* and *Rhizoctonia.*

Management

- Bulbs should be grown in properly drained soil.
- Crop rotation with non host crops should be followed.
- Bulbs should be stored at 35° F (1.6° C) which prevents feeding by the mites during storage.
- Hot water dip (111° F for 1 hour) is also effective against bulb mite (Smith, 1967).
- The predatory mite, *Hypoaspis aculeifer*, when released in the greenhouse feeds at the rate of 1 predator/2 to 5 prey and suppresses the population of the bulb mite to less than 10 individuals per bulb within 6 weeks (Lesna *et al*, 2000).

Thrips

Thrips simplex (Morison).
Liothrips vaneeckei
(Thripidae: Thysanoptera)

Thrips are serious pests of lily bulbs, producing rust coloured sunken areas at the base of the outer scales, which become soft and break off easily. Plants grown from infested bulbs are stunted.

Management

- Dipping the bulbs in hot water at 111° F for 1 hour controls thrips (Pirone *et al,* 1960).

- Fumigation of infested bulbs with methyl bromide.

Lily Beetle

Lilioceris lilii
(Chrysomelidae: Coleoptera)

The lily beetle seems to be mainly confined to some parts of the U.K. and Canada. It was also reported in the U.S.A. as a serious pest in summer 1992 (Casagrande and Livingston, 1995). Both larvae and adults feed on leaves, stems and seed pods of the lily, causing severe damage.

Lily Weevil

Agaspherops nigra
(Curculionidae: Coleoptera)

Lily weevil damages at the grub stage by feeding on the stem below the ground.

Management of Lily Beetle and Weevil

- Lily beetle and weevil can be controlled by spraying with Malathion 50 EC at the rate of 2 ml/l (Smith, 1967) or Carbaryl 50 WP at the rate of 2 g/l of water (Ramesh, 1994).

Other Pests

The stem borer, *Papaipema nebris,* bores into lily stems and feeds on the pith, moving from one plant to another (Smith, 1967). The leaf caterpillar, *Spodoptera litura,* is also recorded on lily (Nair, 1975). Lily bulbs are also prone to attack by rats.

DAHLIA (*DAHLIA VARIABILIS* LINN.)

Aphid

Myzus persicae (Sulz.)
Brachycaudus helichrysi (Kalt.)
Aphis fabae Scopoli.
Macrosiphum euphorbiae Th.
(Aphididae: Hemiptera)

Both nymphs and adults suck the sap from young leaves and flower buds. The affected leaves curl up and present a distorted appearance. The infested buds fail to open and give poor quality of flowers. The aphids may also transmit viral diseases.

Management

- Spray Malathion 50 EC at the rate of 1.5 ml/l or Dimethoate 30 EC at the rate of 1 ml/l of water.

Thrips

Heliothrips haemorrhoidalis (Bouche)
Frankliniella tritici (Fitch)
Thrips tabaci Lind.
Thrips hawaiiensis Morgan
Megaleurothrips distalis (Kirby)
Microcephalothrips abdominalis Crawford.
(Thripidae: Thysanoptera)

Thrips infest the flowers, rasp the surface and feed on the exuding juice. The lower surface of petals turns whitish and withers.

Leaf Hopper

Empoasca fabae
Evacanthus sp.
(Cicadellidae: Hemiptera)

Nymphs and adults suck the sap from leaves and cause discoloration of leaves which appear first along the margin and spread towards the mid vein. The affected plant becomes stunted. *Evacanthus* sp. has been found to cause extensive damage in Himachal Pradesh (Sood and Kakar, 1990).

Management of Thrips and Leaf Hoppers

- Spraying of Malathion 50 EC at the rate of 1.5 ml/l or Methyl Demeton 25 EC at the rate of 1 ml/l or Phosphamidon 85 WSC at the rate of 0.5 ml/l of water is effective to control thrips and leaf hoppers.

Two Spotted Spider Mite

Tetranychus urticae Koch.
(Tetranychidae: Acari)

They cause damage by feeding on the lower surface of leaves, causing white speckling, but as numbers increase, webbing may be spun around the leaves and buds.

Management

- Spray Methyl Demeton 25 EC at the rate of 2 ml/l or wettable sulphur at 3 g/l or Karathane at the rate of 1 ml/l of water.

Bud Worm

Helicoverpa armigera Hub.
(Noctuidae: Lepidoptera)

The larvae feed on flower buds, petals and the reproductive parts of the flower. The attacked flower buds remain unopened and infested flowers give ugly appearance. This pest causes extensive damage to flowers in Ludhiana from the end of March to April and a maximum of seven larvae per plant were recorded (Singh and Sohi, 1990).

Management

- Spraying of Endosulfan 35 EC or Malathion 50 EC at the rate of 2 ml/l of water at bud initiation stage controls this pest.

Other Pests

Adults of the beetle *Phyllotreta downsei* Baly (Chrysomelidae) cause extensive damage to leaves by making innumerable round holes. Sometimes stem and flowers are also attacked by this beetle. The beetles, *Monolepta signata* (Oliv.) (Chrysomelidae), *Mylabris mecilanta, Mylabris pustulata* (Meloidae) and *Oxycetonia albopunctata* (Fab.) (Cetoniidae) feed on flowers (Sood and Kakar, 1990). The maggots of *Chromatomyia horticola* Goureau (Agromyzidae) make serpentine galleries on leaves (Singh and Ipe, 1973). The larvae of *Thysanoplusia orichalcea* (Fab.) (Noctuidae) feed on leaves of *Dahlia* sp. (Sood, 1990). Nymphs and adults of *Attractomorpha crenulata* F.

(Acrididae) feed on flowers and leaves. Its damage has been recorded on *Dahlia* sp. in Himachal Pradesh (Sood and Kakar, 1990).

CYCLAMEN (*CYCLAMEN* SP.)

Cyclamen Mite

Steneotarsonemus pallidus Banks.
(Tarsonemidae: Acari)

The mites are minute, measuring 0.25 mm long, usually found on shoot buds and unexposed leaves, where they are well protected and are rarely seen crawling about exposed leaves. They feed by sucking the leaves and flower buds. The infested leaves are badly deformed, deeply cut, more or less thickened and brittle, curved or twisted. The flower buds are also much deformed or distorted. The life cycle is completed in about two weeks.

Broad Mite or Yellow Mite

Polyphagotarsonemus latus Banks.
(Tarsonemidae: Acari)

Broad mites are minute, broader than the cyclamen mite. Adults are yellowish whereas larvae are creamy white. Males move faster than females. The leaves of infested plants have a glassy or shiny appearance on the undersurface and are curled downwards rather than upwards. The eggs are white, about 1 mm in length, broad, round, with tubercles and whitish. The life cycle is completed within a week.

Management of Cyclamen Mite and Broad Mite

- The predatory phytoseiid mite *Ambyseius ovalis* will check the population of yellow mite (Karuppuchamy *et al,* 1994).
- Spray Dicofol 18.5 EC or Endosulfan 35 EC at the rate of 2 ml/l or Tetradifon 20 WP at the rate of 2 g/l of water (Ramesh, 1994).

Thrips

Heliothrips haemorrhoidalis (Bouche)
Frankliniella occidentalis Pergande.
(Thripidae: Thysanoptera)

Thrips suck the sap from leaves. The infested leaves show irregular brownish red streaks on the upper surface of leaves and roughened brownish spots on the under-surface.

Management

- Weekly release of the predatory mite *Neoseiulus cucumeris* at the rate of 200 or 350 mites m^{-2} resulted in a decline in numbers of *Frankliniella occidentalis*. High inoculative releases of *Neoseiulus cucumeris* early in the flowering cycle followed by frequent low introductions of predators should provide a strong basis for preventive control of *Frankliniella occidentalis* and other thrips species on cyclamen (Courcy Williams, 2001).
- Release of predatory mite *Amblyseius barkeri* and predatory insects *Orius* sp. and *Chrysoperla carnea* controls *Frankliniella occidentalis* (Wilhelm, 1997).
- Spray Dimethoate 30 EC at the rate of 1 ml/l or Fenitrothion 50 EC at the rate of 3.5 ml/l or Malathion 50 EC at the rate of 2 ml/l of water.

Aphid

Aphis gossypii Glover
(Aphididae: Hemiptera)

Nymphs and adults cause damage by sucking the sap from tender shoots and from the undersurface of leaves. In severe infestations, leaves curl, and the growth of the plant is arrested. Young plants gradually dry and die. The affected plants, if alive, get deformed and stunted.

Management

- Spray Endosulfan 35 EC at the rate of 2 ml/l of water.

PESTS OF HERBACEOUS PERENNIALS

ANTHURIUM (*ANTHURIUM* SP.)

Aphid

Myzus circumflexus
(Aphididae: Hemiptera)

Nymphs and adults suck the plant sap, causing yellowing and distortion of leaves and poor growth. They secrete honeydew upon which a black sooty mould often grows.

Management

- Spraying of Dimethoate 30 EC at the rate of 1 ml/l or Phosphamidon 85 WSC at the rate of 0.5 ml/l or Malathion 50 EC at the rate of 2 ml/l of water effectively controls the aphids.
- A formulation containing Pyrethrum extract gave complete control of *Myzus circumflexus* on *Anthurium scherzerianum* plants in the greenhouse, without phytotoxicity (Pignon, 1963).

Thrips

Chaetanaphothrips orchidii (Moulton)
(Thripidae: Thysanoptera)

Thrips suck the sap from leaves and cause mottled effect on foliage and flowers. This life cycle of *Chaetanaphothrips orchidii* on *Anthurium andreanum* is 27.4 days.

Management

- In netted areas in Hawaii, insecticide granules were very effective in decreasing injury to flowers by *Chaetanaphothrips orchidii* for 5 months. Aldicarb 10G and Disulfoton 15G were most effective, followed by Chlorpyriphos 2G and Oxamyl 10 G (Hata and Hara, 1992a).
- As foliar sprays, Bifenthrin 10 WP, Cyfluthrin 2.0 EC and Chlorpyriphos 1.0 ME were also effective in reducing the number of injured flowers. The effectiveness of the spray lasted for 2 weeks (Hata and Hara, 1992a).
- Spraying of Dimethoate 30 EC at the rate of 1 ml/l or Phosphamidon 85 WSC at the rate of 0.5 ml/l also controls thrips.

Whitefly

Bemisia tabaci Guen.
Aleurotulus anthuricola Nakahara.
(Aleurodidae: Hemiptera)

Nymphs and adults suck the sap from leaves. The life cycle of *Aleurotulus anthuricola* on *Anthurium andreanum* is 35.5 days.

Management

- Systemic insecticides like Methomyl 1.8 L or Dimethoate 4.0 EC significantly reduced the number of eggs, nymphs, pupae and adults.
- Contact insecticides like Fenpropathrin 2.4 E was also effective but was not as effective as that of systemic insecticides. Malathion 5.0 EC was only effective against adults and pupae. Bifenthrin 10 WP was effective only against adults (Hata and Hara, 1992b).
- In field trials, Methomyl 1.8 L and Acephate 75 SP significantly reduced the number of nymphs and pupae. The number of eggs was also reduced in Acephate 75 SP treated plots (Hata and Hara, 1992b).

Red Spider Mite

Tetranychus cinnabarinus Bois.
(Tetranychidae: Acari)

Spider mite causes yellowish mottling of the leaves which may become brown and shrivel. A fine web spun on the under-surface of leaves is another sign.

Management

- Spraying the affected plants with Malathion 50 EC at the rate of 4 ml/l or Dicofol 18.5 EC at the rate of 0.8 ml/l of water controls the mite.

Other Insects

Scales and mealybugs also attack anthurium.

GERBERA (*GERBERA JAMESONII* H.)

Greenhouse Whitefly

Trialeurodes vaporariorum (Westwood)

Cotton Whitefly

Bemisia tabaci Guen.
(Aleurodidae: Hemiptera)

The nymphs and adults suck the sap from the underside of the leaves.

Management

- The parasitoid *Encarsia formosa* is used for the control of *Trialeurodes vaporariorum* in greenhouses.
- The use of yellow-orange, Plexiglas sticky traps was also suggested in IPM (Nucifora and Vacante, 1981).
- Application of Pirimiphos-methyl at 3 ml/l or Methamidophos at 1 ml/l effectively controlled whitefly in greenhouses (Baillot, 1977).
- A formulation of potassium salt of fatty acids, namely, MYX 400 I gave control of both the species of whiteflies equivalent to that of Buprofezin and Etofenprox (Pasini *et al*, 1997).

- Spray Monocrotophos 36 WSC at the rate of 2 ml/l or neem oil at the rate of 3 ml/l of water (Anonymous, 1999).

Leaf Miners

Liriomyza trifolii (Burgess)
Liriomyza soncho
(Agromyzidae: Diptera)

The insects bore into the leaf and make irregularly shaped tunnels or blotches which are generally light yellowish tan to brown in colour. As these larvae mature, they fold the leaf together with threads and feed on the inner surface. This pest occurs in greenhouses and in open cultivation.

Management

- Natural immigration of the eulophid *Diglyphus isaea* afforded about 90 per cent parasitism of *Liriomyza trifolii* and effectively protected the crop without artificial control measures.
- Control tests in Sicily showed that combination of yellow trap against adults and aqueous sprays of Pyrazophos against the larvae within the leaf mines were very effective giving 100 per cent mortality (Das and Singh Samanta, 1999).
- Spray Dimethoate 30 EC or Dichlorvos 76 EC or Permethrin at the rate of 1 ml/l of water.
- Spray Chlorpyriphos 20 EC at the rate of 2 ml/l. of water (Anonymous, 1999).

Mites

Polyphagotarsonemus latus (Banks)
Steneotarsonemus pallidus Banks.
(Tarsonemidae: Acari)

In the greenhouse, the development of leaves and flower buds are adversely affected, the flowers become malformed and unsaleable due to the attack by these mites.

Management

- Spray Dicofol 18.5 EC or Tetradifon 20 at the rate of 2 ml/l of water.

Aphids

Myzus persicae (Sulz.)
Aphis fabae Scopoli.
(Aphididae: Hemiptera)

They infest young leaves and cause injury by sucking the sap, resulting in distortion of tissue.

Management

- Spray Malathion 50 EC or Endosulfan 35 EC at the rate of 2 ml/l of water.

Other Insects

The gerbera is also attacked by *Helicoverpa armigera, Memestra brassicae* and *Spodoptera littoralis* (Boisd.) (Noctuidae).

ORCHIDS

The important pests of orchids and their management are discussed.

Scales

Brown Soft Scale

Coccus hesperidum Linn.
Coccus pseudohesperidum
Saissetia coffeae Walker.
Lecanium formicarii L.
Pulvinaria phaiae
(Coccidae: Hemiptera)

Hard Scale

Aspidiotus destructor
Aspidiotus herderae
Aspidiotus nerri

Boisduval Scale

Diaspis boisduvali

Orchid Scale

Furcaspis biformis

Other Scales

Genaparlatoria pseudoaspidiotus
Lepidosaphes mackiana
Lepidosaphes noxia
Lepidosaphes machili
Melanaspis aliena
Parlatoria proteus
Leucaspis cockerelli
Fiorinia fiorinae
Chrysomphalus dictyospermi
Hemiberlesia lataniae Signoret.
Hemiberlesia cyanophylli
(Diaspididae: Hemiptera)

The scales attach themselves to leaves and suck the juice from the cells. Armoured scales (Diaspididae) are amongst the most serious and persistent pests of orchids. They not only extract plant juice but many also secrete honeydew which supports the growth of sooty mould that are cosmetically damaging besides attracting ants. About 100 species of armoured scales have been recorded from orchids and 31 species have been reported from Florida alone (Deckle and Kuitert, 1968). The Boisduval scale is the common armoured scale found on many orchid genera, especially *cattleya* and *cymbidium*. It damages both leaves and pseudobulbs. *Furcaspis biformis* attack *cattleya* orchids while *Genaparlatoria pseudoaspidiotus* was found on *Vanda* and *Lepidosaphes mackiana* and *Lepidosaphes noxia* were observed on *Dendrobium* sp. *Cymbidium* orchids host *Lepidosaphes machili*. The soft scale *Coccus pseudohesperidum* infest *Cattleya* and *Dendrobium* orchids in Europe, the U.S.A. and Hawaii (Yadav and Bose, 1993).

The females of *Aspidiotus destructor* cluster in large numbers on the undersurface of leaves and feed on the plant sap. The leaves as a result turn yellow and in severe cases they wither and dry up; vitality of the plant is drastically reduced (Ramesh, 1994). *Saissetia coffeae* and *Aspidiotus nerri* were reported to attack Cymbidium in Hungary (Ordogh and Takacs, 1983).

Management

- Holttum (1964) recommended the use of root extract of *Derris*, a plant common in Malaysia.
- Spraying of fish oil rosin soap at 1 kg in 67.5 litres of water (Braham and Vatsala, 1981).
- Green lacewing larvae *Chrysoperla carnea* feed on the immature stage of scales. Ladybird beetles *Cryptolaemus montrouzieri* and *Scymnus* sp. are known to attack scale insects.
- Only the crawler stage is susceptible to insecticides. Thus for effective control, frequent application of insecticides is necessary. Spraying of Malathion 50 EC or Methyl Parathion 50 EC or Phosalone 35 EC or Methyl Demeton 25 EC at the rate of 2 ml/l or Fenitrothion 50 EC at the rate of 1 ml/l of water.

Mealy Bugs

Ferrisia virgata Cockerell.
Pseudococcus filamentosus var. *corymbatus.*
Pseudococcus adonidum
Pseudococcus maritimus
Planococcus citri Risso.
Nipaecoccus nipae
(Pseudococcidae: Hemiptera)

Mealybugs cover the stem and leaf, suck the sap, secrete honeydew and attract ants. In large infestations, sooty mould may develop on the leaves. The affected plants lose their vigour.

Management

- Lady bird beetles *Cryptolaemus montrouzieri* and *Scymnus* sp. are known to feed mealybugs.
- Spray Malathion 50 EC at the rate of 2 ml/l or Phosphamidon 85 WSC at the rate of 0.5 ml/l of water.

Two Spotted Spider Mite

Tetranychus urticae Koch.
(Tetranychidae: Acari)

Tenuipalpid Mite

Brevipalpus australis
Brevipalpus cauliformicus
Brevipalpus oncidii
Tenuipalpus pacificus
(Tenuipalpidae: Acari)

Red spider mite *Tetranychus urticae* suck the sap from cells of leaves, leaving them speckled with white. In heavy infestation, webbing occurs on the plants. This is a serious pest on orchids during warm and dry season. *Brevipalpus australis* and *Brevipalpus cauliformicus* were also recorded on many types of orchids. *Brevipalpus oncidii* is a serious pest of *Oncidium* and *Odontoglossum* in England and California. *Tenuipalpus pacificus* is a serious pest of *Phalaenopsis* and other orchids in Australia, Indonesia, the Philippines and the U.S.A.

Management

- The predatory phytoseiid mites, *Amblyseius* sp. and *Phytoseiulus persimilis* will keep the spider mite population in check.
- Spray Dicofol 18.5 EC or Dimethoate 30 EC or Methyl Demeton 25 EC at the rate of 2 ml/l of water.

Thrips

Thrips tabaci Lind.
Anaphothrips orchidaceus
Anaphothrips orchidii
Heliothrips haemorrhoidalis (Bouche)
Hercirothrips femoralis
Frankliniella sp.
(Thripidae: Thysanoptera)

Thrips feed on the foliage or remain hidden in the flower buds. The epidermis is punctured, causing silvery areas on the plant. The flower buds are disfigured and sometimes they even fail to open.

Management

- Spraying with Malathion 50 EC or Methyl Demeton 25 EC at the rate of 2 ml/l of water.

Orchid Weevil

Orchidophilus aterrimus Waterhouse
Cholus sp.
Eucactophagus weissi
(Curculionidae: Coleoptera)

The adults feed on flowers or puncture the flower buds and the grubs bore and feed within the plant tissues. *Orchidophilus aterrimus* is a serious pest of *Dendrobium* orchids.

Management

- Immersion of all susceptible plant materials in chemical dips, regular insecticide applications and regular plant and property inspections controlled *Orchidophilus aterrimus* (Smith and Neal, 2001).
- Spraying with Endosulfan 0.2 per cent emulsion at 3-week intervals for 3 months was also suggested.

Orchid Beetle

Stethopachys formosa Baly.
(Chrysomelidae: Coleoptera)

Orchid beetle is a pest of orchid in eastern Australia.

Management

- Spraying of Bendiocarb at 0.4 g/l, Carbaryl at 1 g/l, Methidathion at 0.5 g/l or Methiocarb at 1.5 g/l effectively controlled the adults of orchid beetle. Dimethoate at 0.3 g/l and Malathion at 0.5 g/l were also effective (Cough and Montgomery, 1994).

Slugs and Snails

The Giant African Snail
Achatina fulica Feussae.
Slug: *Deroceras laeve.*

They are common greenhouse pests where the humidity is high. They feed on the young leaves, roots, flower buds and even open flowers. Slim trails leading to plant injury are indicative of the presence of these pests. Slugs usually hide during the day and become active during night, causing damage by feeding on foliage.

Management

- Apply 15 per cent Metaldehyde dust or 20 per cent Metaldehyde liquid.
- Sprinkle 5 per cent Metaldehyde pellets around the infested plant.
- If the population is low, collect and destroy.

Other Insects

The cattleya fly, *Eurytoma orchidearum* and cattleya midge *Parallelodiplosis cattleyae* attack *Cattleya* orchids and some other genera of orchids. *Eurytoma orchidearum* larvae burrow within flower buds and pseudobulbs. Bugs such as *Tenthecoris orchidearum, Tenthecoris bicolor, Tenthecoris vestitus, Tenthecoris figueiredoi* injure orchids by sucking the sap and cause pale spots on the leaves. The root miner *Agromyza orchidearum* causes wilting of orchids.

REFERENCES

Abraham, A. and Vatsala, P. In: *Introduction to Orchids*. Thiruvananthapuram, India: Tropical Botanic Garden and Research Institute, 1981.

Agarwala, B. K. 'Parasites and predators of rose infesting aphids (Homoptera: Aphididae) in India'. *Entomon.* 8(1): 35-39, 1983.

Agarwala, B. K. and Raychaudhuri, D. N. 'Observation on the population trend of rose aphids, and their hymenopterous parasites in Kalimpong, West Bengal (India)'. *Entomon.* 6(3): 211-214, 1981.

Ali, S. M. 'Coccids (Coccidae: Hemiptera: Insecta) affecting fruit plants in Bihar (India)'. *J. Bombay Nat. Hist. Soc.* 65: 120-137, 1968.

Alverson, D. R. and Gorcuch, C. S. 'Evaluation of chrysanthemum cultivars and insecticides for the control of damage by leaf miner, *Liriomyza trifolii* (Diptera: Agromyzidae)'. *J. Econ. Ent.* 75: 888-891, 1982.

Amutha, S. Bioecology and control of three pyralids *Hendecasis duplifascialis* Hmpsn. *Nausinoe geometralis* (Guen.) and *Margaronia unionalis* Hbn. on *Jasminum* sp. Ph.D. Thesis. University of Madras, Chennai, Tamil Nadu, India, 1994.

Amutha, S. and. David, B. V. 'Screening of insecticides and assessment of economics of chemical control against *Hendicasis duplifascialis* Hmpsn. (Lepidoptera: Pyralidae) on *Jasminum multiflorum* (Burm. F.) Andr'. *Indian J. Envt. and Toxicol.* 5: 69-70, 1995.

Ananthakrishnan, T. N. *Thrips: Biology and Control*. New Delhi, India: Macmillan Company of India, 1973.

Anonymous. In: *Crop Production Techniques of Horticultural Crops*. Coimbatore, Tamil Nadu, India: Dept. of Horticulture and Plantation Crops, Chennai and Tamil Nadu Agrl. University, 1999.

Anonymous. *Package of Practices for Crops of Punjab, Kharif 1990*. Ludhiana, India: PAU, 1990.

Anonymous. *Package of Practices for Crops of Punjab, Rabi 1990-91*. Ludhiana, India: P.A.U. 1991.

Arroyo, M. 'Agriculture'. *Madrid* 32: 283-285, 1963.

Atwal, A. S. In: *Agricultural Pests of India and South-East Asia* . Ludhiana, India: Kalyani Publishers, 1986.

Atwal, A. S. and Dhingra, S. 'Biological studies on rose aphid, *Macrosiphum rosaeiformis* Das'. *Indian J. Ent.* 33: 136-142, 1971.

Ayyar, T. V. R. 'Some South Indian coccids of economic importance'. *J. Bombay Nat. Hist. Soc.* 26(2): 621-628, 1919.

Ayyar, T. V. R. In: *Hand Book of Economic Entomology for South India*. Chennai, India: Government of Madras, 1963.

Bagnall, R. S. 'Brief description of new Thysanoptera'. *12 Ann. Mag. Nat. Hist.*, London. 8(48): 393-400, 1921.

Baillot, F. 'Whitefly control trial on gerbera'. *Horticulture Francaise*, No. 84, p.12, 1977.

Bald, J. G. and Jefferson, R. N. 'Injury to gladioli associated with the root mite, *Rhizoglyphus rhizophagus* (Bks)'. *Plant Disease Rept.*, 36: 435-437, 1952.

Baranowski, T. 'Studies on the biology and control of chrysanthemum gall midge *Diarthonomyia chrysanthemi* (Diptera: Cecidomyiidae)'. *Prace Instytutu Sadownictwa i Kwiaciarstwa w Skierniewicach* B 7: 251-261, 1985.

Beeson, C. F. C. 'Life history and control of *Celosterna scabrator* F. (Coleoptera: Cerambycidae)'. *Indina Fort. Rec.* 15(6): 279-294, 1931.

Bhatt, N. R. 'Carnation'. In: *Commercial Flowers*. (Eds.). Bose, T. K. and Yadav, L. P. Kolkata, India: Naya Prokaşh, pp. 355-416, 1993.

Bhattacharjee, S. K. and De, L. C. In: *Advanced Commercial Floriculture*. Jaipur, India: Aavishkar Publishers, Distributors, Jaipur, 2003.

Bhatti, J. S. 'Taxonomic studies in some *Thripini* (Thysanoptera: Thripidae)'. *Oriental Insects* 3: 373-382, 1969.

Bindra, O. S. and Singh, J. 'Biological observations on the white grub, *Lachnostema consanguinea* Blanchard (Coleoptera: Scarabaeidae) and relative efficacy of different insecticides against adults'. *Indian J. Ent.* 33: 225-227, 1971.

Bose, T. K. and Yadav, L. P. *Commercial Flowers*. Kolkata, India: Naya Prokash, 1993.

Bose, T. K., Maiti, R. G., Dhua, R. S. and Das, P. *Floriculture and Landscaping*. Kolkata, India: Naya Prokash Pub., 1999.

Breaky, E. P. and Batchelor, G. S. 'Phyllody of chrysanthemum and the eriophyid mite, *Paraphytoptus chrysanthemi*'. *Ann. Entomol. Soc. Am.* 43: 452-454, 1950.

Butani, D. K. 'Pests damaging roses in India and their control'. *Pesticides* 8(12): 40-42, 1974.

Butani, D. K. and Dadlani, N. K. 'Pests of ornamental plants in India and their control'. *Pesticides* 17(3): 18-20, 1983.

Cesagrande, R. A. and Livingston, S. 'Lily leaf beetle *Lilioceris lilii*". *Quarterly Bulletin of the North American Lily Society* 49(3): 48-50, 1995.

Chinling, W., Monyu. H., Chinhan, Y. and Yuling, J. 'Toxicity of insecticides for *Scirotothrips dorsalis* Hood (Thysanoptera: Thripidae) on lotus'. *Chinese Journal of Entomology* 19(4): 377-380, 1999.

Cough, N. and Montogomery, B. L. 'Efficacy of insecticides against *Stethopachys formosa* Baly (Coleoptera: Chrysomelidae), a pest of orchids, in eastern Australia'. *Journal of the Australian Entomological Society* 33(4): 351-352, 1994.

Courcy Williams, M. E. De. 'Biological control of thrips on ornamental crops: interactions between the predaceous mite, *Neoseiulus cucumeris* (Acari: Phytoseiidae) and western flower thrips *Frankliniella occidentalis* (Thysanoptera: Thripidae) in cyclamen'. *Biocontrol Science and Technology* 11(1): 41-55, 2001.

Das, B. 'The Aphididae of Lahore'. *Mem. Indian Mus. Calcutta* 6(4): 135-274, 1918.

Das, P. and Singh, Samanta, P. K. 'Gerbera'. In: *Floriculture and Landscaping.* (Eds.). Bose, T. K., Maiti, R. C., Dhua, R. S. and Das, P.). Kolkata: Naya Prokash, pp. 530-547, 1999.

David, B. V. and Kumaraswami, T. In: *Elements of Economic Entomology*. Chennai, India: Popular Book Department, 1975.

David, P. M. M. and Natarajan, S. 'Effect of insecticides on blossom midge caused purple discoloration and drying of flower buds in *Jasminum sambac* Ait. cv. Gundumalli'. *Indian Perfumer* 32: 292-294, 1988.

David, S. K. 'Insects and mites affecting jasmine in Madras State'. *Madras Agric. J.* 45: 146-150, 1958.

David, S. K. 'Notes on South Indian aphids'. *Indian J. Ent.* 19: 289-299, 1957.

David, S. K. and Venugopal, S. 'Life history and habits of the leaf webworm, *Nausinoe geometralis* Guenee. (Lepidoptera: Pyraustidae)'. *J. Bombay Nat. Hist. Soc.* 59: 577-582, 1962.

Deckle, G. W. and Kuitert, L. C. *Fla. Dept. Agric. Bull.* 8: 1-28, 1968.

Denmark, H. A. and Price, J. F. 'Gladiolus thrips'. *Thrips simplex* (DPI Entomology Circular No. 322) No. EENY-36, University of Florida, 2001.

Deol, B. S. Survey of insect pests of ornamental plants particularly flowering shrubs at Ludhiana. M.Sc. Thesis. P.A.U., Ludhiana, 1974.

Dhandapani, N., Gopalan, M. and Sundarababu, P. C. 'Evaluation of insecticides for the control of jasmine bud worm'. *Madras Agric. J.* 76: 50-52, 1989.

Dhingra, S. Biological studies on common rose aphids. M.Sc. Thesis. P.A.U., Ludhiana, 1968.

Dhooria, M. S. 'Two spotted spider mite *Tetranychus urticae*—a serious pest of roses in polyhouses and its control'. *J. Acarol.* 14(1-2): 84-87, 1999.

Dhua, R. S. 'Rose'. In: *Floriculture and Landscaping.* (Eds.). Bose, T. K. *et al.* Kolkata, India: Naya Prokash, pp. 368-405, 1999.

Easwaramoorthy, S., Vadivelu, R. S. and Muthukrishnan, T. S. '*Sycophila* sp. (Eurytomidae: Hymenoptera) a new pest of *Jasminum grandiflorum* Linn'. *Entomon.* 5: 246, 1980.

Ellis, D., McAvoy, R., Ayyash, L. A., Flanagon, M. and Ciomperlik, M. 'Evaluation of *Serangium parcesetosum* (Coleoptera: Coccinellidae) for biological control of silver leaf whitefly *Bemisia argentifoli* on poinsettia'. *Florida Entomologist* 84(2): 215-221, 2001.

Favard, M. P. Une nouvelle maladie: l'acariose des chrysanthemums'. *Compt. Rend.* 41: 572-573, 1955.

Field, R. P. and Hoy, M. A. 'Biological control of spider mites on greenhouse roses'. *Calif. Agric.* 38(3-4): 29-32, 1984.

Fletcher, T. B. 'Pests of garden plants. Roses'. Proc. *Second. Ent. Mtg.* Pusa, pp. 264-266, 1917.

Fransen, J. J., Boogaard, M. and Tolsma, J. 'The minute pirate bug *Orius insidiosus* (Say) as a predator of western flower thrips *Frankliniella occidentalis* in chrysanthemum, rose and Saintpaulia'. *Bulletin SROP* 16(8): 73-97, 1993.

Gardner, W. A., Oetting, B. D. and Storey, G. K. 'Scheduling of *Verticillium lecanii* and Benomyl applications to maintain aphid (Homoptera: Aphididae) control on chrysanthemums in greenhouses'. *J. Econ. Ent.* 77(2): 514-518, 1984.

Gaud, S. M., Bennett, F. D. and Franqui, R. A. 'New records and notes on whiteflies (Homoptera : Aleyrodidae) from Puerto Rico'. *J. Agric. University of Puerto Rico* 75: 297-299, 1991.

Geissler, K. and Schliephake, E. 'Possibilities of application of insect pathogenic viruses as biological pesticides'. *Gesunde Pflanzen* 43(3): 92-95, 1991.

Ghosh, S. K. 'Prospects of employing predatory mites in the management of phytophagous mites'. In: *Biodiversity and Management of Phytophagous Mites.* (Eds.). Rabindra, R. J., Karuppuchamy, P., Umapathy, G., Chinnaiah, C. and Balasubramani, V.). Coimbatore, India: Tamil Nadu Agrl. University, pp. 158-166, 2001.

Ghosh, S. K., Jayanth, K. P. and Nethi, K. P. 'Predatory potential of an indigenous mite *A. tetranychivorus* on *T. urticae* under lab and polyhouse conditions'. Abstract No. 181. Proceedings of the *Entomocongress* 2000 held on 5-8 November, 2000 at Thiruvananthapuram, Kerala, India, 2000.

Glover, P. M. Rep. Department of Entomology. *Indian Lac Res. Inst.* 13-25, 1934-35.

Gokkes, M. 'Physical barriers in greenhouses'. *Phytoparasitica* 27(1): 76, 1999.

Govindarajan, R., David, B. V., Srinivasan, P. M., and Subramaniam, T. R. '*Aleurotrachelus* sp. A new whitefly pest of *Jasminum auriculatum* and its control'. *South Indian Hort.* 22: 24-25, 1974.

Gunasekaran, V. Studies on bio-ecology of jasmine pest complex. M.Sc. (Ag.) Thesis. Tamil Nadu Agricultural University, Coimbatore, Tamil Nadu, India, 1989.

Hameed, S. F. 'New records of insect pests on wild roses in Himachal Pradesh'. *Indian J. Ent.* 40(4): 455-456, 1978.

Hassan, S. A., Albert, R., Bigler, S., Bilaisinger, P., Bogenschutz, H., Boller, E., Burn, J., Chiverton, P., Edwards, P., Englert, W. D., Huang, P., Inglesfield, C., Naton, E., Oomen, P. A., Overmeer, W. P. J., Rieckmann, W., Samace-Peterson, L., Staubli, J. A. L., Tuset, J., Viggiani, G. and Vanwetswinkel, G. 'Results of the third joint pesticide testing programme by the IOBC/WPRS Working Group "Pesticides and beneficial organisms." *J. Appl. Entomol.* 103: 92-107, 1987.

Hata, T. Y. and Hara, A. H. 'Anthurium thrips, *Chaetanaphothrips orchidii* (Moulton): Biology and insecticidal control on Hawaiian anthuriums'. *Tropical Pest Management* 38(3): 230-233, 1992a.

Hata, T. Y. and Hara, A. H. 'Anthurium whitefly, *Aleurotulus anthuricola* Nakahara: Biology and control in Hawaii'. *Tropical Pest Management* 38(2): 152-154, 1992b.

Heinz, K. M. and Parella, M. P. 'Biological control of insect pests on greenhouse marigolds'. *Environ. Entomol.* 19(4): 825-835, 1990.

Helyer, N. L. and Wardlow, L. R. 'Aphid control on chrysanthemum using frequent, low dose application of *Verticillium lecanii*". *Bulletin SROP* 10(2): 62-65, 1987.

Herbert, H. J., Smith, R. F. and McRae, K. B. 'Evaluation of non insecticidal methods to reduce damage to chrysanthemums by the leaf miner *Liriomyza trifolii*. *Can. Entomol.* 116: 1259-1266, 1984.

Holttum, R. E. In: *Revised Flora of Malaya*. Third Edition. Singapore: Government Printing Office, 1964.

Isabel, J. N. Studies on the management of major pests of Jasmine (*Jasminum sambac* Ait.) cv Gundumalli with special reference to *Nausinoe geometralis* Guenee. (Pyralidae: Lepidoptera). M.Sc. (Ag.) Thesis. Tamil Nadu Agricultural University, Coimbatore, Tamil Nadu, India, 1996.

Isabel, J. N., Renuka, S. and Rajabaskar, D. 'Evaluation of certain plant products for the control of jasmine leaf webworm, *Nausinoe geometralis* Guenee'. In: *National Symposium on 'Pest Management Strategies: Current Trends and Future Prospects'*. p. 29, 2001a.

Isabel, J. N., Renuka, S. and Rajabaskar, D. 'Stage preference and predatory potential of *Chrysoperla carnea* against jasmine leaf webworm *Nausinoe geometralis* Guenee'. In: *National Symposium on Pest Management Strategies: Current Trends and Future Prospects'*. p. 25, 2001b.

Jefferson, R. N., Bald, J. G., Morishita, F. S. and Close, D. H. 'Effect of vapum on *Rhizoglyphus* mites gladiolus soil diseases'. *J. Econ. Ent.* 49: 584-589, 1956.

Jesudasan, R. W. A. and David, B. V. 'Taxonomic studies on Indian Aleyrodidae (Insects: Homoptera)'. *Oriental Insects* 25: 231-234, 1991.

Jones, V. P. and Parella, M. P. 'The movement and dispersal of *Liriomyza trifolii* in a chrysanthemum greenhouse'. *Ann. Appl. Biol.* 109: 33-39, 1986.

Jose, F., Nachiappan, R. M. and Selvarayan, V. 'Management of jasmine eriophyid mite *Aceria jasmini* (Acari: Eriophyidae)'. *J. Acarol.* 14: 41-44, 1999a.

Jose, F., Nachiappan, R. M. and Selvarayan, V. 'Seasonal incidence of jasmine eriophyid mite, *Aceria jasmini* (Acari: Eriophyidae) and its influence on the foliar biochemical factors'. *J. Acarol.* 14: 45-46, 1999b.

Kalshoven, L. G. E. 'Pests of Crops in Indonesia'. Vanhoeve, Jakarta, Indonesia: P.I. Ichtiar Baru, 1950.

Karuppuchamy, P. and Mohanasundaram, M. 'Bioecology and control of chilli muranai mite *Polyphagotarsonemus latus* (Banks) (Tarsonemidae: Acari)'. *Indian J. Pl. Prot.* 15: 1-4, 1987.

Karuppuchamy, P., Balasubramanian, G., Sundarababu, P. C. and Gopalan, M. 'A potential predator of chilli mite, *Polyphagotarsonemus latus* (Banks) (Tarsonemidae: Acari)'. *Madras Agric. J.* 81: 552-553, 1994.

Karuppuchamy, P., Dhandapani, N., Sundarababu, P. C. and Mohanasundaram, M. 'Efficacy of certain acaricides for the control of carmine spider mite, *Tetranychus cinnabarinus* (Boisduval) on jasmine'. *Madras Agric. J.* 73: 236-239, 1986.

Kauppuchamy, P. 'Life stages and biology of different groups of mites'. In: *Biodiversity and Management of Phytophagous Mites*. (Eds.). Rabindra, R. J., Karuppuchamy, P., Umapathy, G., Chinnaiah, C. and Balasubramani, V. Coimbatore, India: Tamil Nadu Agrl. University, Coimbatore, pp. 118-123, 2001.

Khan, N. H. and Abedi, Z. H. 'Notes on the biology and control of the lily moth, *Brithys crini* Fab'. *J. Bombay Nat. Hist. Soc.* 54: 785-788, 1957.

Knodel, M., Lyons, R. E. and Poe, S. E. 'Photoperiod affects chrysanthemum host plant selection by leaf miners (Diptera: Agromyzidae)'. *HortScience* 20: 708-710, 1985.

Krishnamoorthy, A. 'Release and recoveries of an exotic predatory mite *Phytoseiulus persimilis*'. *Entomon.* 13: 95-97, 1988.

Ladieu, M. S. and Helyer, N. L. 'Pyrazophos: A fungicide with insecticidal properties including activity against chrysanthemum leaf miner *Phytomyza syngenesiae*'. *Ann. Appl. Biol.* 102: 275-279, 1983.

Lakshminarayana, K. V. 'Notes on the Thysanoptera collected during Western and Southern Indian survey, 1962, with a review of Thysanoptera complex of the hosts'. *J. Bombay Nat. Hist. Soc.* 65: 348-368, 1968.

Larew, H. G. and Locke, J. C. 'Repellency and toxicity of a horticultural oil against whiteflies on chrysanthemum'. *HortScience* 25(11): 1406-1407, 1990.

Larew, H. G., Montz, J. K., Webb, R. E. and Warthen, J. D. '*Liriomyza trifolii* (Burgess) (Diptera: Agromyzidae) control on chrysanthemum by neem seed kernel extract applied to soil'. *J. Econ. Ent.* 78(1): 80-84, 1985.

Lesna, I., Conin, C. G. M., Sabelis, M. W., Straalen, N. and Van, M. 'Biological control of the bulb mite, *Rhizoglyphus robini* by the predatory mite, *Hypoaspis aculeifer* on lilies: predator-prey dynamics in the soil, under greenhouse and field conditions'. *Biocontrol Science and Technology* 10(2): 179-193, 2000.

Lindquist, R. K. 'New greenhouse pests with particular reference to the leaf miner *Liriomyza trifolii*'. In: Proc. 10th International Congress Plant Protection, Brighton, England, pp. 1087-1094, 1983.

Lindquist, R. K., Casey, M. L., Helyer, N. and Scopes ,N. E. A. 'Leaf miners on glasshouse chrysanthemum: Control of *Chromatomyia syngenesiae* and *Liriomyza trifolii*'. *Journal of Agricultural Entomology* 1(3): 256-263, 1984.

Magie, R. O. and Poe, S. L. 'The world of the Gladiolus, NAGC, U.S.A., pp. 155-181, 1972.

Mallik, B., Onkaruppa, S. and Harish Kumar, H. Management of the spider mite *Tetranychus urticae* Koch. on rose using phytoseiid predator *Amblyseius longispinosus* in polyhouse'. *Pest Management in Horticultural Ecosystem* 4: 46-48, 1998.

Mallikarjunappa, S., Mishra, R. K., Mithyantha, M. S. and Sivasankaran, K. 'Bio-efficiency of pesticides against the eriophyid mite infesting jasmine'. *Curr. Res.* 20: 242-243, 1991.

Mansour, F. A. and Ascher, K. R. S. 'Effect of neem (*Azadirachta indica*) seed kernel extract from different solvents on the carmine spider mite, *Tetranychus cinnabarinus*'. *Phytoparasitica* 11: 177-185, 1983.

McComie, L. D., Gosine, S. and Siew, P. 'The effect of *Cryptolaemus montrouzieri* (Mulsant) on the hibiscus mealybug *Maconellicoccus hirsutus* (Green) on hibiscus plants in Trinidad'. *Tropical Fruits Newsletter* No. 23: 7-10, 1997.

McMahon, R. W., Lindquist, R. K., Baith, B. D., Makin, T. L. and Casey, M. L. 'Comparison of commercially and locally produced *Encarsia formosa* Gohan on the control of sweet potato whitefly on poinsettias'. *Hort. Technology* 4(3): 295-298, 1994.

Medina, H. R. M. and Paris, F. M. 'Evaluation of physical and chemical methods of control for the chrysanthemum leaf miner *Liriomyza trifolii* (Burgess) *Revista Colombiana de Entomologia* 11(1): 36-38, 1985.

Mohan, N. J. and Udayagiri, S. 'Pest management in medicinal and aromatic crops' . *Annual Report*, IIHR, Bangalore, 1985.

Moorhouse, E. R., Gillespie, A. T. and Charnley, A. K. 'The progress and prospects for the control of the black vine weevil, *Otiorhynchus sulcatus* by entomogenous fungi'. Proc. and Abst. *V International Colloquium on Invertebrate Pathology and Microbial Control,* Adelaide, Australia, 20-24 August, 1990, pp. 381-385, 1990.

Morison, G. D. 'On a collection of Thysanoptera from South Australia'. *Bull. Entomol. Res.* 21: 9-14, 1930.

Moulton, D. and Steinweden, J. B. 'A new *Taeniothrips* on gladiolus'. *Can. Entomol.* 43: 20-21, 1931.

Nair, C. P. R. and Nair, M. R. G. K. Studies on the biology of the lace wing *Corythauma ayyari* Drake, a pest of jasmine'. *Agric. Res. J. Kerala* 12: 172-173, 1974.

Nair, M. R. G. K. In: *Insects and Mites of Crops in India.* New Delhi, India: I.C.A.R., 1975.

Nandakumar, C., Reghunath, P. and Sheela, K. R. 'On the control of red scale *Aonidiella aurantii* Muskll. on rose'. *Entomon* 13(3-4): 275-277, 1988.

Nayar, K. K., Ananthakrishnan, T. N. and David, B. V. In: *General and Applied Entomology.* New Delhi, India: Tata McGraw Hill Publishing Company Ltd., 1976.

Nelson, S. J., Venugopal, M. S., Janarthanan, R. and Natarajan, S. 'Efficacy of certain plant products against jasmine bud worm, *Hendecasis duplifascialis* Hmp. (Pyraustidae : Lepidoptera)'. *Indian Perfumer* 37: 236-239, 1993.

Nucifora, A. and Vecante, V. 'The integrated control of *Trialeurodes vaporariorum* on gerbera'. *Culture Protette* 10(3): 33-36, 1981.

Olson, D. L. and Oetting, R. D. 'The efficacy of myco-insecticides of *Beauveria bassiana* against silver leaf whitefly on poinsettia'. *Journal of Agricultural and Urban Entomology* 16(3): 179-185, 1999.

Ordogh, G. and Takacs, A. 'The efficacy of Pyrotox against larvae of scale insects (*Pseudococcus maritimus, Saissetia coffeae, Aspidiotus nerii*) damaging potted ornamentals'. *Novenyvedelem* 19: 417-419, 1983.

Osborne, L. S. and Landa, Z. 'Utilisation of entomogenous fungus, *Paecilomyces fumosoroseus* against sweet potato whitefly, *Bemisia tabaci'*.. *Bulletin OILB/SROP* 17(3): 201-206, 1994.

Pandolfo, F. M. 'Chemical control test against the red spider mite (*Tetranychus urticae* Koch) in carnation in the greenhouse'. *Informatore Fitopatologica* 34: 46-48, 1984.

Parella, M. P. and Jones, V. P. 'Coping with leaf miner crisis'. *Calif. Agric.* 38(9): 17-19, 1984.

Parella, M. P., Jones, V. P., Youngman, R. R. and Lebeck, L. M. 'Effect of leaf mining and leaf stippling of *Liriomyza* sp. on photosynthetic rates of chrysanthemum'. *Ann. Entomol. Soc. Am.* 78: 90-93, 1985.

Pasini, C., D'Aquila, F. and Gandolfo, M. 'Gerbera: salts of fatty acids in controlling aleurodids'. *Colture Protette* 26(4): 73-76, 1997.

Patel, R. B. 'Biology of rose thrips (*Retithrips syriacus* Mayet) and chemical control of pest complex of rose'. *Thesis Abst.* 15(3): 288-289, 1988.

Patil, S. P. and Veeresh, G. K. 'The larva of *Apogonial rauca* Fab. (Coleoptera: Scarabidae)'. *J. Maharashtra Agric. Univ.* 9(1): 111-112, 1984.

Perumal, R. S., Lakshmanan, P., Subramaniam, T. R. and Santhanaraman, T. 'Occurrence of the white grub *Holotrichia* sp. as a new pest of jasmine and nerium'. *Madras Agric. J.* 58: 519, 1971.

Pignon, C. *Pyreth. Post D.* 7: 33-36, 1963.

Pirone, P. P., Dodge, B. O. and Rickett, H. W. In: *Diseases and Pests of Ornamental Plants.* London, U.K.: Constable and Co. Ltd., 1960.

Prabhakar Rao, V. 'Icerine scale insects records from the Orient'. *Indian J. Ent.* 12: 39-66, 1950.

Prasad, D., Singh, K. M., Singh, R. N. and Methu, D. N. 'A new predator of new pest of jasmine in Delhi'. *Bull. Ent.* 24: 140-141, 1983.

Raghavarao, J., Ragarao, P. V. and Rao, B. H. K. 'Pests of ornamental trees, shrubs and lawns'. In: *Entomology in India*. Entomological Society of India, pp. 127-147, 1964.

Ramachandra Rao, Y. and Cherian, M. C. 'The fluted scale *Icerya purchasi* (Mask) as a pest of wattle in South India and its control by biological method'. *Madras Agric. J.* 32(3-7): 40-43, 1944.

Ramakers, P. M. J. 'Possibilities of biological control of *Thrips tabaci* Lind. in glasshouses'. *Medet. Fac. Landbouww. Rijksuniv. Gent.* 43(2): 463-469, 1978.

Ramakrishnan, T. A. *The Indian Rose Annual* 3: 64-70, 1984.

Ramesh, P. In: *Pests of Floriculture Crops and Their Control*. New Delhi, India: Kalyani Publishers, 1994.

Rami Reddy, P. V., Janakiram, T. and Rao, T. M. 'Evaluation of chrysanthemum germplasm for the incidence of thrips, *Microcephalothrips abdominalis* (Crawford)'. In: *IPM in Horticultural Crops: Emerging Trends in the New Millennium*. (Eds.). Abraham Verghese and Parvatha Reddy, P. Bangalore, India: IIHR, pp. 119-120, 2001.

Randhawa, G. S. and Mukhopadhyay, A. In: *Floriculture in India*. Mumbai, India: Allied Publishers Ltd., 1998.

Rani, B. J. and Mohan, N. J. 'Pest management in ornamental, medicinal and aromatic crops'. *Annual Report*. Bangalore, India: IIHR, 1994.

Rao, V. T., Perraju, A. and Rao, P. V. R. '*Contarinia maculipennis* Felt. occurrence as jasmine bud midge in South India'. *Andhra Agric. J.* 2: 208-209, 1955.

Rao, V. T., Perraju, A. and Rao, P. V. R. 'The jasmine fly'. *Andhra Agric. J.* 1: 313-315, 1954.

Reddy, A. S., Rao, B. H. K. and Wilson, Y. 'Chemical control of jasmine pests'. *South Indian Hort.* 26: 25-27, 1978.

Regupathy, A., Palanisamy, S., Chandramohan, N. and Gunathilagaraj, K. In: *A Guide on Crop Pests*. Coimbatore, Tamil Nadu, India: Sooriya Desktop Publishers, 1994.

Rice Mahr, S. E., Cloyd, R. A., Mahr, D. L. and Sadof, C. S. 'In: *Biological Control of Insects and Other Pests of Greenhouse Crops*'. University of Wisconsin, U.S.A.: North Central Regional Publication 581, 2001.

Sachan, J. N. and Srivastava, B. P. 'Biology of lily moth, *Polytela gloriosa* Fabr. (Lepidoptera: Noctuidae)'. *Indian J. Ent.* 27(2): 127-129, 1965.

Sandhu, G. S. and Shukla, G. K. 'Chemical control of jasmine leaf webworm'. *Pestology* 8: 17-19, 1984.

Sell, P. and Kuo-Sell, H. L. 'Well tried and new beneficial organisms, practical knowledge obtained in biological pest control in ornamental cultivation under glass and possibilities for its development'. *Deutscher Gartenbau* 43(42): 2548-2553, 1989.

Sharma, P. and Bhattacharjee, S. K. 'Plant protection in ornamental crops'. *AICRP on Floriculture*. New Delhi, India: ICAR Technical Bull. No. 19, 2002.

Shukla, G. K. and Sandhu, G. S. 'New records of natural enemies of jasmine leaf webworm in India'. *Bull. Ent.* 24: 135-136, 1983.

Singh, J. and Sohi, A. S. '*Helicoverpa* (*Heliothis*) *armigera* (Hub.), a pest of ornamentals in Punjab, India'. 23rd Inst. Hort. Congress held at Firenze (Italy), August 27–September 1, 1990.

Singh, R. and Singh, A. '*Aphis gossypii* Glover, an insect vector of carnation (*Dianthus caryophyllus* L.) mottle virus—a new record'. *Curr. Sci.* 58(16): 9, 1989.

Singh, S. and Ipe, M. In: *The Agromyzidae from India*. Agra, India: St. John's College, 1973.

Singh, S. P. '*Heliothis armigera* (Hub.) as a pest of ornamental plants in Simla Hills'. *Indian J. Pl. Prot.* 11 (1-2): 95, 1985.

Smith, E. S. C. and Neal, M. J. 'Eradication of the orchid weevil *Orchidophilus aterrimus* (Waterhouse) (Coleoptera: Curculionidae) from the NT'. *Technical Bulletin.* Dept. of Primary Industries and Fisheries, Northern Territory of Australia, No. **288**: 27-29, 2001.

Smith, F. F. 'Controlling insects of flowers'. *Agricultural Information Bulletin,* No. 237, U.S. Dept. of Agriculture, 1967.

Sohi, A. S. and Mann, J. S. 'Biology and host plants of Typhlocybinae (Homoptera: Cicadellidae) in India'. Proc. 2nd *Int. Workshop on Leaf Hoppers and Plant Hoppers of Economic Importance,* held at Provo, Utah, 28 July-1 August, 1986. (Eds.). Wilson, M. R. and Nault, L. R. London, U.K.: C.E. 1987: 275-281, 1986.

Sohi, A. S. and Singh, D. 'Insect pests of ornamental plants'. In: *Advances in Horticulture,* vol. 12. *Ornamental Plants.* (Eds.). Chada, K. L. and Bhattacharjee, S. K. New Delhi, India: Malhotra Publishing House, pp. 913-961, 1995.

Sohi, A. S., Shnemar, M. S. and Mann, J. S. 'Typhlocybine (Cicadellidae, Typhlocybinae) fauna of the perennial plants of Punjab, India'. *Zool. Orient.* 5(1-2): 10-20, 1988.

Sood, A. K. Insect pests of ornamental plants in Himachal Pradesh. M.Sc. Thesis. Dr. Y. S. Parmar University of Horticulture & Forestry, Solan. *Abstract* 16(1): 29-30, 1990.

Sood, A. K. and Kakar, K. L. 'Record of insect and non-insect pests of ornamental plants from Himachal Pradesh'. *J. Insect Sci.* 3(2): 141-145, 1990.

Srinivasan, P. M., Govindarajan, R. and Paul, A. V. N. 'Control of the rose scale *Lindingaspis rossi* (Mask.) by granular insecticides'. *South Indian Hort.* 22(3-4): 124, 1974.

Steel, H. V. Thrips investigation. Some common Thysanoptera in Australia'. Commonwealth of Australia: CSIR, Pamphlet, pp. 54-59, 1935.

Sundararaj, T. S., Seemanthani, B. and Varadhan, K. M. 'A 'Mullai', *Jasminum auriculatum* selection highly resistant to gall mite'. *Madras Agric. J.* 54: 599-601, 1967.

Sunderland, K. D., Chambers, R. J., Helyer, N. L. and Sopp, P. I. 'Integrated pest management of greenhouse crops in Northern Europe'. *Horticultural Reviews* 13: 1-66, 1992.

Tewari, G. C. 'Seasonal activity of *Myllocerus* weevils damaging roses'. *Bull. Ent.* 24(2): 116-120, 1983.

Tewari, G. C. and Mohan, N. J. 'Studies on the pests of important ornamental and medicinal plants'. *Annual Report.* Bangalore, India: IIHR, pp. 74-75, 1981.

Trehan, K. N. 'Biological observations on *Trypanea amoena* Frfld. on marigold'. *Indian J. Ent.* 8: 107, 1947.

Uda, A., Yase, J., Yamnaka, M. and Fujii, H. 'Control of leaf beetle *Aulacophora nigripennis* attacking carnation (*Dianthus caryophyllus*) using some *Dianthus* as trap plants'. *Journal of the Japanese Society for Horticultural Science* 70(6): 753-759, 2001.

Udayagiri, S. 'First record of *Liriomyza compositella* Spenser and *L. brassicae* (Riley) (Diptera: Agromyzidae) as a pest of ornamental crops in South India'. *Entomon.* 12(2): 85-88, 1987.

Ulson, D. L. and Oetting, R. D. 'The efficacy of micro-insecticides of *Beauveria bassiana* against silver leaf white fly on poinsettia'. *Journal of Agricultural and Urban Entomology* 16(3): 179-185, 1999.

Umapathy, G. and Rajendran, B. 'Management of the Eriophyid mite, *Aceria jasmini* Chan. (Eriophyidae: Acari) on jasmine'. *Madras Agric. J.* 86: 83-86, 1999.

Valunj, A. R., Pawar, S. A., Kadam, U. K. and Khaire, V. M. A new acaricide AC-303-630 (Chlorfenapyr 10 EC) against carnation mite, *Tetranychus cinnabarinus* Biosd'. *Pest Management in Horticultural Ecosystems* 5(1): 21-23, 1999.

Verghese, A. and Prasad, V. G. 'Note on the record of insect pests of rose and *Hibiscus* from Bangalore'. *Indian J. Hort.* 40(1-2): 131-133, 1983.

Verma, G. S. and Singh, P. D. *Nat. Sem. Prodn. Tech. Comm. Flower Crops*, Coimbatore, pp. 101-103, 1980.

Vyas, H. N. 'Pest complex of ornamental plants—rose, *Rosa indica* Linn. and jasmine, *Jasminum sambac* Linn'. *Crop Research* (Hissar) 16: 251-252, 1998.

Wang, C. L. 'Chemical control of insects of chrysanthemum cut flowers'. *J. Agric. Res. China* 31(4): 339-346, 1982.

Wardlow, L. R. 'Biological pest control on ornamentals'. United Kingdom Summary *Bull. SROP/WPRS* 12(3): 23-25, 1989.

Wilhelm, C. 'Beneficials in the cyclamen crop take over the securing of quality'. *TASPO Gartenbaumagazin* 6(2): 16, 1997.

Winstone, A. A. and David, B. V. 'Effect of chitin inhibitor diflubenzuron on jasmine bud worm *Hendecasis duplifascialis* Hampn. (Pyraustidae: Lepidoptera) pest management'. *FIPPAT Entl. Ser.* 1: 9-17, 1986a.

Winstone, A. A. and David, B. V. 'Effect of chitin inhibitor diflubenzuron on jasmine bud worm *Hendecasis duplifascialis* Hmpsn. (Pyraustidae: Lepidptera) pest management'. *FIPPAT Entl. Ser.* 1: 18-20, 1986b.

Woets, J. 'Integrated control in vegetables under glass in the Netherlands'. *Bulletin SROP/WPRS* 4: 26-31, 1973.

Yadav, L. P. and Bose, T. K. 'Orchids'. In: *Commercial Flowers.* (Eds.). Bose, T. K. and Yadav, L. P. Kolkata, India: Naya Prokash, pp. 151-265, 1993.

Yase, J. 'Control of cotton bud worm, *Helicoverpa armigera* by YFLs in carnation greenhouse'. *Agrochemicals Japan* 78: 10-12, 2001.

—oo(O)oo—

10

DISEASES OF ORNAMENTAL PLANTS AND THEIR MANAGEMENT

PRATIBHA SHARMA

INTRODUCTION

INDIA grows a large number of flowering plants under varied climatic and soil conditions during different seasons. A large number of ornamentals, including annual and biennial herbaceous plants, bulbous plants, foliage plants, flowering shrubs and trees are grown in India for decorative purposes. The different crops are cultivated for their attractive flowers, foliage, fragrance, hedging, rockeries and window boxes, besides use as cut flowers. Information on diseases of various ornamental crops is scattered in many research papers, reviews, bulletins and books. The present article comprises an effort of collection of important information in one article, the information on aetiology, symptomology, epidemiology and management of diseases affecting these crops. This chapter compiles the various important diseases of annuals, biennials and perennials used commercially. Many ornamentals are multiplied vegetatively, and hence, disseminated through planting material. So far seed certification has not been introduced for ornamental plants in our country. The disease-causing organisms are carried by sees from one locality to another. Growers hardly adopt any control measures because of their ignorance about the disease and lack of information on control measures. This review article is an attempt to provide up to date information on diseases affecting important ornamental plants. Many diseases produce characteristic symptoms that make it possible to find out the cause with a fair degree of certainty. Accordingly, the various diseases have been categorised based on the symptoms produced by each of them. Information on time of appearance, intensity of infection, favourable condition of disease development, host range and varietal reaction are also incorporated wherever possible. In order to make it useful for scientific workers, the morphology of the causal organism is also given. Currently accepted scientific names of different pathogens and their host plants have been given. Control measures sugge•ted against different diseases are based on existing knowledge and research of bioagents. Hot water treatments, solarisation, resistant varieties are also being included in management of most of the diseases, based on the availability of the information.

ANNUAL AND BIENNIAL FLOWERS

ALTHEA (Hollyhock)

Althaea rosea; Family: Malvaceae

Fungal Diseases

Anthracnose and Seedling Blight (*Colletotrichum malvacearum* A. Barun. & Casp.)

Symptoms: The disease is very destructive in warm humid weather. Seedlings are comparatively susceptible. It also infects the lower part of the stem and roots. Black blotches are formed on veins, leaf blades, petioles and stems. Severely infected leaves curl and dry up. The disease is commonly observed in gardens.

Management: Collection and burning the diseased leaves and plants is recommended. Humid conditions should be avoided. Bavistin (0.15 per cent) or Zineb (0.25 per cent) may be sprayed as a remedial measure.

Rust (*Puccinia malvacearum* Mont.)

Symptoms: Leaves, stems, bracts and other parts are infected resulting in small raised, dark brown pustules. A dense growth of pustules may destroy large portions of the leaf which later dries out. In humid weather the pustules are ash-grey. The disease is commonly being observed in Himachal Pradesh and Bangalore, and becomes severe in humid weather.

Management: Sulfex (0.2 per cent) or Karathane (0.025 per cent) spray is recommended. The infected seedlings and plant debris should be destroyed.

Cercospora Leaf-Spot or Stem-Blemish (*Cercospora nebulosa* Sacc.)

Symptoms: The infection initiates as minute spots on the leaves, which later grow bigger and appear circular to sub-circular, light to dark and 1-3.5 mm in diameter. The central portions dry up and fall off. The spots present on the stem are oblong, blackish-grey, initially separate, later coalescing and forming irregular dark brown streaks (Crous and Brown, 1996). Infection is serious during humid weather. Another species, namely *Cercospora altheina*, is also reported to attack this host.

Management: Bavistin (0.1 per cent) or copper oxychloride (0.3 5) spray is recommended. Destruction of debris and avoiding excessive irrigation of beds is advisable.

Powdery Mildew (*Erysiphe cichoracearum* DC.)

Symptoms: Whitish floury patches appear on the leaves, stem bracts and sepals. Severely infected leaves turn yellow and dry up. The plants are also attacked by *Leveillula taurica*. Powdery mildew is a common and widely distributed disease.

Management: Wettable sulphur (0.2 per cent) spray is recommended.

Myrothecium Leaf Spot (*Myrothecium roridum* Tode ex. Fr.)

Symptoms: Circular to irregular, brownish spots appear. Diseased tissues are dry and brittle with sporodochia. The disease becomes serious during humid and rainy weather.

Grey Mould (*Botrytis cinerea* Pers.)

Symptoms: The pathogen attacks flowers under humid weather and spoils them.

Management: Spraying with Bavistin (0.1 per cent), following good sanitation, avoiding humid conditions, and removal and destruction of diseased flowers are helpful in checking disease spread.

ALYSSUM (Madwort)

Alyssum maritimum (*Lobularia maritima*); Family: Cruciferae.

Fungal Diseases

Powdery Mildew (*Erysiphe polygoni* DC.)

Symptoms: Whitish floury patches are noticed on the leaves, turning them yellow. The disease is not very common.

Management: Spraying with wettable sulphur (0.2 per cent) or Karathane (0.025 per cent) checks the disease.

White Rust [*Albugo candida* (Pers.) Kuntze]

Symptoms: The infection is noticed as raised on the underside of leaves, which release powdery whitish spores. The infected leaves turn yellow and dry. The disease is uncommon, and is favoured by humid weather. The fungus perpetuates through zoospores, hence disposal of plant debris is very essential.

Management: Remove and destroy diseased plants and all debris. Remove surrounding weeds. Sprays of Captafol, Carbendazim (0.2 per cent) are effective to control this disease.

Foot Rot and Stem Rot [*Sclerotinia sclerotiorum* (Lib.) de Bary and other soil-borne pathogens]

Symptoms: The infection appears as brown, depressed spots at the collar region which may girdle the stem. The infected plant dies after wilting. Numerous sclerotia are formed in diseased tissue which help the fungus to perpetuate. The disease is common and widespread.

Management: Use of Bavistin and Captan gives some control. Diseased plants may be rouged out and destroyed. Use of disease-free seed in disease-free soil is also advocated.

AMARANTH (*Amaranthus*)

Amaranthus sp; Family: Amaranthaceae

Fungal Diseases

Damping Off (*Pythium myriotylum*)

Symptoms: Common symptoms of seedlings are observed at pre-emergence and post emergence stages.

Management: Avoid stagnation of water. Use treated seed with Captan or Carbendazim so as to prevent decaying of seed and seedlings.

White Blister *Albugo bliti* (Biv.-Bern.) Kuntze

Symptoms: Whitish pustules develop on the leaves which, on maturity, turn reddish brown. Severely infected plants remain stunted and distorted.

Management: Spray of copper oxychloride (0.3 per cent) reduces the disease.

Downy Mildew [*Peronospora tarinosa* (Fr. ex. Fr.)]

Symptoms: The disease is serious in humid and cool areas. Growth of pathogen is observed on the lower surface of the leaves alone with yellowish patches on the upper surface. Under favourable conditions, leaves drop off and cause severe losses.

Management: Spraying of crop with copper oxychloride (0.3 per cent) helps in controlling the disease.

ANTIRRHINUM (Snapdragon)

Antirrhinum majus; Family: Scrophulariaceae

Fungal Diseases

Anthracnose [*Colletotrichum fuscum* Laubert and *Colletotrichum gloeosporiodes* (Penz.) & Sacc.]

Symptoms: The disease is most serious under glasshouse conditions and in humid weather. Plant is infected at all stages of growth. Infection spots on older stems are sunken, oblong, pale yellowish-green to grey with narrow brown borders. On leaves, it produces brownish, irregular spots. Several spots may girdle the plant at the base. The disease is serious during the rainy season. The fungal acervuli later appear on the spots as minute black dots, which contain a number of long setae (Weidemann *et al,* 1988).

Management: Spraying with Bavistin (0.1 per cent), Bordeaux mixture (1 per cent) or Zineb (0.2 per cent), collecting and destroying the diseased debris, and avoiding overhead watering are recommended.

Blight (*Phyllosticta antirrhini* Syd.)

Symptoms: Cream-coloured or light brown, large, circular, zonate spots sometimes surrounded by haloes appear on the leaves. On the stem the infection spots are ashy-grey which later turn brown. If the spots encircle the stem, the plants show wilting. Pycnidia are epiphllous, central and dark. Spores are hyaline and 1-celled. Plants of all ages are susceptible. The disease is more serious in summer at temperatures above 16° C (Sohi, 1984).

Management: Same as for anthracnose.

Rust (*Puccinia antirrhini* Dietl & Holw.)

Symptoms: It is one of the most serious diseases. It attacks leaves, young stems and calyxes. Brown-coloured pustules develop on the underside of leaves. Infected plants bloom prematurely, bear only small flowers or they may be killed when the infection takes place very early. The disease is favoured by temperature below 21° C. It attacks the plants at all stages of growth and is confined to hosts of this family. Alpha and Beta races of the pathogen have been reported (Aitken *et al,* 1989).

Management: Spraying with wettable sulphur (0.25) or Karathane (0.025 per cent) before disease appearance and repeating the spray at periodic intervals, and avoiding crowding of plants are remedial in disease control.

Powdery Mildew (*Oidium* Link ex Fr.)

Symptoms: The disease is serious only during the late stages of plant growth. It adversely affects seed production.

Management: The disease can be checked by spraying wettable sulphur and Karathane at 0.2 per cent.

Downy Mildew (*Peronospora antirrhini* Schroet.)

Symptoms: It is very common on seedlings and produces light brown, downy growth on the lower surface of leaves, with the corresponding upper surface turning yellow (Paul *et al,* 1991).

Management: It can be controlled by spraying Bordeux mixture (0.1 per cent) or Dithane M-45 (0.2 per cent) at fortnightly intervals. Paulus *et al* (1983) controlled downy mildew successfully by spraying Metalaxyl at the rate of 0.75 pints/100 gl water and Manzate 80 W at 1.5 lb/100 gl water.

Crown Rot (*Sclerotium rolfsii* Sacc.)

Symptoms: Typical symptoms appear as in other cases and the disease is serious in sandy-loam soil (Merlo *et al*, 1986).

Management: Raising the nursery in disease-free soil from healthy seed, dipping the seedlings in Bavistin (0.1 per cent) for a few hours before transplanting, drenching the soil with Captan (0.2 per cent) at the regions where seedlings are to be transplanted, and raking the soil frequently when the plants are growing, give good control.

Collar Rot and Leaf Spot (*Myrothecium roridum* Tode ex Fr.)

Symptoms: The pathogen produces sunken, dark brown, cracked cankers at the collar region. Eventually the infected plants wilt and die. Dark brown necrotic spots also appear on the leaves. Sporodochia are cushion-shaped, black with white margins. They develop on the rotten bark at ground level and on the lower surface of diseased leaf spots.

Management: Raising the nursery in disease-free soil from healthy seed, dipping the seedlings in Bavistin (0.1 per cent) for a few hours before transplanting, drenching the soil with Captan (0.2 per cent) at the regions where seedlings are to be transplanted, and raking the soil frequently when the plants are growing, give good control.

BELLIS (Daisy or English Daisy)

Bellis perennis; Family: Compositae

Fungal Diseases

Leaf Spot [*Alternaria tenuis* Nees = *Alternaria alternata* (Fr.) Keissler]

Symptoms: The fungus produces dark brown, circular or oval spots on both the leaf surfaces. The spots initiate mostly from the leaf margins and result in browning of the apical leaf portion

Management: Spray of Indofil M-45 (0.25 per cent) can cure the problem.

No insects reported in this flower of major significance.

CALENDULA (Pot Marigold)

Calendula officinalis; Family: Compositae

Fungal Diseases

Leaf Spot (*Cercospora calenduliae* Sacc.)

Symptoms: This is probably the most serious disease of this host. Dark brown, irregular spots, which run together, appear. Adjacent spots coalesce. Severe infection kills the seedlings when they are more than a month old. Younger plants are not attacked. The infection and disease development is favoured by humid weather .The disease has been reported to cause losses of up to 50 per cent.

Management: Spraying with Bavistin (0.1 per cent) or copper oxychloride (0.3 per cent) or Zineb (0.2 per cent) is recommended. Also, the plant debris should be disposed of.

Powdery Mildew (*Erysiphe cichoracearum* DC. and *Erysiphe polygoni* DC.)

Symptoms: It produces whitish patches on leaves. In severe cases the stem and flower buds are also attacked. Diseased leaves turn yellow, curl and dry up. Infection is more under excessively moist and shade conditions. The mycelium is ectophytic with erect conidiophores.

Management: Spraying with Karathane (0.025 per cent) or wettable sulphur (0.2 per cent) and avoiding overcrowding and shading of plants are recommended.

Smut (*Entyloma calendulae*)

Symptoms: Pale yellow spots, which turn brown to black, 1/4 inch are diameter is noticed on leaves.

Management: All diseased leaves should be collected and burnt. Also, it is advisable to avoid excessive watering, change beds frequently and spray with Zineb (0.2 per cent) for disease control. Copper oxychloride (0.2 per cent) spray checks the disease.

Leaf Spot (*Colletotrichum gloeosporiodes* Penzig)

Symptoms: The infection is noticed as brown irregular spots on leaves, which later turn greyish. Diseased tissues fall off in certain cases. Acervuli typical of the species are produced later in the season with pink spore masses and dark setae.

Management: Spraying with Bavistin (0.1 per cent) or Zineb (0.2 per cent), destroying diseased debris, and avoiding humid conditions are recommended. Copper oxychloride or Mancozeb (0.2 per cent) or Carbendazim (0.05 per cent) at weekly intervals reduces the incidence of disease.

Alternaria Leaf Spot (*Alternaria calendulae*)

Symptoms: Infection spots are similar to *Cercospora* leaf spots. These later turn greyish with concentric rings. Infection is also noticed on stem, inflorescence and seeds.

Management: Spraying Zineb (0.2 per cent) or Dithane M-45 (0.2 per cent), using disease-free seed, burning the plant debris and following long rotation are recommended.

Flower Blight (*Phomopsis* Sacc.)

Symptoms: The pathogen attacks ray florets and peduncles resulting in dark brown spots or lesions, which later turn greyish. The flower heads dry. Infection is carried through seed and diseased debris.

Management: Spraying with Indofil M-45 (0.25 per cent), Carbendazim (0.2 per cent) prevents disease.

Viral and Mycoplasmal Diseases

Cucumber-mosaic virus, tomato spotted-wilt virus and aster yellows attack *Calendula.*

Management: Infected plants should be rouged out and destroyed. Spraying Malathion or Metasystox or Sevin 1 ml/litre of water is recommended to control sucking insects responsible for spread of these diseases.

CALLISTEPHUS (China Aster)

Callistephus chinensis (Linn.) Nees syn. *Aster chinensis* Linn. (China Aster); Family: Compositae

Fungal Diseases

Anthracnose (*Colletotrichum gloeosporiodes* Penzig)

Symptoms: It produces dark brown to black irregular spots or blotches on leaves and stem. Acervuli are dark, sub-epidermal, punctiform setose, spores are 1-celled, hyaline and guttulate.

Management: A periodic spray of Indofil M-45 (0.2 per cent) helps in reducing the diseases.

Stemphylium Leaf Spot (*Stemphylium callistephae*)

Symptoms: Brown, nearly circular, concentrically zonate spots with dark margins appear on leaves, bracts, petals and stems. It adversely affects the quality of flowers.

Management: Spraying with Indofil M-45 (0.2 per cent), and using balanced fertilisers help in disease control.

Stem Canker and Flower Blight (*Phomopsis callistephi*)

Symptoms: Depressed, dark brown lesions appear on the stem, which may girdle it, resulting in the death of the plant. The infection is also noticed on ray florets as dark brown or irregular spots, which later turn greyish. The disease symptoms are also observed on peduncles as brownish to greyish spots resulting in drying.

Management: It is advisable to avoid injury to the plants during cultural operations, destroy diseased debris and use disease-free seed.

Grey Mould Blight (*Botrytis cinerea* Pers.)

Symptoms: It produces grey-mould symptoms on flowers in cloudy and humid weather. In addition, it also attacks the flower buds and blossoms producing blight symptoms.

Management: Good sanitation, removal and burning of diseased flowers and affected plants, avoidance of overcrowding of plants and overhead watering and spraying with Zineb and Maneb (0.2 per cent) are recommended.

Alternaria Leaf Spot [*Alternaria alternata* (Fr.) Keissler syn. *Alternaria tenuis* Nees]

Symptoms: Infection first appears as minute dots which later form large spots. The spots are amphigenous, circular to irregular, up to 10 mm in diameter with concentric rings, and greyish to dark brown, giving a blighted appearance to the foliage.

Alternaria tenuissima (Fr.) Wiltsh. has also been reported on *Callistephus chinensis*. It produces pale brown necrotic spots with a dirty yellow margin on the leaves. Flower petals show water-soaked, irregular, brown lesions. The disease is seed-borne. The fungus also infects *Chrysanthemum, Dahlia, Gerbera*, Sunflower, Nasturtium and *Zinnia*. The disease is widespread and common.

Management: *Alternaria* leaf spot diseases can be controlled by spraying Indofil M-45 (0.2 per cent).

Wilt [*Fusarium oxysporum* f. sp. *callistephi* (beach) Snyd. & Hans.]

Symptoms: It is one of the most serious diseases. Plants wilt, wither and die out at any stage from seedling to full bloom. Older plants are often stunted, with a one-sided development and a brown discoloration of the vascular bundles. Sometimes all lower leaves wilt and the base of the stem blackens, often with a mass of pink spores at ground level. Plants in full bloom may suddenly droop. The fungus produces macro- and micro-conidia and chlamydospores. It is seed-borne and persists in the soil for many years.

Management: Use of sterilised soil for nursery beds, treatment of seed before planting with mercuric chloride solution (1:1,000), Captan, Thiram or any biological agent (*Trichoderma*), use of healthy seeds and avoidance of planting in disease-sick soil, are the control measures.

Root Rot and Foot Rot [*Phytophthora erythroseptica, Phytophthora nicotianae* var. *parasitica* (Dastur) Waterh., *Rhizoctonia solani* Kuhn, *Rhizoctonia bataticola* (Taub.) Butler, *Sclerotium rolfsii* Sacc. and *Pythium* Pringsh.]

Management: Both these diseases are serious at times and need the following proper control measures as discussed earlier. Infected plants turn pale and later wilt.

Aster Yellows

Symptoms: The first sign of infection is the pale yellowish tinge of the leaves. Usually adventitious shoots develop more or less like those of witches' broom. The blooms that may develop to any extent are off-colour, the rays being usually yellowish green. The plants do not die, like those infected with wilt and root rot, but continue to be the source of the disease for neighbouring plants.

Management: The causal organism is transmitted from plant to plant by the aster leaf hopper *Macrosteles fascifrons*. Grow the plants in screened enclosures to keep away the insects. Spray of Diazinon, Malathion or Metasystox R at early stage of disease development helps in managing the crop.

***CENTAUREA* (Cornflower)**

Centaurea cyanus: Other common names: Bacherlor's button, Ragged robbin: Family: Compositae

Fungal Diseases

Leaf Spot (*Alternaria tenuis* Nees)

Symptoms: Dark brown spots of indefinite shape with distinct margins start appearing generally from the leaf tip. Disease attack is mostly detected around flowering time.

Management: Spraying with Dithane M-45 (0.2 per cent), burning of debris, and using disease-free seeds are recommended.

Downy Mildew (*Bremia lactucae* Regel)

Symptoms: The fungus usually produces pale greenish or reddish irregular spots on the upper side of the leaves, while the undersides may be covered with a mouldy growth. Affected leaves collapse and the attacked parts die. The infection is severe on young plants in humid weather. The disease is not very common.

Management: Spraying with copper fungicides helps check the disease. Crowding of plants and excessive watering should be avoided. Infected plants and diseased debris should be removed and destroyed.

Stem Rots (*Phytophthora cactorum* (Leb. & Cohn.) Schroet.; *Sclerotium sclerotiorum* (Lib.) de Bary; *Pellicularia filamentosa* (Pat.) Rogers]

Symptoms: *Centaurea cyanus* is highly prone to infection by these fungi. At times appreciable mortality is noticed. *Fusarium oxysporum* f. *callistephi* also attacks this host producing wilt symptoms.

Management: Use of disease-free soil and seed, better sanitation, and disposal of debris is helpful in minimising mortality. Also, following 2-3 years rotation is advisable.

Foot Rot [*Fusarium solani* (Mart.) Sacc.]

Symptoms: The disease becomes apparent when the plants are about to flower. Water-soaked, dark brown, depressed lesions appear at the collar region leading to drying of foliage and wilting of plants.

Management: This disease can be checked by avoiding waterlogged conditions, use of disease-free seed, and by following proper rotation.

CHRYSANTHEMUM (*Chrysanthemum segetum*); *Chrysanthemum coronarium* (Crown Daisy) and *Chrysanthemum carinatum* (Tricoloured chrysanthemum). Family: Compositae

Fungal Diseases

Basal Stem Rots (*Pythium* and *Rhizoctonia* sp.)

Symptoms: Rotting of cuttings in rooting beds is most common and serious and the fungus may also enter the established plants through the wounds caused by pinching, etc. Sudden wilting of infected parts is the main feature.

High humidity and temperature periods account for this disease.

Rhizoctonia: May be distinguished by the following

- The decay originates near the soil surface rather than in the root system.
- The diseased area has a brown coloration.
- Strands of brownish mycelium can be seen with a hand lens.
- Where plants are crowded, such as in a propagating bed or container, the fungus can form a brown coloured web over the plants.
- The expression of the disease is often rapid and rots mature plants upward from an infection at soil line.

Pythium: May be distinguished by the following symptoms:

- Rotting may start at the root tips.
- Does not produce a characteristic colour, roots will be brown and black depending on the stage of decomposition. Rapid blackening of stem on actively growing plants is not uncommon.
- May destroy the root system, causing a slow retardation in growth.
- In almost all cases, the root system has stood in water at some time. Field observation reveals both *Pythium* and *Rhizoctonia* attacking the plant at the same time. Spread of the organisms is through the transfer of plants, tools and soil. Next to poor soil aeration and inadequate drainage, high salt concentrations are known to predispose the plants to infection and increase the severity of the disease expression. The most severe symptoms of the disease are usually found on beds having an uneven soil surface permitting surface water pockets. Under cover, the greatest incidence will be observed in that part of the ground bed where surface run-off flooding was permitted. In both the cases, the excess of water seals the soil and provides conditions favourable for both the diseases.

Management: Pre-planting sterilisation of soil with Formalin (0.2 per cent) and regular drenching with Thiram controls the spread of infection to neighbouring cuttings.

Phoma Root Rot (*Phoma* sp.)

Symptoms: The disease is first noticed on lower leaves, which show chlorosis and later wilt, and then spread upwards.

Management: Pre-planting soil drench with Nabam (1.5 g/l) gives excellent control.

Stem Rot and Wilt (*Fusarium oxysporum* Schlechtend ex. Fr.)

Symptoms: Disease develops only after the appearance of flower buds although infection might have occurred even in rooting beds. Stem near soil becomes dark brown and dries. Lower leaves turn yellow and plant wilts permanently afterwards.

Management: Soil application of Dithane M-45 (0.2 per cent) controls the disease.

Grey Mould: (*Botrytis cinerea* Pers. ex. Fr.)

Symptoms: The fungus can attack the plant at any stage from rooting bed to flowers in transit. First girdling of stem takes place, which results in death of upper part. Leaf infection starts from margin and proceeds towards centre and base, showing semi-circular bands. Flower infection starts with brown water soaked spots on lower petals.

Management: Good aeration, more planting distance and spraying with Captaf (1 g/l) or Thiram (3 g/l) provides good control.

Septoria Leaf Spot (*Septoria obesa* Sacc. and *Septoria chrysanthemella* Sacc.)

Symptoms: Most common and serious disease, produces brown spots on leaves that grow in size and number in humid conditions. Leaves turn yellow and die afterwards. Of the 38 chrysanthemum cultivars assessed for their disease reaction under field condition, the cultivars 'Alandi Local', 'Varsha', 'Mohini', 'Harvest House', 'Sarad Shobha', 'Meera' and 'Kirti' were found resistant, while local cv 'Zipri' was must susceptible (M.P.K.V., Pune).

Management: Control measures include picking and destroying leaves, providing dry and airy environment and spraying Bavistin 0.1 per cent. M.P.K.V., Pune, and P.A.U., Ludhiana, had reported that Dithane, M-45 (0.2 per cent) six sprays to be given at 15-day intervals starting from first incidence of disease was effective. A new leaf spot of chrysanthemum caused by *Phoma chrysanthemicola* has been reported at Ludhiana in severe form under field conditions (P.A.U.). For the control of leaf spot disease of chrysanthemum, it was also recommended to spray the crop with 0.2 per cent copper-oxychloride or 0.2 per cent Mancozeb with sticker given time at intervals of 10-days during June to October (M.P.K.V., Pune) (Sharma and Bhattacharjee, 2002).

Chrysanthemum Ray Blight (*Didymella chrysanthemi*)

Symptoms: It affects all above ground parts but is most severe on the shoots and on the flowers which develop reddish petal spots on light coloured cultivars and brownish spots on darker coloured types. The lesions spread rapidly and the flowers collapse with the ensuing decay. On shoots, the disease often begins with the rotting of terminal buds while the leaves are affected by patches of brown black decay. When large lesions form at the base of the stem, the plants may develop small, yellow and often shrivelled leaves.

Management: Destroy diseased plants promptly and avoid high humidity in greenhouses.

Spray at least three times at weekly intervals with Zineb or Thiram (0.2 per cent) following the appearance of symptoms. Where the disease is suspected or prevalent, dip unrooted cuttings in Benomyl suspension.

The incidence of blight disease of chrysanthemum was low (5.83 per cent) under polyhouse conditions than under open field conditions (38.4 per cent) reported from M.P.K.V., Pune.

Powdery Mildew (*Erysiphe cichoracearum* DC)

Symptoms: White powdery patches appear on the upper surface and on tender shoots.

Management: Keeping dry environment and spraying with Carbendazim controls the disease.

Rust (*Puccinia chrysanthemi* Roze)

Symptoms: The pustules appear as blister-like swellings, which break open releasing masses of powdery uredospores. Severely infected plants are stunted and do not bloom properly.

Management: Steam sterilisation of soil and pots, keeping humidity below 75 per cent, cutting and burning infested plants and spray of Mancozeb are some of the control measures.

Verticillium Wilt (*Verticillium albo-atrum* Reinke and Berth and *Verticillium dahliae* Kleb.)

Symptoms: Symptoms appear at the flower bed initiation and the plants wilt suddenly.

Management: Selection of healthy stock plants, soil sterilisation and spraying with Bavistin have been recommended for preventing its spread.

Ray Specks (*Erwinia chrysanthemi*)

Symptoms: Wilting of one or more branches during sunny days and recovery at night is an initial symptom. Later on, stem tips turn brown, brittle and collapse. Stem becomes hollow with brownish streaks extending up to base. Cuttings either remain stunted or die.

Management: Soil sterilisation, using disease free cuttings and avoiding contamination during pinching are some of the precautions and use of Streptomycin in rooting beds is advised to control it.

Bacterial Leaf Spot

Symptoms: Circular elliptical leaf spots appear, increase and coalesce to form large necrotic areas. The disease appears first on basal leaves and progresses upward to flower buds, which turn dark and die later.

Management: Use of disease free cuttings and spraying in rainy season with copper sulphate controls the problem.

Viral Diseases

As many as 20 viral diseases are reported on this crop. In all, stunted plants, curling of leaves and rosette formation on terminal growth are the main symptoms. Thrips and aphids, etc. transmit the virus so they should be controlled in time.

Chrysanthemum Stunt (Chrysanthemum Stunt Virus)

Symptoms: Overall reduction in plant size. Pale foliage with margins failing to enlarge giving stiff appearance. Red and bronze flowers, often bleached, may open prematurely.

Management: As the virus is transmitted through the wounds the tools used to cut or pinch affected plants should not be used on healthy plants without proper sterilisation. Use cuttings from healthy plants.

Combination Spray for Chrysanthemums

The following combination sprays will control all the more prevalent fungus diseases and insect pests of outdoor chrysanthemums:

Sevin, 50 per cent wettable powder	2 tablespoonfuls
Malathion, 25 per cent WP	4 tablespoonfuls
Zineb, 65 per cent WP	1 tablespoonful
Kelthane, 35 per cent WP	1½ teaspoonfuls
Water	4 litres

Mix the four ingredients dry, then add enough water to make a very thin paste. Pour this mixture into the spray tank, preferably through a fine screen or cheesecloth, add water and stir.

Biological Control

Chrysanthemum is largely damaged by pests like aphids, leaf eating caterpillars of moths and butterflies, etc. Moths and butterflies lay their eggs on leaves and newly emerging buds causing irreparable damage. The insecticides sprayed on the plants fail to kill them. The growth of the plant is also greatly affected by environmental pollution like toxic gases, smoke, heavy dewdrops, smog, dirt, etc. It is not practicable to spray chemicals and insecticides directly on flowers and buds. Use translucent paper bags or polythene bags to cover the buds and flowers. It protects the buds and flowers from moths and butterflies and does not allow them to lay eggs on them. It also saves the buds and flowers from dirt, dust, smoke, dew, toxic gases and other sources of pollution.

COSMOS (Mexican Aster)

Cosmos bipinnatus, Cosmos bipinnatus Cav., Cosmos. Family: Compositae

Fungal Diseases

Stem Blight or Canker (*Diaporthe stewarti* Harrison)

Symptoms: It is a disease of mature plants. The pathogen attacks the stem and branches forming brown spots, which girdle them. The portions beyond these cankers wilt and die. Affected stems become weak and break. The disease is not common.

Management: The diseased plants should be removed and burnt.

Leaf Spot (*Cercospora* Fres. and *Septoria* Sacc.)

Symptoms: These produce dark brown irregular spots. Adjacent spots coalesce giving a blighted appearance. The diseases are serious during rainy weather.

Management: Bavistin (0.1 per cent) or Dithane Z-78 (0.2 per cent) may be sprayed.

Powdery Mildew (*Erysiphe cichoracearum* DC.)

Symptoms: Typical symptoms appear on leaves and young branches. The disease is common.

Management: Spraying the crop with Dinocap (0.02 per cent) or Carbendazim (0.05 per cent) (Neibisch and Kelling, 1986).

Dieback (*Phoma herbarum* West)

Symptoms: The infection is noticed on twigs and stems forming cankers, which result in dieback symptoms and eventual death of plants.

Management: Spray of Bavistin (0.2 per cent) is recommended.

Bacterial Wilt (*Pseudomonas solanacearum* E. F. Smith)

Symptoms: It produces typical wilt symptoms resulting in the death of plants.

Management: The infected plants should be rouged out, and 3-4 years rotation may be followed.

CARNATION (*Dianthus caryophyllus* Linn.)

Family: Caryophyllaceae

Fungal Diseases

Wilt (*Fusarium oxysporum* f. sp. *dianthi* (Beach) Snyd & Hansen)

Symptoms: This disease is one of the major problems in carnations causing losses to the commercial growers. It is favoured by warm temperature and can cause very high percentage of mortality (Fletcher and Martin, 1972). Chief diagnostic symptoms are abnormal growth and stunting of young shoots. Leaves become yellowish and the stems are soft so as to be easily crushed. The entire plant wilts and collapses in a short time after the attack. The part of the stem in contact with the soil shows shredding of bark while the wood beneath remains firm. If pulled, the plant breaks off easily.

Fusarium wilts are very difficult to manage. The fact that even a single infection of a plant by one spore is sufficient to introduce the pathogen into the plant in which it then grows and spreads internally makes prevention of infection and subsequent control with surface fungicides practically impossible. Also, Fusarium sp. can survive in the soil for a long time and this makes control through crop rotation and other cultural practices ineffective. Management practices of Fusarium wilt in ornamental crops include many factors, which, though not successful in completely eliminating the pathogen, are helpful in reducing the pathogen population and disease severity.

A complete disease management programme against Fusarium wilts is based on:

- Pathogen-free stock;
- Soil free of pathogens; and
- Sanitation.

Pathogen-Free Stock

For vegetatively propagated ornamental crops, the use of clean, pathogen free propagules (cuttings, divisions, runners, corms, bulbs, buds and grafts) is of utmost importance. Since the Fusarium pathogen occurs in the vascular elements it may be already infected when removed from parent. Clean vegetative propagules can be obtained through the cultured cuttings, apical meristems, single cell cultures and indexing procedures.

Pathogen-Free Soil

The population of pathogens in the soil can be reduced by:

- Clean cultivation—direct destruction of crop refuse, debris and infested plant material;
- Soil treatment with fumigants;
- Pasteurising the soil with steam; and
- In specific climatic conditions, by soil solarisation.

Sanitation

Soil Treatment with Fumigants: Soil fumigation can greatly alleviate disease losses. Methyl bromide, Chloropicrin, a mixture of both, Vapam, etc., are general fungicides used for fumigation of nurseries or in high profit crops like greenhouse ornamentals and vegetables. Ethylene dibromide (EDB) and DD mixture control nematodes while predisposing crops to the wilt fungi.

Carnations are generally grown as monocultures, sometimes consecutively for more than 15 years. Fumigation with methyl bromide (800 kg/ha) has become a common practice to reduce the initial inoculum of *Fusarium oxysporum* f. sp. *dianthi*. This treatment, however, does not completely eradicate the pathogen from fumigated soils, especially in deeper layers.

In glasshouses, soil sterilisation by steam pasturing is a standard practice to eliminate inoculum before planting, to prevent Fusarium as well as other diseases due to soil-borne pathogens. Soil solarisation using clear transparent polyethylene film (0.1 mm thick) for 30 days gave satisfactory control of Fusarium wilt (Hardy and Sivasithamparam, 1985).

This disease can be prevented by drenching of soil around the plants base with Carbendazim (0.1 per cent) or Bavistin (0.15 per cent). Spraying of plants with Dithane M-45 (0.1 per cent) + Bavistin (0.1 per cent) at fortnightly intervals was also reported to be effective.

Alternaria Leaf Spot and Blight (*Alternaria dianthi* Stevens and Hall)

Symptoms: This is the most common and serious foliar disease of carnations. Small and purple coloured spots appear on the leaves. Under moist conditions these enlarge to form leaf spots up to one cm in diameter. The margins of the lesions are usually purple in colour and the centres grey-brown, and black spores may be present in this area. Several spots may merge resulting in the death of the leaf. The pathogen also causes basal rot of cuttings. This fungus also affects the stems, usually at the nodes, where it enters through the wounds or growth cracks producing black lesions.

Management: Destruction of diseased debris and Dithane M-45 (0.2 per cent) spraying are recommended. It was recommended by AICRP centre, that fungicides like Blitox (0.3 per cent), Foltaf (0.2 per cent), Captaf (0.3 per cent) and Indofil M-45 (0.2 per cent) were effective to reduce the leaf spot infection in a significant manner.

Rhizoctonia Stem Rot

This disease is caused by *Rhizoctonia solani* and affects the plants at the soil line and they die within a week or so. Grow the crop in raised beds treated with 0.5 per cent Formaline. Drench the infected plants either with Carbendazim 0.1 per cent or Thiram 0.2 per cent or Dithane M-45 0.2 per cent.

Grey Mould

Botrytis cinerea is the most common fungus in carnation production. The biggest problem is often the attack on the flowers. In the greenhouse, it happens when there is high humidity. Reduce the humidity.

Viral Diseases

Carnations are subject to many viral diseases, the most common ones being streak, mosaic, mottle, ring spot, etched ring and vein mottle. Use virus-free planting material raised through shoot tip culture.

Streak (Carnation Streak Virus)

Symptoms: Yellow or red spots and streaks occur parallel to the leaf veins. The lower leaves may turn yellow and die.

Mosaic (Carnation Mosaic Virus)

Light green mottling and irregular to elongate blotches occur on leaves. Conspicuous breaking of flower colour also occurs. The green peach aphid, *Myzus persicae,* transmits the disease.

Mottle (Carnation Mottle Virus)

The virus is very common in commercial cultivars in which it produces either faint leaf mottling or no symptoms at all. It causes breaking of flower colour in some of the varieties.

Ring Spot (Carnation Ring Spot Virus)

Irregular grey or yellow spots appear on the leaves. The leaf margins are wavy. The green peach aphid also transmits the virus.

Management: The viruses can be controlled by treating the plants or infected cuttings at 38° C for two months. Less heat tolerant varieties can be subjected to 30° C and then treated at 38° C.

HELIANTHUS (Sunflower)

Helianthus annus; Family: Compositae

Fungal Diseases

Leaf Spot (*Cercospora helianthicola* Chup & Viegas)

Symptoms: Blackish-brown, circular to irregular spots, surrounded by broad, yellowish irregular areas appear. Adjacent spots coalesce resulting in the death of the leaves.

Management: Spray of Bavistin (0.2 per cent) prevents the disease.

Powdery Mildew (*Erysiphe cichoracearum* DC.)

Symptoms: Whitish powdery spots are noticed on the lower side of leaves, which turn yellow, curl and dry out. Severely attacked plants remain stunted and do not flower. The disease is very widespread.

Management: Wettable sulphur (0.2 per cent) or Karathane (0.025 per cent) spray is recommended.

Rust (*Puccinia helianthi* Schw.)

Symptoms: Numerous brown pustules develop on the underside of leaves, resulting in their curling and drying out. Uredia are irregular, scattered, cinnamon to brownish and up to 1 mm in diameter. The disease is common and widespread.

Management: The carryover of the disease mostly takes place through crop debris, which should be destroyed after harvest. Maneb, Zineb or Mancozeb (0.2 per cent) spray gives good protection. Cultivation of resistant varieties is also advocated.

Downy Mildew [*Plasmopara halstedii* (Farlow) Berl. & de Toni]

Symptoms: The fungus enters the young seedlings through root hairs and moves upward into the stem and leaves within one week after seedling emergence. The seedlings become stunted, abnormally thickened and their cotyledons show downward curling. The leaves show pronounced chlorotic mottling on their upper surface along the veins. The seedlings become resistant with age. Primary infection takes place through oospores and progresses systematically on seedlings. The fungus perpetuates through the oospores in the soil. The disease is favoured by low temperature and high humidity. Losses in different varieties depend upon the time of attack and the disease intensity.

Management: Dithane M-45 (0.2 per cent) or Deconil (0.2 per cent) spray, following proper rotation (2-3 years), destruction of diseased debris, and use of resistant varieties and disease-free seed are recommended.

Stem Rot [*Sclerotinia sclerotiorum* (Lib.) de Bary; *Macrophomina phaseoli* (Maubl.) Ashby].

Symptoms: The plants are infected when they are full grown by *Sclerotinia sclerotiorum*, whereas *Macrophomina phaseoli* may attack at any time. The infected plants die showing root-rot, foot-rot and stem-rot symptoms. Numerous dark, irregular sclerotia are formed inside the hollow stem. A thick mycelial growth develops on the stem near the ground. The disease is widespread and destructive.

Management: Infected plants may be removed and destroyed. Following 2-3 years rotation, avoiding close planting, use of disease-free seed and balanced fertilisation offer good control. Soil application of PCNB against *Macrophomina phaseoli* and spraying with Bavistin (0.1 per cent) reduce disease incidence appreciably.

Leaf Spot [*Alternaria helianthi* (Hansf.) Tubaki & Nishihara]

Symptoms: It appears initially in the form of small, scattered brown spots during the rainy season. These increase in size, number and coalesce. Mature lesions are irregular, 1-2.5 cm in diameter, dark brown with pale margins and yellow haloes. Linear necrotic lesions are formed on the petioles. The disease is widespread.

Management: Dithane M-45 (0.2 per cent) or Zineb (0.2 per cent) spray and use of disease-free seeds are recommended.

Southern Blight (*Sclerotium rolfsii* Sacc.)

Symptoms: It is serious in sandy loam soils. The infected plants show wilting followed by death. Mustard seed-like sclerotia are noticed near the foot region of infected plants. The disease is widespread.

Management: Same as for stem-rot.

Septoria Leaf Spot (*Septoria helianthi* Ell. & Kell)

Symptoms: The infection first appears at the cotyledon stage as small necrotic spots, 1-2 mm in diameter. These spots later become angular or sub-circular (up to 2 cm in diameter). Adjacent spots coalesce to form irregular lesions. The tissues surrounding these spots lose chlorophyll and turn light green or yellow.

Management: As the fungus is seed-borne, it is advisable to use disease-free seed and spray with Bavistin (0.1 per cent) or Zineb (0.2 per cent).

Wilt (*Fusarium oxysporum* Fr. and *Verticillium dahliae* Kleb.)

Symptoms: *Fusarium oxysporum* is serious on common varieties and produces typical symptoms on grown-up plants. The disease caused by *Verticillium dahliae* is also called leaf-mottle. It produces leaf mottling, inter-veinal chlorosis and necrosis on young plants. The infection first appears on the lowermost leaf and spreads further, ultimately killing the plant. *Verticillium dahliae* is seed-borne.

Management: Both these diseases can be kept under check by the use of disease-free seed and by following 3-4 years rotation along with cultivation of relatively resistant varieties.

Viral Diseases

Symptoms: Mosaic: Mosaic patterns, ring spots and chlorotic spots appear. It is host-specific and is transmitted through sap. The disease is common and widespread.

Management: Diseased plants should be rouged out.

IBERIS (Candytuft)

Iberis umbellata and *Iberis amara*; Family: Cruciferae

Fungal Diseases

Leaf Spot [*Alternaria brassicicola* (Schw.) Wiltsh.]

Symptoms: Dark blackish, elongated, circular or irregular spots with indistinct margins appear on both surfaces of leaves, stems and inflorescence.

Management: Dithane M-45 (0.2 per cent) or Difolatan (0.3 per cent) may be sprayed.

Powdery Mildew (*Erysiphe polygoni* DC.)

Symptoms: Whitish floury patches are noticed on the leaves turning them yellow.

Management: Spraying with wettable sulphur (0.2 per cent) or Karathane (0.025 per cent) checks the disease.

Foot Rot and Stem Rot [*Sclerotinia sclerotiorum* (Lib.) de Bary and other soil-borne pathogens]

Symptoms: The infection appears as brown, depressed spots at the collar region, which may girdle the stem. The infected plant dies after wilting. Numerous sclerotia are formed in diseased tissue, which help the fungus to perpetuate. The disease is common and widespread.

Management: Soil application of Bavistin and Captan gives some control. Diseased plants may be rouged out and destroyed. Use of disease-free seed in disease-free soil is also advocated.

Damping-Off (*Rhizoctonia solani* Kuhn and other soil-borne pathogens)

Symptoms: The disease is serious in the nursery and is responsible for both pre- and post-emergence mortality. Pre- and post-emergence damping-off is also caused by *Pythium ultimum*. The disease is serious under humid weather conditions.

Management: Same as in other cases.

BALSAM (*Impatiens balsamina* Linn.)

Fungal Diseases

Downy Mildew (*Peronospora obducens* Schroet.)

Symptoms: The disease is generally noticed in the hills of Solan, Kasauli and Mussoorie during the rainy season and is quite serious at Solan (Himachal Pradesh). It attacks all the aerial parts, including flowers, fruits and seeds but the symptoms are most conspicuous on the leaves (Paul et al, 1991). Numerous brownish, circular to irregular spots which are more prominent on the under-surface of leaves appear and are covered with white cottony growth. The upper surface of spots is at first yellowish and then dark brown. Leaves of all ages are susceptible but the infection first appears on the lower leaves and proceeds upwards. Severely diseased leaves turn yellow and are shed. Diseased plants die before flowering under favourable weather conditions (high humidity and 25-28° C temperature).

Management: Disposal of debris and spraying with Dithane M-45 (0.2 per cent) or Difolatan (0.2 per cent) keeps the disease under check.

Powdery Mildew [*Erysiphe cichoracearum* DC. ex Merat.; *Erysiphe polygoni* DC.; *Leveillula taurica* (Lev.) Arn. and *Oidium* Sacc.]

Symptoms: The infection is noticed on all the aerial parts of the plant and produces whitish floury-patches resulting in yellowing of leaves and defoliation. Severely infected plants do not flower and die. The disease is favoured by humid weather and moderate temperature, and is serious in south India and at mid-hill elevations in north India. The disease is common in the late season.

Management: Spraying with wettable sulphur (0.2 per cent) or Karathane (0.025 per cent) or Bavistin (0.1 per cent), and avoiding shade and dense planting are recommended.

Leaf Spot [*Drechslera australiensis* (Bugnic.) Subram. & Jain]

Symptoms: Small dark brown to black, circular to irregular, amphigenous spots, 2-3 mm in diameter initially appear and later coalesce forming necrotic lesions. Very often leaf tips and margins also dry.

Management: Spraying with Dithane M-45 (0.2 per cent) or Zineb (0.2 per cent), and destroying plant debris at the end of the season help in controlling the disease.

Alternaria Leaf Spot (*Alternaria tenuis* Nees)

Symptoms: It produces spots similar to those of *Drechslera*. The disease is widespread.

Management: Use of disease-free seed and spraying with Indofil M-45 (0.2 per cent) are recommended.

Ascochyta Leaf Spot (*Ascochyta impatientis*)

Symptoms: Large pale brown blotches with paler margins are formed, ultimately resulting in the death of leaves. Pycnidia are black or brown, globose to subglobose, erumpent and ostiolate; conidia are 2-celled and hyaline.

Management: Spray of Indofil M-45 (0.2 per cent) is recommended.

Cercospora Leaf Spot [*Cercospora fukushlana* (Mats.) Yamamota syn. *Cercospora balsaminae* Kel. & Sw.]

Symptoms: Spots are circular to sub-circular, at first brown then with dingy grey centre and brown margins. The fungus has been collected from Bangalore. The disease is common during rainy weather.

Management: Zineb (0.2 per cent) or Bavistin (0.1 per cent) may be sprayed.

Stem Rot, Foot Rot or Collar Rot [*Macrophomina phaseoli* (Maubl.) Ashby]

Symptoms: Dark brown necrotic lesions running parallel to the stem appear from the foot region and proceed upwards resulting in the wilting and eventual death of plants. Affected tissues later turn ochraceous. Numerous, dark, hard, irregular bodies which are the sclerotia appear on the diseased tissues. The fungus perpetuates through sclerotia in the soil and has a very wide host range. The disease is serious under stress conditions and kills the plant at any stage of growth including the seedlings (Castello and Peterson, 1990).

Management: To avoid build-up of the inoculum in the soil, diseased plants may be removed and burnt. Proper watering is required to avoid stress conditions. Soil application of Bavistin is also advocated in the nursery. Dipping of seedlings in Bavistin or Benomyl solution (0.1 per cent) for a few hours before transplanting, and treatment of seeds with these fungicides give good control.

Damping-Off [*Rhizoctonia solani* Kuhn and *Phytophthora nicotianae* var. *parasitica* (Dastur) Waterh.]

Symptoms: Both these pathogens are responsible for pre- and post-emergence damping-off of seedlings and death of established seedlings. It is serious when the beds are kept too wet or waterlogged. Both the pathogens are widespread, serious, have a wide host range and perpetuate in the soil as saprophytes.

Management: Raising the nursery in disease-free soil, avoiding excessive watering and over-crowding of seedlings, and drenching the nursery with Captan (0.2 per cent) are helpful in disease control. Use of Metalaxyl and Fosetyl-Al are effective against both the pathogens (Benson, 1995).

Foot Rot (*Rhizoctonia solani* Kuhn)

Symptoms: Wilting and collapse of seedlings, wilting, yellowing and desiccation of leaves of older plants and disintegration of the collar region are seen.

Management: Dipping of seedlings in Bavistin (0.1 per cent) helps in preventing disease.

LATHYRUS (Sweet Pea)

Lathyrus odoratus; Family: Leguminosae

Fungal Diseases

Anthracnose, Blossom and Shoot Blight [*Glomerella cingulata* (Stonem.) Spauld. & Schrenk, *conidia* stage; *Colletotrichum gloeosporiodes* Penzig]

Symptoms: The disease is very destructive. Initially the leaves, flowers and shoots are disfigured with whitish spots, which spread very fast and eventually the infected parts wither and dry up. The leaves and stems turn dark brown and die.

Management: Spray of Captan (0.2 per cent) and Bavistin (0.2 per cent) at regular intervals helps in controlling the disease.

Powdery Mildew [*Microsphaera alni* (Wallr.) Salm., conidial stage; *Oidium* Link ex Fr.]

Symptoms: The fungus produces whitish floury patches on both sides of the leaves. These later turn brownish and dry off.

Management: Spray of Sulfex (0.3 per cent) and Bavistin (0.2 per cent) at regular intervals helps in controlling the disease.

Downy Mildew (*Peronospora trifoliorum* de Bary)

Symptoms: Yellow translucent spots appear on the upper side of the leaves with greyish mouldy growth on the lower surface.

Management: The fungus perpetuates through oospores; hence destruction of debris and proper crop rotation are essential for disease control.

Other Leaf Spot [*Colletotrichum pisi* Pat.; *Isariopsis griseloa* Sacc.; *Phyllosticta orobella*; *Mycosphaerella pinodes* (Berk. & Blox.); Vestergr and *Ramularia deusta* var. *odorati*]

None of these is common and serious. Spraying fungicides against other diseases will control some of these leaf-spot diseases also.

Viral Diseases

- Mosaic (pea-mosaic virus): It attacks the seedlings and causes curling of leaves and dwarfing of plants along with mosaic symptoms. The very few flowers produced by the infected plants show discoloured streaks.
- Infected plants should be destroyed. Malathion spray is recommended for controlling aphid. Planting near bean and other leguminous crops should be avoided.
- Tomato spotted-wilt. Purplish spots appear on the stem and leaves following leaf-mottle symptoms.

Management: Spraying with Malathion, avoiding susceptible plants like dahlia, tomato and potato in the vicinity of the crop and rouging out diseased plants are recommended measures for disease control.

MATTHIOLA (Stock)

Matthiola incana R. Br.(Stock); Other common name: Gillyflower; Family: Cruciferae

Fungal Diseases

Leaf Spot (*Alternaria raphani* Groves & Skolko)

Symptoms: Pale greyish-green depressed spots up to 3 cm in diameter appear. These later become blackish with concentric zones, mostly on the lower leaves. The disease is common and widespread.

Management: Copper oxychloride or Dithane M-45 (0.2 per cent) spray is recommended. Also, diseased leaves should be collected and destroyed.

White Rust [*Albugo candida* (Pers.) Kuntze]

Symptoms: White raised pustules are formed on the underside of lower leaves. These give rise to powdery-spores. Oospores are produced late in the season.

Management: It is advisable to remove and destroy all diseased plants, plant parts and their debris, spray with copper oxychloride (0.2 per cent), and follow a 2-year rotation.

Damping-Off [*Rhizoctonia solani* Kuhn; *Rhizoctonia bataticola* (Taub.) Butler; *Phytophthora* de Bary and *Pythium* Pringsh]

Symptoms: Roots and basal parts of plants turn black and eventually die. In older plants initially wilting is noticed leading to their death. The disease is common and widespread.

Management: Use of disease-free soil for raising the nursery, and drenching the nursery with formaldehyde (1 per cent) helps in disease control.

Downy Mildew [*Peronospora parasitica* (Pers.) de Bary]

Symptoms: It produces typical symptoms. Tender stems, leaves and flower parts are attacked. Affected plants remain stunted.

Management: Same as for leaf spot.

Wilt (*Fusarium oxysporum* f. *matthioli* Baker)

Symptoms: Typical wilt symptoms result. Seedlings show vein clearing and stunted growth.

Management: Same as for other wilt diseases. It is seed-borne hence special care is needed to use healthy seed.

Stem Rot and Wilt (*Sclerotinia sclerotiorum*)

Symptoms: The initial symptoms are drooping and yellowing of leaves, which is followed by complete wilting of plants. Infected plants can be pulled out easily since there is a total disintegration of tissues in the collar region. Mycelial patches and dark irregular sclerotia are also noticed in the diseased tissues. The fungus has a very wide host range and perpetuates through sclerotia. It has smaller sclerotia, asci and ascospores than *Sclerotinia sclerotiorum.*

Viral Diseases

Mosaic: Stocks are also commonly infected by Anemone-mosaic virus and CMV just as anemones are. The predominant symptom is flower breaking which is sometimes accompanied by leaf distortion and mottling.

PETUNIA

Petunia hybrida; Family: Solanaceae

Fungal Diseases

Leaf Blotch [*Cercospora petuniae* (Saito) Muller & Chupp]

Symptoms: Circular to irregular, dark brown spots appear which later turn greyish. Numerous spots coalesce resulting in leaf blotching followed by their death. The disease is serious in humid weather. The fungus also attacks other species of *Petunia.*

Management: Spraying with Bavistin (0.1 per cent) or Dithane M-45 (0.2 per cent), destroying the debris, and avoiding crowding of plants and excessive irrigation are measures to check disease spread.

Leaf Spot (*Ascophyta petuniae)*

Symptoms: The infection spots resemble those of *Cercospora petuniae.* Two-celled, oval to elongated hyaline spores are produced in pycnidia.

Management: Same as for *Cercospora* leaf spot.

Powdery Mildew (*Oidium* Link ex Fr.)

Symptoms: Whitish floury patches develop on the leaves, which become yellow and dry out. Severe infection kills the plant.

Management: Spraying with Sulfex (0.2 per cent) or Bavistin (0.1 per cent) and avoiding shading of plants is recommended.

Phyllosticta Leaf Spot (*Phyllosticta petuniae* Speg.)

Symptoms: Infection is noticed as small, circular to irregular spots starting usually from the leaf tip. These become salmon to chestnut or greyish. It also attacks other species of *Petunia.*

Management: Same as for other leaf spots.

Crown Rot (*Phytophthora parasitica* Dast.)

Symptoms: The fungus attacks the leaves, branches and stems producing water-soaked, brownish spots leading to the death of plants.

Management: Spraying with Difolatan (0.3 per cent) or Dithane M-45 (0.2 per cent) or copper oxychloride (0.3 per cent), avoiding water stagnation by judicious irrigation, following 2 years rotation and destroying the debris are recommended.

Bacterial Wilt (*Pseudomonas solanacearum* E. F. Smith)

Symptoms: The disease is very serious and kills the plant at any stage of growth.

Management: Planting in disease-sick plots should be avoided, treating the soil with bleaching powder (neutral) before sowing and following crop rotation with non-host crops are recommended.

Damping-Off [*Pythium aphanidermatum* (Eds.) Fitz. and *Rhizoctonia solani* Kuhn]

Symptoms: Typical symptoms appear in the nursery before and after seedling emergence.

Management: Same as in other cases of damping-off.

Viral Diseases

Symptoms: *Petunia* is attacked by tobacco mosaic (TMV), cucumber mosaic (CMV), tobacco ring spot, aster yellows, tomato spotted wilt, beet curly top and potato virus X diseases. In case of TMV infection, the diseased plants show stunted growth with reduction in leaf size. Mottling of leaves with vein-banding symptoms is also common on the lower leaves. TMV is transmitted by sap and *Aphis gossypii.* CMV is transmitted by *Myzus persicae.*

Management: To reduce the incidence of viral diseases, it is essential to spray with insecticides at periodic intervals to check secondary spread. Use of healthy seedlings is also desirable.

Aster Yellows

This disease has also been reported from petunias.

Management: Remove infected plants. Spray of Malathion or Sevin helps management of disease.

TAGETES (Marigold)

African Marigold (*Tagetes erecta*); French Marigold (*Tagetes patula*); Family: Compositae

Fungal Diseases

Leaf Spot (*Alternaria tagetica* Shome & Mustafee)

Symptoms: It produces small, blackish-brown, circular spots initially, which spread fast and become irregular later and cover large areas. The disease is common and has been reported on *Tagetes* sp. from marigold growing areas. *Tagetes erecta* and *Tagetes patula* is also attacked by *Alternaria zinniae* Pape which produces leaf-blight or marginal-blight symptoms. The infection appears as minute spots, which later enlarge, coalesce to form large irregular spots or blotches that are amphigenous and light brown to dark brown.

Management: Dithane M-45 (0.2 per cent) or copper oxychloride (0.2 per cent) spray may be used.

Flower Bud Rot [*Alternaria alternata* (Fr.) Keissler and *Alternaria dianthi* Stevens & Hall]

Symptoms: This disease is serious in humid climate. The infection appears on young unopened buds as dark brown spots. These buds do not open properly and the flower quality (shape and size) is adversely affected (Dhiman and Arora, 1990).

Management: Same as for the previous disease.

Cercospora Leaf Spot (*Cercospora thunbergiae* Boedijn)

Symptoms: It produces brown, circular-oblong spots, 0.1-0.5 cm in diameter surrounded by blackish-brown margins.

Septoria Leaf Spot (*Septoria tageticola*)

Symptoms: Oval to irregular, smoky-grey to black spots, interspersed with minute, black fruiting bodies appear. Infection starts from the lower leaves and progresses upwards. Varieties of *Tagetes erecta* are susceptible, whereas those of *Tagetes patula* are either resistant or immune.

Management: Bavistin (0.1 per cent) or Dithane M-45 (0.2 per cent) may be sprayed as a remedial measure.

Viral Diseases

Cucumber mosaic virus (CMV) produces mottling symptoms on the leaves leading to reduction in leaf size and distortion. Infected plants are stunted and produce few low-grade flowers. Transmission is by sap and by *Aphis craccivora, Aphis gossypii* and *Myzus persicae*. The virus has a very wide host range.

TROPAEOLUM (Nasturtium)

Tropaeolum majus Linn.; Family: Tropaeolaceae

Fungal Diseases

Leaf Spot [*Acroconidiella tropaeli* (Bond) Linquist]

Symptoms: Tiny red spots appear which later become sub-circular, brownish blotches 1-15 cm in diameter due to coalescence of adjacent spots (Hashimoto *et al,* 1987). Central portions shrivel with a thin brownish-yellow margin. Lesions develop especially on the underside of leaves, in moist weather.

Management: Use of healthy seed and spraying with Captan (0.2 per cent), Zineb (0.2 per cent) help in disease control.

Leaf Spot (*Cercospora papavericola* Chupp)

Symptoms: It is a common disease and produces well defined spherical to oval, brown spots with greyish centres 1-7 mm in diameter. Adjacent spots coalesce forming large irregular patches. It is favoured by humid weather.

Management: Spraying with Bavistin (0.1 per cent) helps in controlling the disease.

VIOLA (Pansy)

Other common names: Viola, Tufted Pansy; *Viola tricolor hortensis;* Family: Violaceae

Fungal Diseases

Scab (*Sphaceloma violae* Jenk)

Symptoms: The infection initially causes minute, water soaked dots on leaves in the inter-veinal areas. These spots later turn light yellowish, brown ashy, whitish-ashy and given scabby appearance. The spots are circular and surrounded by a deep green border. Later the spots give shot hole effect due to falling of diseased areas. Stem infection causing girdling of the infected parts results in the death of the parts above the points of attack. The lesions are generally linear to elliptical. On petioles and sepals these are round, circular to irregular while those on capsules are often raised.

Management: Treating them with Carbendazim (3 g/Kg) checks the disease. Sprays of Mancozeb (0.2) as soon as the disease appears are recommended.

Downy Mildew [*Bremiella megasperma* (Berl.) Wils. and *Peronospora viola* de Bary]

Symptoms: Light greyish felt-like fungal growth appears on the under-surface of the leaves giving yellowish discoloration on the upper surface. The leaves near the soil when attacked rot away rapidly. The whole plant may die and droop when stem is infected without showing any mildew symptoms on the leaves.

Management: Destroy infected crop debris and avoid damp conditions in the beds. Mancozeb is effective in the management of the disease.

Leaf Spots [*Alternaria tenuissima* (Kunze ex Fr.) Wiltshire; *Ascochyta violicola* M. Alpine; *Ascochyta violae* Sacc. & Speg.; *Cercospora granuliformis* Ellis & Holw.; *Cercospora violae* Sacc.; *Cylindrosporum viola* Sacc.; *Sphaceloma violae* Jenk.; *Marsonina violae* (Pass.) Magnus; *P. nigrescens* Bonar and Cook; *Septoria viola* Westend]

Symptoms: There are various foliar pathogens causing leaf spots in pansy. The symptoms are almost similar to other leaf spots except for minor differences due to pathogen variations. When more than one leaf spot attacks the crop simultaneously or depending upon favourable conditions, the disease becomes more severe, causing leaf defoliation and losses in quality and number of flowers.

Management: Sprays of Mancozeb (0.25 per cent) at weekly intervals checks the disease spread.

Root Rot (*Fusarium oxysporum* Schlenchtend ex Fr.; *Rhizoctonia solani* Kuhn.; *Pythium ultimum* Trow)

Symptoms: The root infections by these fungi either cause wilting of plants or disintegration of roots or collar region, plants die and fall off. In nursery, some of the pathogens cause damping-off symptoms. The disease appears in patches. Nurseries may be raised in disease-free and sterilised soils.

Management: Soil drenching with Carbendazim or Captan is also effective in controlling the disease. Biocontrol with *Trichoderma viride* and *Gliocladium virens* has been found effective against *Rhizoctonia solani* and *Pythium ultimum* (Lewis *et al*, 1995).

Rust [*Puccinia violae* (Schumach.) DC]

Symptoms: The disease first appears as pale green spots. Then orange coloured aecia develop on the opposite side containing aeciospores. Aecia also develop on petioles and stems. Late in the season, brown uredia containing urediospores develop, which help in the rapid spread of the disease during the growing season. Later, dark brown to black teliosori containing teleutospores appear which, on germination, release basidiospores. The basidiospores on infection cause pale green spots on the under-surface of fresh leaves. The disease also attacks wild hosts.

Management: Spray of Zineb (0.2 per cent) at weekly intervals can be effective.

Smut [*Urocystis violae* (Sowerby) Fisch. V. Waldh]

Symptoms: The pathogen attacks both cultivated and wild species of pansy. All plant parts including leaves, calyx, petioles, flower stalks and offshoots are subject to attack. On leaf stalks, more or less elongated blisters develop due to infection and cause considerable deformity. On breaking open, the pustules release a large number of dark spores, which are held in small sporeballs.

Management: Destruction of infected debris and crop rotation of 2-3 years reduces the disease (Catson and Luttrell, 1988).

Crown and Stem Rot [*Myrothecium roridium* Tode ex Fr.; *Sclerotium rolfsii* Sacc.; *Sclerotium rolfsii* sacc. *var. delphinum* (Welch) Boerema & Hamers]

Symptoms: The disease attacks the plant at soil level, resulting in drying and brittleness of the infected tissues. The diseased tissues later bear abundant black spore masses. Infected leaves turn purplish black in the spots or streaks. These pathogens also attack hollyhock. Crop rotation and sanitation practices reduce the disease incidence.

Under favourable weather conditions, powdery mildew [*Sphaerotheca humuli* (DC.)] grey mould (*Botrytis cinerea* Pers, ex. Fr.) are also serious in pansy. Though bacterial diseases are not serious on pansy, sporadic occurrence of *Pseudomonas viridiflava* (Burkholder) Dowson has been reported.

Viral Diseases

Curly Top

Symptoms: The disease causes dwarfing and rosetting of shoots and reduction in flower size. Diseased plants bear flowers for a shorter period and die prematurely. The virus is transmitted by leafhopper.

CMV

Symptoms: The disease appears in the form of clearing of veins and veinlets, dwarfing and yellowing of plants. Young plants die prematurely. Flowers show colour-breaking streaks on the flowers, which later turn white. The disease is transmitted.

Management: General control measures should be followed, namely, rouging of infected plants, spraying lower beds with Malathion or Rogor (1 ml/l) for managing the vectors.

Physiological Diseases

Oedema, Wart Disease

Corky or wart-like growths 1/8 inch long develop on the leaves and flower stalks and cause these parts to become dry and brittle. Insect punctures are often followed by such growths. Ordinarily, improper water relations lead to oedema; either a soil kept too wet by over-mulching or excessive humidity of the air due to faulty lighting conditions. Admit more air to the greenhouse and stir the soil to make it more porous. Avoid rapid changes of conditions, however.

ZINNIA *(Zinnia elegans)*

Other common names: Youth and Old Age; Family: Compositae

Fungal Diseases

Leaf Spot (*Cercospora zinniae* Ell. & Mart.)

Symptoms: It produces numerous, small, conspicuous, circular, white to greyish spots with reddish-brown margins. In severe cases of infection, the leaves turn brownish and dry off. Fruiting is amphigenous but more on the upper surface.

Management: Bavistin (0.1 per cent) or Zineb (0.2 per cent) spray is recommended. It is advisable to collect and destroy diseased debris, use healthy seed and avoid crowding of plants and excessive watering.

Phyllosticta Leaf Spot (*Phyllosticta zinniae* P. Brun.)

Symptoms: It produces small, reddish-brown spots, which later turn greyish, coalesce with one another and form irregular patches. Severe infection results in leaf drying. It also attacks flowers and seeds (Wu and Yang, 1992).

Management: Spraying the crop with Mancozeb (0.2 per cent), Carbendazim (0.05 per cent) or copper oxychloride (0.2 per cent) reduces the disease spread (Wu & Chou, 1995; Yadahalli *et al*, 1994).

Powdery Mildew (*Erysiphe cichoracearum* DC.)

Symptoms: The infection is noticed as dirty-white floury irregular patches on the leaves, stems and inflorescences.

Management: Spraying with wettable sulphur (0.2 per cent) or Karathane (0.025 per cent), and avoiding shade are recommended.

Alternaria Leaf Spot or Blight (*Alternaria zinniae* Pape)

Symptoms: Infection starts as minute spots, which later enlarge and coalesce to form large circular to irregular, light to dark brown blotches and whitish patches. Severely infected leaves dry off.

Management: Treating the seed with Captan or Thiram, destroying plant debris, following a 2-year rotation, spraying the plants with Maneb, Zineb or copper oxychloride (0.2 per cent) and raising the seed crop in disease-free areas are the recommended measures.

Twig Dieback or Blight (*Phomopsis zinniae*)

Symptoms: The infection is noticed in the form of dieback of flowering branches resulting in their death (including the flower). Leaves are also infected, producing dark brown, irregular necrotic lesions. Infected tissues turn greyish with numerous, dark, pinhead-like pycnidia.

Management: Disease-free seed, and Zineb (0.2 per cent) or copper oxychloride (0.3 per cent) or Bavistin (0.1 per cent) spray are recommended. Also, diseased debris should be collected and burnt.

Southern Blight (*Sclerotium rolfsii* Sacc.)

Symptoms: The disease is serious in sandy-loam soils. The fungus attacks several plants belonging to different families and lives in the soil as a saprophyte. Among ornamentals, it attacks Ageratum, Amaryllis, Azalea, Caladium, Calendula, Canna, Carnation, Cosmos, China Aster, Chrysanthemum, Dahlia, Delphinium, Gladiolus, Hollyhock, Iris, Jasmine, Verbena, Lily, Lupine, Marigold, Narcissus, Orchid, Phlox, Rose, Scabiosa, Sedum, Sweet Pea, Tulip, Violet, Zinnia and many others. Initially it forms white wefts of mycelium at the base of the stem, which may come over the ground as well in wet weather. Numerous mustard seed-like reddish-tan or brown sclerotia are produced around the stem. The plant is killed at the ground level leading to wilting and death.

Management: The disease can be kept under check by application of more organic matter (FYM) or nitrogenous fertilisers. Periodic raking of soil brings down disease incidence. Drenching of soil with formaldehyde, or Captan is also helpful. Rotation with non-host crops (cereals) helps in minimising disease incidence.

Viral Diseases

Symptoms

Mosaic: It causes mottling of leaves. Affected plants are pale and produce few flowers. Transmission through sap and also *Aphis gossypii* and *Myzus persicae*. It produces systemic mosaic symptoms on tobacco, petunia, *Cucumis sativus* and bitter gourd.

Leaf Curl: Thickening of veins and veinlets, curling of leaves, dwarfing of plants, distorted flowers and sterility symptoms are produced. It is transmitted by the white fly.

Management: Remove and destroy infected plants. Spray with Malathion at regular intervals for controlling sucking insects, which spread viruses.

BULBOUS CROPS

AMARYLLIS (*Amaryllis belladonna*)

Other common name: Belladonna Lily; Family: Amaryllidaceae

Fungal Diseases

Red Spot [*Stagonospora curtisii* (Berk.) Sacc]

Symptoms: The tips of young shoots develop a red-brown, typically scorched appearance soon after they emerge. Flower stalks are severely stunted and distorted. It causes bright red or vermilion spots, which become soft, brown and sunken as they enlarge. In humid conditions fungal mycelium may develop. The centre of the lesions always retains red colour at the edges. Leaf scorching is also frequent in some areas. The disease has been recorded in Iraq, the U.S.A. and New Zealand (Smith 1929; Dingley and Brian, 1956; Tarbeih *et al,* 1980).

Management: Use of Benomyl (0.2 per cent) is recommended against this disease. Infected bulbs should be discarded. Diseased leaves should be removed and burnt. Plants should be sprayed with Zineb (0.2 per cent) or Bordeaux mixture with a sticker as soon as the disease appears and the spray may be repeated at fortnightly intervals until blooming.

Viral Diseases

Mosaic

Symptoms and Management: The leaves show light green to yellow mottling. The plant vigour and flower quality is adversely affected. Rouging out and destroying all affected plants and using disease free planting material is recommended.

ANEMONE (Windflower)

Other common names: Poppy Anemone; *Anemone coronaria*; Family: Ranunculaceae

Fungal Diseases

Leaf Spots

A large number of leaf spotting fungi occur on both cultivated and wild species of *Anemone*. Among the most common are *Phyllosticta anemones, Phyllosticta anemonicola, Cercospora pulsatillae, Cercosporella filiformis, Didymaria didyma, Ramularia ranunculi, Septoria anemones* and *Septoria punicea*.

Management: Valuable plants can be protected by periodic applications of copper fungicides.

Rhizome Rot (*Pellicularia rolfsii* and *Sclerotium delphinii*)

A decay of the rhizomes followed by death of the plant may be caused by these soil-infesting fungi.

Management: Dig up and destroy diseased plants as well as the soil immediately surrounding them. Treat remainder of soil with heat or chemicals.

Collar Rot (*Botrytis cinerea*)

Occasionally, this fungus results in extensive destruction of plants by rotting the crowns.

Management: Same as for rhizome rot.

Spot Disease (*Synchytrium anemones*)

The flowers become discoloured in spots, dwarfed and distorted, and fall. The leaves and stems also are spotted.

Management: Discard badly infected plants or remove spotted flowers and leaves.

Downy Mildew (*Plasmopara pygmaea*)

Symptoms: The fine white mildew covers the under surface of the leaves. Plants become discoloured. Stamens abort.

Management: The same as for the preceding disease.

Rust (*Tranzschelia pruni-spinosae*)

Symptoms: This is a systemic disease, which stimulates the anemone to abnormal development early in the spring. The alternate host is *Prunus.*

Management: Infected plants should be dug up and destroyed. This rust rarely occurs on anemone if cherry and other species of *Prunus* are not grown in the vicinity.

Other Rusts

Several other species of rust infest anemone. *Tranzschelia fusca* causes a thickening and paling of the leaves and crown; brown pustules develop on the under sides of the leaves. This fungus winters in the rhizome. Other species are *Tranzschelia cohaesa, Tranzschelia suffusca, Tranzschelia tucsonensis* and *Puccinia recondita* var. *agropyri*.

Smut (*Urocytis anemones*)

Symptoms: Dark brown powdery pustules appear on swollen regions of the leaf stalks and leaf blades. They break the epidermis and scatter their spores, which are united in balls. This smut winters in the rootstalk.

Management: Diseased leaves should be taken off and placed in the trashcan. Use only sound rootstalks from disease-free plants for propagation.

Virus Diseases

A virus disease, mosaic, has been reported on Japanese anemones. Infected plants should be rouged out and destroyed.

BEGONIA (*Begonia)*

Family: Begoniaceae

Fungal Diseases

Anthracnose [*Colletotrichum dematium* (Press. ex Fr.) Grove]

Symptoms: It produces conspicuous, circular to irregular and marginal to scattered, whitish-buff necrotic lesions with blackish-tinged margins on the leaves. Diseased tissues fall off. Infection is favoured by humid cool weather. The fungus is very destructive during the rainy season.

Management: Spraying with Bavistin (0.1 per cent) or Zineb (0.2 per cent), and avoiding excessive watering help in controlling the disease.

Leaf Spot (*Colletotrichum gloeosporiodes* Penzig)

Symptoms: Small, circular necrotic lesions, whitish to pale buff with dark brown halo appear on leaves. Diseased tissues fall off.

Management: Same as for Anthracnose.

Blotch or Grey Mould (*Botrytis cinerea* Pers. ex Fr.)

Symptoms: It attacks flowers and leaves producing dark brown, irregular blotch symptoms. The disease is serious under humid and closed atmosphere. The infected portions turn brown and are covered by a greyish mouldy growth. The quality of flowers and foliage is spoiled.

Management: Removal and destruction of infected plants or plant parts, proper ventilation and circulation of air and spraying with Benomyl or Bavistin (0.1 per cent) or Zineb (0.2 per cent) help in controlling disease spread.

Powdery Mildew (*Erysiphe cichoracearum* DC. and *Oidium begoniae*)

Symptoms: It produces whitish, floury patches on leaves and flower stalks adversely affecting the plant vigour. The disease is common in shade and closed atmosphere.

Management: Spraying with Karathane (0.025 per cent) or Sulfex (0.2 per cent) and avoiding shading of plants check the disease from spreading.

Bacterial Leaf Spot (*Xanthomonas begoniae*)

Symptoms: Diseased spots are at first small and blister-like. These enlarge later and coalesce with one another forming large lesions. Serious infection results in the poor growth or death of plants. The disease is widespread.

Management: Removal and destruction of diseased plants and leaves, avoiding injury and excessive watering, spraying with copper oxychloride (0.2 per cent) or Actidione (500 ppm) and growing plants by giving proper spacing are the recommended control measures.

Damping-Off (*Rhizoctonia solani* Kuhn and *Pythium debaryanum* Hessee)

Symptoms: The disease is serious during humid weather. Seedlings and young plants are attacked resulting in damping-off or rotting of the seedlings. It also rots the leaf blades. The disease is widespread.

Management: Use of disease-free soil, judicious watering, proper spacing of plants and drenching the beds either with Captan (0.2 per cent) or Thiram (0.2 per cent) or Bordeaux mixture (1.0 per cent) help in controlling the disease.

Viral Diseases

Symptoms: *Begonia rex* is attacked by tomato spotted-wilt and Aster Yellow (mycoplasma-like organism, MLO) diseases which cause mottling, stunting and ring-spots. These diseases are not common.

Management: Diseased plants should be destroyed as soon as noticed. Spraying the plants with insecticides, such as Rogor and Nuvocron, controls thrips, thereby avoiding further transmission. Use of healthy seedlings is recommended. Cucumber-mosaic virus also infects bedding begonias resulting in mottling symptoms. It is transmitted by *Myzus persicae.*

Physiological Disease

Corky Scab

This disease, also known as oedema, appears as light brown corky growths on the undersides of the leaves and along the stems. It is thought to be due to high humidity and over-watering in cloudy weather.

CONVALLARIA (Lily of the Valley)

Convallaria majalis; Family: Liliaceae

Fungal Diseases

Stem Rot (*Botrytis paeoniae*)

Symptoms: This disease appears in early summer in the garden or propagating beds on plants 2 or 3 years old. It is characterised by arrested growth and, under very humid conditions, by the decay of the inflorescence. In the greenhouse, peduncles and blossoms are attacked as the buds open. Small round yellowish or greyish specks first appear on stalks and leaves, and later develop into dark brown sunken spots. Decayed spots appear on the stalks near the attachment of the leaves, and from them the disease spreads upward into the leaves and downward into the lower parts of the stem. The pulpy tissues of the stem are entirely destroyed so that the plant breaks over and the upper parts die. The same thing happens to infect flower stalks. The underground parts are not attacked. New growths are stunted by the lack of stored nourishment. The fungus develops most rapidly in humid greenhouses with poor aeration. It may spread over a cluster of flowers even after the plants have been marketed if they have been kept moist under paper overnight. The black sclerotia, which usually live through the winter, are formed on the lower part of the stem.

Management: Since this fungus also attacks peonies, *Convallaria* should not be grown near them; avoid working on this host after handling diseased peonies. Do not handle *Convallaria* while the plants are wet, for this spreads the fungus to clean plants. Avoid excessive dosages of high nitrogenous fertilisers. Space the plants well apart. Remove all decayed foliage in the fall. Destroy all diseased plants as soon as possible. Reduce moisture in the greenhouse and provide more air. Replace infested soil with fresh soil or steam-pasteurise the beds before replanting. For propagating, avoid the use of pips taken from diseased plants.

Anthracnose (*Gloeosporium convallariae*)

Symptoms: Circular or oval brown spots with purplish-red margins, up to more than half inch or more in diameter, may develop on the leaves. The diseased tissues drop out and expose the water-conducting vessels. Lesions occur also on leaf and flower stalks. The foliage dies prematurely. Young plants are dwarfed because of poorly developed root systems.

Management: Remove and destroy all diseased foliage in the fall. Avoid propagating from infected plantings. Change the location of the plantings. Spray repeatedly with a copper fungicide, beginning as soon as the young shoots appear. Avoid propagating with pips taken from diseased plantings.

Leaf Spots (*Kabatiella microsticta* and *Phylloticta* sp.)

Symptoms: The former causes brown spots bordered by purple-brown margins; the latter causes more or less circular, small, brown spots.

Management: If the stem rot disease is controlled, leaf spots will not be serious.

Other Diseases

A leaf blotch caused by *Ascochyta majalis,* whose perfect stage is *Mycosphaerella convallaria,* and a crown rot by *Pellicularia rolfsii* are two other fungus diseases which occur occasionally.

Physiological Disease

If conditions are abnormally moist, especially where the soil is heavy clay, plants may fail to develop properly in propagation. Blossoms open imperfectly or not at all; they soon become greyish-brown and dry up.

Avoid an excess of nitrogenous fertiliser and provide good storage conditions for the pips.

CROCUS

Other common names: Saffron, Kesar, *Crocus capsularis*; Family: Iridaceae

Bacterial Disease

Scab (*Pseudomonas marginata*)

Symptoms: This bacterial disease is important on gladiolus also. Occasionally it occurs on imported crocus stocks. On leaf surface blackish or reddish spots or specks appear which gradually enlarge and become circular, slightly raised with dark margins. The rot proceeds rapidly leading to toppling over of the plant at or near soil levels with typical necrotes. Plants also have wilted appearance, whereas bulbs are affected by spots on their surface. Yellow, water soaked, circular and shallow lesions develop on corm surface, which slowly deepen and become dark brown to black in colour. The lesions often ooze gum like exudation.

Management: Avoid overhead watering to the plants. Soak cormels and corms in 0.5 per cent formaldehyde (38 per cent Formalin) for 2 hours, then in 200 ppm Streptomycin for 2 hours just before planting.

A mixture containing 60 per cent Thiram and 15 per cent Dieldrin dusted on the corm prior to planting gives excellent control. Dipping of corms in 0.2 per cent suspension of Thiram for 15 minutes also checked the infection.

Corm treatment with 0.1 per cent phenyl mercury acetate for 30 minutes combined with a one year crop rotation, which includes cultivation of resistant plants or a two-month fallow during July and August, is also helpful.

Fungal Diseases

Dry or Neck Rot (*Stromatinia gladioli, Sclerotinia gladioli*)

Symptoms: It is also known as root rot caused by *Stromatinia gladioli* or *Sclerotinia gladioli* and is seen on the stored corms as small, dark, more or less superficial spots or lesions which can also produce collar rot, killing the plants or its delayed attack may only harm the new corms for carrying the disease to the next season. Dry rot is a widespread disease attacking gladioli plants in the field and corms in storage. It is more severe during humid conditions in the field where premature yellowing occurs, ultimately causing death of the plants. The leaves turn brown from the tips downwards and at the leaves they carry neck rot, but corms remain attached to the stem firmly. Later, the sheaths become shredded and covered with numerous round and black resting bodies just large enough to see. These tiny bodies and the absence of grey spore masses differentiate this disease from Botrytis stem rot. Newly rotted neck tissues are yellow brown with mouldy odour. Diseased corms show numerous round, black and small lesions (first reddish brown) varying from merely a dot to 0.6 cm in diameter with slightly raised edges which may coalesce forming an irregular area. Often the small lesions are found in the ring along the leaf scars and then black resting bodies are seen on the decayed leaves and tunics. These spots on the corms are more or less circular and sunken but can be distinguished from scab since dry rot does not produce shining or gummy exudation. In moist storage conditions, these black lesions run together to make the corms mummified but in the dry season it is not so serious. The tunics of these infected corms look snuffy and turn brittle. This is a soil-borne fungus and its longevity in the soil has been found to be 10 years or so.

Management: Corms may be cured by soaking them in cold water for 24 hours and then for 30 minutes in hot water at 54.5° C.

Dip corms in Dicholoran or Dichofluanid (300 g a.i./100 kg corms in 100 litres of water) and spray it in the furrows before planting of corms.

Dip the corms in Procymidone (aucyline) at 0.3 per cent or treat the soil at 0.5-r/m in furrows before planting.

Storage Rot (*Penicillium gladioli*)

Symptoms: In storage, the corms are severely affected due to the attack of certain storage fungi. Infection usually starts at places where the husks are adherent to the corm. A number of infection rings are formed leading to corm decay. Ultimately, the lesions are sunken and reddish brown, covered with greenish mycelial growth. Numerous light brown minute sclerotia appear on the diseased tissues, which help the fungus to perpetuate. Ultimately, the corm is mummified.

Management: Corms should be cured properly and handled carefully to avoid injury. Damp storage conditions should be avoided.

Curing of corm at 85° F for 8-10 days.

Treat corm/corms with Vitavax 300 at 0.1-1 per cent or treat corm/cormels with mercuric compound (inorganic and organic), namely Ceresan wet at 0.1 per cent solution for half an hour and thereafter dry them before planting.

CYCLAMEN (Sowbroad)

Cyclamen persicum giganteum; Family: Primulaceae

Bacterial Disease

Tuber Rot (*Erwinia carotovora*)

The causal bacterium enters the cyclamen tuber via wounds and causes a soft, wet decay.

Management: Avoid wounding of tubers. Plant them in steam-pasteurised soil.

Fungal Diseases

Blight (*Botrytis cinerea*)

Symptoms: Buds and leaves may be rotted and flower petals spotted by this fungus. Another fungus, *Glomerella cingulata*, causes leaf and bud blight.

Management: Avoid overcrowding of plants. Pick off and destroy spotted leaves as soon as noticed. Mist-spray periodically with Benlate.

Stunt (*Ramularia cyclaminicola*)

Symptoms: Plants are stunted with the flowers remaining below the small, sometimes yellow, leaf blades. Reddish-brown necrotic areas may be seen when infected tubers are sliced open. The fungus is systemic, invading the shoots and roots.

Management: Discard diseased plants. Start seed from healthy plants in new or pasteurised soil.

DAHLIA (*Dahlia variabilis*)

Family: Compositae

Fungal Diseases

Powdery Mildew (*Erysiphe cichoracearum* DC)

Symptoms: The leaves and young stem at first develop circular spots, which gradually increase in size, and ultimately the entire foliage is covered with a white powdery growth. Generally warm humid conditions favour the growth of this disease.

Management: To check the disease, the plants should be dusted fortnightly with sulphur dust, or sprayed with wettable sulphur, Dinocap, Benomyl and Carbendazim (Pernez et al, 1994).

Stem Rot [*Sclerotinia sclerotiorum* (lib.) de Bary and *Sclerotium rolfsii* Sacc.]

Symptoms: The disease is soil borne in nature. The stem rot fungus attacks the main stem and branches near the base of the plant. A white mould may completely encircle the stem. Plants growing in heavy wet soil infected with this fungus ultimately wilt and die (Chaudhary *et al*, 1982; Jones, 1991).

Management: To avoid the occurrence of this disease it is advisable to change the growing area. Wide spacing and good drainage will help to keep the disease in check. Affected plants should be uprooted and burnt. For *Sclerotinia rolfsii* control, the tubers are treated in Benomyl (0.1 per cent) for 30 minutes after harvest and before planting. Drenching of soil with Quintozene (0.2 per cent) is recommended to reduce the infection.

Wilt (*Verticillum alboatrum* and *Fusarium* sp.)

Symptoms: The plants wilt and die because of the blockage or destruction of water-conducting vessels, or death of the living tissue due to toxins produced by the fungus. Wilted plants should be destroyed and only healthy tubers should be used. The only characteristic symptom is a brown or black discoloration seen in the conducting elements of the stem. This discoloration can be traced both upwards in the plant and downwards towards the roots but if it is confined to the basal parts it is more likely to be caused by root damage than by a wilt.

Management: Wilt diseases are very difficult to tackle in gardens. When plants are lightly infected, pack the bases of the stems with damp peat, which may promote healthy new root growth. After removal of diseased plants, thoroughly disinfect greenhouse or garden soil, all pots, benches, tools and raise the crop in sterilised compost.

Smut (*Entyloma calandulae* f. sp. *dahliae*)

Symptoms: Circular or elliptical pale yellow-brown spots arise on the leaves, normally after seedling stage, which later enlarge, darken and merge but usually retain a yellow margin. Darker areas within the lesions can sometimes be seen in a strong light. It appears first on the lower leaves, then spreads upwards and large areas of leaf tissue may ultimately die.

Management: Burn diseased foliage and remove any leaves adhering to the tubers before storing. Do not grow dahlias in contaminated soil for at least five years. Three or four fortnightly sprays of Captan (0.2 per cent) may check the disease on growing plants, if applied early in the season.

Bacterial Diseases

Crown Gall (*Agrobacterium tumefaciens*)

Symptoms: Usually irregular but sometimes more or less spherical, knotty swellings can arise on almost any part of the affected plants but occur most typically at the junction of root and stem. Infected plants become stunted and the shoots spindly.

Management: Dipping roots at time of planting in Streptomycin sulphate solution can prevent the disease. Roots and crowns of plant with tumours should be destroyed and changing the location of planting will also help to manage the disease.

Bacterial Wilt (*Pseudomonas solanacearum*)

Symptoms: Infected plants usually droop and wilt rather suddenly. The conducting system will be found plugged with yellowish masses of bacteria, which ooze out when the stems are cut. The wet soft rot of the stems near the soil is characteristic and distinguishes this wilt from that caused by the *Verticillium* and the *Fusarium* wilt fungi.

Management: Destroy all wilted plants; be sure that no infected parts are thrown into the compost pile. This bacterium, like the wilt fungi, winters in the soil. Heavily infested soil must be pasteurised with steam. The best way to avoid this disease is to practise rotation, as the farmer does to avoid soil-borne diseases.

Bacteriosis (*Erwinia carotovora* var. *carotovora*)

Symptoms: This disease appears as a browning and softening of the stem. The layers of pith that line the hollow stem are moist and blackish with the rot, which extends into the bark. If one uses a compound microscope, great masses of the large bacteria can be seen swimming about. The rot gives off a foul odour. This disease also affects the tubers.

Control: Special methods for the control of this disease have not as yet been found. Destroy the affected parts of the plant so as to prevent the spread.

Mosaic

Symptoms: Mosaic caused by *Marmor dahliae* is one of the serious diseases reported for dahlias. The leaves become mottled and develop pale green bands along the midribs and larger secondary veins. Yellowish or pale green vein banding is the most common symptom. It is also accompanied with twisting or crumpling of leaves. It is transmitted by aphids, most notably *Myzus persicae*.

Management: Growing of cuttings in closed environment. Careful selection and multiplication from symptom-less plants is essential.

Spotted Wilt

Symptoms: The common symptoms are mottling of leaves, stunting and ring spot symptoms. Disease spreads by thrips.

Management: Spraying of 0.1 per cent Malathion or Metacid to control insect vectors is recommended. Using virus-free tubers or cuttings can also check the disease.

FREESIA (*Freesia refracta*)

Family: Iridaceae

Fungal Diseases

Wilt and Corm Rot (*Fusarium oxysporum* Schlecht.)

Symptoms: Infected plants become yellow, then gradually wilt, die and collapse. The fungus invades the main roots, the laterals and the feeding rootlets, which turn pink and then dark brown on rooting. From these the infection spreads to the corms. The lesions on the corms are deep pink. The tissues of the diseased corms first become yellowish, and as the decay progresses, they turn dark brown. In the advanced stage, the vascular bundles turn brown surrounded by a pink discoloration. Ultimately, the affected tissues become chalky-white, and the dried corm is a hard, dry, friable mass, which crumbles under light pressure. The disease is not common.

Management: Use of disease-free soil and healthy planting material, following 3-4 years rotation, and dipping the corms in Benomyl or Bavistin (0.1 per cent) for half and hour before planting are recommended.

GERBERA (Transvaal Daisy)

Gerbera jamesonii; Family: Compositae

Fungal Diseases

Foot Rot and Root Rot (*Phytophthora* de Bary; *Pythium* Pringsh; *Rhizoctonia solani* Kuhn)

Symptoms: These types of maladies are caused by soil borne pathogens either singly or in combination. The plants are killed in the nursery stage known as damping-off and also at various stages of crop growth. These pathogens are favoured by water-logged conditions and high humidity.

Management: Seed treatment with Captan, use of healthy seeds so as to get healthy nursery, avoidance of water-logged conditions, good sanitation and proper application of fertilisers are general control measures.

Anthracnose (*Colletotrichum gloeosporiodes* Penz.)

Symptoms: It produces circular, scattered, reddish-brown spots, which coalesce with one another in moist weather and involve large areas, thereby resulting in withering, rolling and drying of leaves.

Management: Spraying with Bavistin (0.1 per cent) gives effective control. Excessive watering or overcrowding of plants should be avoided. Infected debris should be burnt in time.

Blossom Blight (*Botrytis cinerea*)

Symptoms: In cool moist weather or prolonged drizzling rains the disease is encouraged. Light brown irregular water soaked areas appear on flower stalks, which enlarge and coalesce producing distinct depressed lesions. Ultimately the infection spreads on the entire flower. The stalk of the base is girdled which causes drooping and death of plants.

Management: Deep planting, bad drainage and poor ventilation predispose the plants to infection and should be avoided. Infected plants should be removed and destroyed. Seed treatment with 0.1 per cent Benlate or 0.2 per cent Thiram helps in preventing the disease.

Powdery Mildew (*Erysiphae cichoracearum* DC and *Oidium erysiphoides* f. sp. *erysiphoides* Link ex. Fr.)

Symptoms: The affected parts are covered with white, floury, mycelial growth. The diseased areas get dried and fall.

Management: Spraying with Bavistin (0.1 per cent), Sulfex (0.3 per cent), Karathane (0.5 per cent), effectively controls powdery mildew.

Bacterial Blight

Symptoms: The disease is characterised by small to large circular or irregular, brownish black leaf spots with or without concentric rings.

Management: Use of Spreptocycline (0.01 per cent) was found effective.

Tobacco Rattle Virus

Symptoms: Characterised by yellow or black annolated leaf spots on the foliage.

Management: Soil sterilisation to control the nematode vectors, for example *Trichodorus* sp., and application of Aldicarb or Furadon are recommended.

GLADIOLUS

Other common name: Sword Lily; Family: Iridaceae

Fungal Diseases

Fusarium Wilt, Yellows, Corm Rot [*Fusarium oxysporum* f. sp. *gladioli* (Massee) Snyder & Hansen]

Symptoms: The characteristic symptom is inter-veinal leaf tip yellowing, which extends down the leaf and the whole leaf gradually turns brown and becomes narrow. The other common symptoms include stunting, curving, arching and bending of the leaves. On advancement of infection, the plant suddenly wilts or turns yellow and dies prematurely. The leaf infection is usually basal and associated with corm rot.

Roots arising from corms may show brown lesions. Corm rot in storage is distinct, appears dry and is very often restricted to the corm base. But in the field it is characterised by soft rot, invaded by other organisms and frequently associated with brown, wiry roots. The centre of the bulbs turns black and rots completely. Lesions on corms are reddish-brown with well-defined margins, round to oval, somewhat depressed leading to hard shrunken mummified corms. Singh (1969) reported this disease from India. The disease has also been reported from Pakistan, Italy, Germany, Russia and China (Mirza and Shakir, 1991; Infantino and Rumino, 1993; Chen *et al,* 1994). Heavy crop loss to the tune of 30 per cent was estimated in Germany and 60 to 80 per cent in Russia (Bruhn, 1955; Protsenko, 1958; Vlasova and Shitan, 1974).

Management

Resistant Varieties: Cultivars like Albana, Apricot Glow, Souvenir, Hoppman's Glory, Sylvia, Dhiraj, White Friendship and White Prosperity are reported to be resistant while Australian Fair and Mansore are tolerant to the disease (McCullough, 1944; Georgia and Peikova, 1976; Sezgin *et al*, 1983; Chandra *et al*, 1985; Bajaj *et al*, 1989). Use of resistant cultivars together with soil drench of Carbendazim 4 days after planting and three times at 10-day intervals is recommended (IARI, New Delhi). In gladiolus, out of 11 cultivars tested against the disease, the cultivars 'IARI Set-1', 'Psittacinus Hylo' and 'Sueliton' were found resistant (M.P.K.V., Pune) (Sharma and Bhattacharjee, 2002).

Cultural Practices: Crop rotation, sanitation, early lifting and proper curing of corms at 29.5-30° C for about one week and hot water dips of the corms reduce the disease. Diseased corms should be discarded, infected and late blooming plants should by rouged from the fields as early as possible. The cormels floating in water before the hot water treatment should also be discarded and removal of corm scales (cataphylls) (Woltz *et al*, 1978; Magie and Poe, 1972; Katan, 1980). Adjusting the soil pH to 6.7-7.0 by using nitrogen 80-90 per cent from nitrate sources has great impact on lowering the inoculum potential (Groen, 1990). Adequate aeration in storage, raised beds, improved water drainage and incorporation of organic mulches in soil helps in preventing corm rot.

Hot Water Treatment: Hot water treatment of cormels at 53-55° C for 30 minutes + a soak or dry dressing with Sumislox (promycidone) is helpful. The daughter bulbs also carry slight to moderate infection from mother bulbs (Harrison, 1962; Grimal'skya, 1978). The cormels are more tolerant to hot water treatment than corms (Magie, 1968). Gladiolus bulbs treated for 1 hour. at 50° C have been reported to be effective by Slogtern *et al*, 1974. The combined effect of heat and fungicides has been found to be effective for gladiolus corms by Valaskova (1977).

Use of Chemicals: Dipping of corms in one per cent potassium permanganate before storage or steeping in 0.2 per cent solution for four hours and dusting with Captan gives good control. Captan, Chlorothalonil, Benomyl, Thiabendazole and Zineb also provide effective and excellent protection. For *Fusarium* wilt and rot of corms of gladiolus, the AICRP centres of Pune, Ludhiana and Srinagar had reported that pre-planting and pre-storage treatment of gladiolus corms with Carbendazim (Bavistin) 0.1 per cent and Thiram (0.3 per cent) among systemic and Mancozeb among non-systemic fungicides were effective.

Pre-sowing dip treatment of corms with Emisan-6 (0.2 per cent) or dipping the corms and cormels twice, that is, before storage and before planting for 15 minutes in the fungicidal solution followed by drenching twice at monthly intervals with Captan 0.3 per cent or Thiram 0.3 per cent was effective against *Fusarium* wilt of gladiolus (M.P.K.V., Pune and P.A.U., Ludhiana). (Sharma and Bhattacharjee, 2002).

Scale and Neck Rot (*Pseudomonas gladioli*)

Symptoms: The infection is first noticed as small, rounded, red-brown spots on the leaves which later darken. The basal portion of the leaf may rot. Bacterial ooze comes out through craters. The bacterium perpetuates in the soil and can also attack Freesia, Crocus and other related plants.

Management: Use of healthy planting material, following 2-3 years rotation, and rouging out and destroying diseased plants are remedial measures.

Storage Rot

The fungus *Penicillium gladioli* is the most common cause of this disease. Other fungi reported to be responsible for storage rot are species of *Aspergillus, Rhizopus, Alternaria*, etc. Infection occurs through injuries.

Symptoms: The disease mostly appears in the form of black, brown, greenish or yellowish mouldy growth or the corms may rot and emit a foul smell.

Management: As infection occurs through injuries, wounding of corms at the time of digging or during handing should be avoided. The corm dip treatment and the curing technique as described under 'Wilt' should be followed. Damp storage conditions should also be avoided. High temperature (more than 5° C in the cold store) also needs to be avoided, as it could lead to rapid rotting of corms under humid conditions.

For the control of storage rot of gladiolus corms, it was recommended by M.P.K.V., Pune Centre, to dip the corms in Mancozeb 0.2 per cent, or Thiram 0.3 per cent or Benomyl at 0.2 per cent for 15 minutes after harvest, dry them in shade and store in cold storage (Sharma and Bhattacharjee, 2002).

Dry or Neck Rot [*Stromatinia gladioli* or *Sclerotinia gladioli* (Drayt.) Whetz.]

Symptoms: It is also known as root rot, dry rot or neck rot and its prevalence has been recorded from Argentina, Greece, Portugal, the Netherlands, South Africa and many other countries (Alippe, 1966; Grigariu, 1976; Dias *et al*, 1981; Groen, 1981; Darvas, 1984). The disease was first recorded by Drayton (1928) who, in 1971, renamed it as *Stromatinia gladioli* on the base of sclerotial characters. The disease is seen on the stored corms as small, dark, more or less superficial spots or lesions which can also produce collar rot, killing the plants or its delayed attack may only harm the new corm for carrying the disease to the next season. Dry rot is a widespread disease attacking gladioli plants in the field and corms in storage. It is more severe during humid conditions in the field where premature yellowing occurs, ultimately causing death of the plants. The leaves turn brown from the tips, dip downwards at the leaves they carry (neck rot), but corms remain attached to the stem firmly. Later, sheaths become shredded and covered with numerous round and black resting bodies just large enough to see. These tiny bodies and the absence of grey spore masses differentiate this disease from the *Botrytis* stem rot. Newly rotted neck tissues are yellow brown with mouldy odour. Diseased corms show numerous round black and small lesions (first reddish brown) varying from merely a dot to 0.6 cm in diameter with slightly raised edges (Magie, 1957) which may coalesce forming an irregular area. Often the small lesions are found in the ring along the leaf scars and then black resting bodies are seen on the decayed leaves and tunics. These spots on the corms are more or less circular and sunken but can be distinguished from scab since dry rot does not produce shining or gummy exudation. In moist storage conditions, these black lesions run together to make the corms mummified but in dry season it is not so serious. The tunics of these infected corms look stuffy and turn brittle. This is a soil borne fungus and its longevity in the soil has been found to be 10 years or so.

Management: Corms can be cured by dipping them first in cold water for 24 hours and then for 30 minutes in hot water at 54.5° C. Use of 0.1 per cent potassium permanganate or 0.2 per cent Bavistin for 20 minutes and spraying with copper solution twice or thrice during the growing season at fortnightly intervals has been recommended. Dipping corms in Dicloran (300 g a.i./100 kg corms in 100 litres of water) and spraying in furrows before planting of corm is useful.

DISEASES OF LEAF AND FLOWER

Botrytis blight and Flower rot (*Botrytis gladioli* Timmermanns)

Symptoms: It is destructive to leaves and flowers. Flower shippers sustain the greatest losses when flowers develop Botrytis rot after packing due to incipient infections.

Botrytis survives in corms, soil and in plant debris. Disease epidemics erupt during cool, wet weather conditions. Leaves of gladioli are most susceptible to infection after frost injury. Sclerotia can be formed on all plants as well as on the corms. They have a dark colour, are flat and of valuable size (1-9 mm), and employ a part in spreading the disease in a flowering season under and above the ground. They can produce spores, which are spread by the winds. The corm infections arise especially under highly moist conditions during the storage of the corms. The other leaf infections manifest themselves after prolonged wet periods in which the crop remains moist too long. Such conditions especially prevail in a dense crop and in stuffy greenhouses.

The lightest symptoms on the corm consist of little black-brown spots which, as a rule, are found mainly at the top of the corm around the stalk base. Brown stains of variable shape and size may also occur on the surface. In case of severe infection, the corms become soft all over and reddish brown in colour. Sometimes the outside shows white fungus threads (mould) and black sclerotia. When the corms dry up, they will shrivel and become hard. The infections of the plant above the ground can occur at the soil level, where the tissue is brown and rotted, so that the plant becomes yellow and sometimes topples. The disease below ground tissue reveals large black sclerotia. Otherwise, the corm will be sound and well rooted. The symptoms on the leaves consist initially of light, afterwards of dark brown round spots. Later in the season, large and dead grey brown patches appear on the tissues; germinating spores may cause colourless watery spots on the flower.

Management

- Keep the relative humidity low (maximum 80 per cent) and maintain good air circulation.
- Do not plant too densely.
- Remove clearly infected plants with their corms.
- When growing in greenhouses, ventilate well and if necessary, dry the crop by heating water in the morning so that the crop will be dry before nightfall.
- After harvesting the flowers, remove the remaining plants and corms as soon as possible.
- Effective fungicides: Vinlozolin, Iprodione and Benlate. Sprays of dithiocarbamate fungicides, Daconil, Captan and Dichloran have been used with success (Holloman and Young, 1951; Nagler, 1965). Cyprex and Mancozeb are applied in checking the disease of flowers.

Leaf Spots (*Curvularia trifolii* f sp. *gladioli* Parmela and Luttrell)

Symptoms: Brown oval spots on petals, young leaves with small clumps of black spores near centre of larger spots are formed.

This is also a soil-borne disease pathogen. It is destructive to cormel planting and may develop on flowers during warm humid weather. The disease attacks young leaves, survives in corms and in soil and is carried on seed. Infection from infested soil causes rot of leaves below soil. The disease acts as a 'damping-off' of young cormel and seed-linked plants. Small plants may be killed without any sign of aboveground infection.

Management: Maneb is effective in protecting leaves and stem against *Botrytis*, but open flowers and the bases of buds are not well protected. Maneb is less effective against *Curvularia* but is the best fungicide for both diseases. (Maneb + zinc ion formulations) are sold under various labels such as Manzate 200 and Dithane M-45. Maneb spray should be applied 1-3 times per week, the number of applications depending upon the weather/season.

As soon as the disease appears, sprays should be applied every other day, especially during wet weather. Critical periods for infection with *Botrytis* are during cool weather after freeze or frost injury; the critical period for *Curvularia* disease is during warm weather. It is important to spray regularly before a freeze or frost occurs, whether *Botrytis* is present or not. Both diseases are more difficult to control after infection is established.

Stemphylium Leaf Spots (*Stemphylium* Wallr.)

Symptoms: The fungus attacks mature leaves of certain cultivars in cool and warm weather. Although leaf spots caused by this fungus may be mistaken for those caused by *Botrytis*, they may be distinguished by size. *Botrytis* spots vary greatly in size, the smallest being visible only on the upper leaf surface. *Stemphylium* spots are usually more uniform and visible on both sides of the leaf.

Management: Spray of Maneb or Dyrene (0.2 per cent) controls the disease.

Leaf Spots (*Septoria gladioli* Pass.)

Symptoms: It is of minor importance and is easily controlled. Small, black spore bodies may be seen near the centres of older spots.

Management: Weekly spraying of Maneb (0.2 per cent) controls the disease.

Bacterial Leaf Spots (*Pseudomonas marginata, Xanthomonas germisudans*)

Symptoms: They may damage leaves in warm rainy weather. Plants that are crowded, especially those from cormels in planting stokes, are most affected. Spraying does not protect very well; sanitation is helpful in reducing spread of bacteria. Overhead irrigation, use a fine spray to avoid the large drops, which drive bacteria into leaf pores. Irrigation early in the morning before light causes the leaf pores or stomata to open.

Management: Cut flowers for use or break out old flower heads only when foliage is dry, also avoid working among plants that are wet. If a cutting instrument is used, sterilise it, generally in 70 per cent alcohol. Avoid use of excess N fertiliser, which results in lush growth. Corms to be dipped in mercuric chloride (1:1000) before planting and storage.

Viral Diseases

Several viruses are known to infect gladiolus. Most of them are transmitted by aphids or nematodes. Aphids transmit mostly mosaic type viruses (for example, cucumber mosaic and bean yellow mosaic virus), whereas soil-living microorganisms the ring spot viruses (for example, tobacco ring spot and tomato ring spot virus). Aster Yellows is reported to be transmitted by leaf hoppers.

Management: Use of virus-free planting material is required to raise healthy plants. Meristem tip culture is known to eliminate most viruses from infected plants and hence of value where whole of the planting stock is infected. Healthy planting stock must be maintained under insect-proof structures or by controlling insect vectors with suitable insecticides. Genetic resistance of the host to the major viruses is a long-term approach to the problem.

Non-parasitic disease symptoms are caused by unfavourable weather conditions, faulty nutrition, injurious chemicals and polluted air. A bud sheath burn or crooked stems may result when the young spike is under a water stem and wilting. A wilting exposes the young bud sheaths to the sun and wind, causing brown tips. Older leaves exposed directly to the sunlight may have small rusty brown spots due to the action of the sun, or water drops over a film of spray chemicals. The bright morning sunlight causes yellow horizontal bands on young leaves and on the short sheathing leaves. This occurs most often on moderately cool mornings on newly exposed leaf tissues near the soil surface. The bends often indicate the daily growth of the leaves. The yellow tissue gradually turns green in most cases.

Holding the cut flowers in water containing one or more parts per million of fluorides causes break down and discoloration of petal edges. The mineral content of some well water may also injure the florets. Petal edges may also be injured by dusts of certain chlorinated hydrocarbon pesticides applied to buds just before openings. Physiological flower bud rot and toppling of the flower heads were found to be caused by calcium deficiency.

Leaves and flowers may be injured by spray chemical herbicides and air pollution. Gladioli are extremely sensitive to certain fluorine compounds whether present in the air, oil or dusted on plants. Fluorinated drinking water contains enough fluoride to damage florets opened from tight buds on cut spikes. The insecticide Cryolite or Kryolide, is a fluoride, which causes severe leaf, tip burn and scorch when sprayed or dusted on gladiolus. Phosphate fertilisers normally contain a fluoride band when applied to acid soils (below pH 5); the fluorides may be absorbed by the roots and cause sleeve leaf scorch. Other factors causing leaf tip burn include roots injuries due to flooding, excessive fertilisers and nematodes.

Tip Burn

It is believed that it occurs due to high levels of aerial fluorides in the atmosphere.

Symptoms: This physiological disorder results in discoloration and drying up of tips of leaves of gladiolus. Some of the saprophytic fungi also start invading the necrotised tissue.

Management: It is recommended that a spray of Blitox 50 WP (0.2 per cent) should be given at initiation of symptoms and repeated once more if required. There is not much to worry about this malady, as it does not cause serious economic loss to the grower.

HYACINTH (*Hyacinthus orientalis*)

Family: Liliaceae

Fungal Diseases

Black Slime [*Sclerotinia bulborum* (walker) Rehmn]

Symptoms: The leaves can be easily pulled out from decayed bases of the bulbs. The bulbs rot from the outer scales first and slowly the infection spreads to the inner fleshy scales. They turn dark grey in colour and remain attached to the root plates. Under moist weather, the white fluffy mycelium of the fungus along with sclerotia formation starts appearing on the bulbs. As a result, the heavily infected bulbs shrivel into a black dry mess, which disintegrates teh bulbs completely.

Management: Cultural remedies like removal of infected plants from the field and infested soil is recommended. Sowing of diseased bulbs should be avoided. Infection is sometimes prevented by disinfecting bulbs for 15 minutes prior to planting in a mixture of 0.5 per cent phenol derivative and 1.5 per cent Zineb, Maneb or in a phenol derivative plus Captafol that resulted in slightly higher yield and better disease control than disinfecting in mercury compounds.

Rust (*Uromyces muscari* var. *hyacinthi*)

Symptoms: This fungus occasionally attacks hyacinth leaves, bursting out in masses of dark brown teliospores.

Management: Thus far the disease has not proved to be serious enough to warrant control measures.

Bacterial Diseases

Soft Rot (*Erwinia carotovora* var. *carotovora* (Jones) and *Erwinia rhapontici* (millard) Burkholder)

Symptoms: The leaf tips turn yellow, shrivelled and plant growth stops at flowering time; affected flowers open irregularly and fall off. The bulb scales near the soil level usually rot leading the inflorescence to topple over. Rotting progresses rapidly to leaves, scales and reaches the bulbs, which appear soft, dirty white and emit foul smell within a few days after infection. The rooting may be generated both ways, either nose downward or from the root plate upward (Sellwood and Lelliot, 1978).

Management: Destruction of the infected bulbs and sterilisation of the soil with Fundazol, Cuprocin and Orthophaltan prior to planting gave good control. The application of an anti-microbial substance, Decamethoxin, has been reported to be effective. Growing of resistant or fairly resistant varieties such as Bismarc, Delft, Blau, Miozotis, Amsterdam, and Queen of the Pinks is recommended (Bakrdzhiev, 1979). Destruction of the infected bulbs and sterilisation of the soil with Fundazol, Cuprocin and Orthophaltan prior to planting gave good control. The application of an antimicrobial substance, Decamethoxin, was also reported to be effective (Kabsna *et al,* 1975).

Yellow Disease [*Xanthomonas hyacinthi* (Wakker) Vauterin *et al*]

Symptoms: Rotting of the bulbs either before or soon after planting results in diseased plants, which fail to flower properly. Leaves which get infected via natural openings or wounding, produce long, water soaked dark stripes. The outer leaves often develop dark edges. Bacteria, if it reaches the root plate, disperses rapidly in the vascular system throughout the bulb. When cut open, large number of small, scattered yellow spots become visible, which are arranged in concentric circles. These spots, on squeezing, ooze out a yellow slimy liquid containing bacteria, which enter into the scales from the foliage which is initially confined to the vascular strand of the leaves and the scales.

Management: Heavily damaged bulbs during storage should be removed. Keep bulbs for 2 weeks at 30° C and then for 3 days at 44° C after storing them for 4 to 6 weeks at 30° C. The temperature should be raised to 44° C as early as possible. The protection induced in this way may last for three years. Microwave heating and new chemical treatments have also advised against yellow disease of hyacinth (Boer, 1972). Selection of resistant varieties is a must to eliminate infection (Ejik *et al*, 1976).

Viral Diseases

Mosaic (Ornithogalum mosaic virus): Leaves show blue-green mottling, streaking, and blotching, often accompanied by reduction and withering of flowers. The virus has been transferred from hyacinth to *Ornithogalum thyrsoides* by the potato aphid, *Macrosiphum euphorbiae*. Bulb plants belonging to the genera Lachenalia and Galtonia are also susceptible to this disease.

Management: Same as for mosaic of *Iris*.

IRIS (Iris, several species)

Other common name: Flag; Family: Iridaceae

Fungal Diseases

Bulb Rot or Blue Mould *(Penicillium corymbiferum* Westl).

Symptoms: The disease is chiefly manifested in storage. It spreads either from the base or from relatively large lesions at the site of the bulb. Symptoms are quite variable on bulbs, the fleshy scales beneath the tunic develop white to greenish grey rot, covering the whole bulb. The tunic itself turns hard and chalky. Plants emerging from infected bulbs may get stunted with limited root growth, turn chlorotic, lack flowers and die prematurely.

The fungus gets associated with the musk rot at the leaf and stem bases. Pathogens often enter through wounds created by root initiation process or by any other agency in heavily contaminated soils. High relative humidity and 17° C temperature results in heavy infection within 3 days in wounded bulbs. Ethylene or Ethephone treatment used for early forcing performance in iris cultivars does not predispose iris bulbs to rots. Prolonged cool, moist conditions, early and very late digging, predispose bulbs for rapid infection. Environmental factors like warm temperature and high humidity, on the other hand, decreases the disease severity. Moist weather favours disease spread and if bulbous irises are left in the ground for two consecutive years, the disease potential is increased to a greater extent (Ferrari, 1974).

Management: The disease can be controlled by preventing mechanical damage to the bulbs. Manipulation of storage temperature and humidity requirements and adequate ventilation during storage is essential to keep bulbs healthy. A very early or late digging and planting of diseased bulbs should be avoided because early lifting leads to more than 30 per cent loss of yield. Sometimes fungicidal dips with Thiram and Benomyl prior to planting are suggested. Sterilisation of the bulbs in one per cent Formalin before forcing, reduced fungal infection (6 per cent) as compared to 13 per cent infection in untreated lots. Late planting, 3 to 4 years crop rotation and cultural methods to promote good air flow are desirable for commercial production of bulbous irises (Pirone, 1978).

Crown Rot (*Sclerotium rolfsii* Sacc. and *Sclerotium delphinii* Weleh)

Symptoms: The pathogen attacks the base of the leaves and flower stalks and induces rot that causes the leaves to die back. The crown portion is coated with a white fan-like growth of the fungal mycelium containing numerous sclerotial bodies. Warm and moist storage conditions are favourable for spread of the pathogen in storage. The disease is common on rhizomatous iris.

Management: Since the disease is soil-borne, the control of the disease is difficult and expensive. The infected stocks should be discarded. Removal of infected plants and contaminated soil surrounding the diseased bulb up to a distance of a foot helps in managing the disease spread. Early digging of bulbs and sun drying immediately before storage is effective. In addition, deep ploughing and three to four years' rotation is helpful in lowering the inoculum. Chloropicrin as a soil fumigant and Terrachlor (PCNB) or Quintozene as dust applicant have been recommended. Application of *Trichoderma harzianum* either as a bulb coating or broadcast treating soil with Quintozene or soil solarisation prior to planting are reported to be effective in reducing the incidence up to 93 per cent and increasing the yield by 35 to 41 per cent.

Ink Disease [*Drechslera iridis* (Oud) Iris]

Symptoms: The initial symptoms begin as minute spots and streaks and enlarge to dark reddish brown lesions with brown to black masses of spores. Infected leaves, stems and flowers collapse and appear blighted. Infected debris and bulbs, to a limited extent, act as a source of primary infection.

Management: To eliminate infection, burning of straw, removing diseased bulbs and older leaves, weeding operation, deep ploughing, 3-4 year rotation, raking up and hot water plus Formalin treatment of bulbs are essential to achieve a good control of the disease. Regular spraying during leaf spotting phase with dithiocarbamate fungicides is helpful in reducing the severity of crop infection.

Leaf Spot [*Mycospharella macrospora* (Bleb.) Jousted]

Symptoms: Infection is visible only on leaves of rhizomatous irises but in bulbous iris leaves, stem and flowers are infected. Early diagnosis can be made by looking for the first symptom that appears on both surfaces of the leaf as minute brown spots surrounded by a dark green water soaked halo, which later turns yellow. Lesions enlarge and assume an oval shape bordered by a red brown zone and greyish centre.

Management: To prevent infection, removal and destruction of the older leaves prior to spring has been recommended. Spraying with Mancozeb, Bordeaux mixture and Zineb is quite effective. Late planting, three to four years' crop rotation and cultural methods to promote good airflow are desirable for commercial production of bulbous irises.

Bacterial Diseases

Soft Rot [*Erwinia carotovora* sub. sp. *carotovora* (Jones) Bergey *et al*]

Symptoms: The soaked lesions on the leaf blades develop into large irregular spots. With the progress of the disease, sheaths and rhizomes rot completely and result in shoot collapse. The affected rhizomes also exhibit a wet, soft decay and emit foul smell. Bacteria gain entry into the plant through wounds and insect injury. The pathogen is gram negative and soil borne in nature.

Management: Removal and destruction of crop debris can check the spread. The infected areas on the rhizomes can be removed with the help of a knife but the knife should be disinfected each time by dipping or wiping it with 70 per cent denatured alcohol. Shallow planting, careful handling is advisable while shady places or poorly drained sites should be avoided. The control of iris borer is also necessary to eliminate the soft rot disease.

LILIUM (Lily)

(*Lilium* species and hybrids); Family: Liliaceae

Fungal Diseases

Foot Rot (*Phytophthora cactorum*)

Symptoms: Infected plants turn violet brown at the base of the stem and then the infection moves upwards. The leaves become yellow acropetally. The severely infested plants fall over due to the collapse of the stem.

Management: Good drainage is very important for prevention of the disease. Cultivation in wet weather should be avoided. The affected plants should be removed from the site and destroyed. Hot water treatment of bulbs for 2 hours at 39° C followed by a 30 minute dip in 0.2 per cent Benomyl eliminates infection and stimulates plant growth.

Fusarium Scale Rot (*Fusarium oxysporum* Schlechtend ex. fr. f. sp. *Liliae*)

Symptoms: The initial characteristic symptoms are foliage yellowing and wilting. Although the bulbs of these plants may appear healthy, the roots develop a reddish coloured decay at their tips. The plants remain stunted with yellow foliage and extensively rotted scales. In early infection, the scales develop brown to reddish necrotic areas which gradually enlarge; the infected scales later fall away or shatter from the basal plate.

Management: Some cultivars of *Lilium hansoni, Lilium sargentiae* and *Lilium maximoviczii* have been reported resistant. Disease-resistant plants to *Fusarium* by using fusaric acid as a selective agent on pollens of various lily cultivars of known resistance to the pathogen were obtained. A high level of resistance is noticed in Asiatic hybrids and to a lesser extent in cultivars of *Lilium longiflorum* and moderate resistance is observed in oriental hybrids.

Removal of infected plants and soil treatment with Formalin, Chlopicrin or stem sterilisation have shown good disease control. Use of pathogen-free stock in combination of fungicides, hot water and surface sterilisation of scales used for bulblet production.

Hot water treatment of bulbs for 2 hours at 39° C followed by a 30-minute dip in Benomyl eliminates infection and stimulates plant growth. Addition of mycorrhizal fungi increases plant growth and offsets the detrimental impact of root pathogen.

Botrytis Blight [*Botrytis elliptica* (Berk.) Cooke and *Botrytis cinerea* Pers. ex. Fr.]

Symptoms: Occasionally, the growing points of the shoots are attacked and dieback occurs with no further growth. The first symptom of the disease initially appears as small, circular or elongated, sunken spots somewhat water-soaked. The centre becomes reddish brown at first and is gradually surrounded by a yellowish zone and becomes pale brown on enlargement (Dimock and Tammen, 1967). Flower buds thus produced are shrivelled, distorted or disfigured, depending on the severity.

Management: Use of disease-free stock and resistant plant material such as *Lilium pyrenacium, Lilium regale* and *Lilium willmottiae* are recommended for healthy flower production (McHorter, 1963). Removal of all old foliage and flowers from field is necessary to eliminate the inoculum potential.

For many years two per cent Bordeaux mixed with a spreader stick have served as a good control. Captan, Zineb, Thiram and other dithiocarbamate fungicides, which leave fewer deposits also reduce the disease. Foliar sprays of Carbendazim plus Diethofenacol gave complete control of *Botrytis elliptica* while Cyproconazole, Fenpropimorph and Chlorothalonil fungicides have been reported to be effective. Fungicides such as Captafol, Chlorothalonil, Febram, Mancozeb, Triforine, Zineb and a mixture of thiophanate methyl and Mancozeb effectively control the disease. Smoky pellet formulations (Procymidone 30 per cent + dichlofluanid dust) have been reported to be effective against this disease as compared to wettable powders in controlling the disease (Oh *et al*, 1991).

Bacterial Diseases

Soft Rot (*Bacillus lilii, Pectobacterium carotovorum*)

Symptoms: The characteristic symptoms are soft and wet decay of bulbs.

Management: All diseased bulbs should be discarded. Healthy bulbs should be planted in well-drained soil. Care should be taken for avoiding injury to the bulbs while lifting from soil.

Viral Diseases

Lilium, particularly, are extremely susceptible to infection by viruses, especially those transmitted by aphids. Gardeners often find lily stocks have a much shorter working life than irises and narcissi before they succumb to the build up of viruses. The reason may be due to the fact that the aphids colonise lilies and thus introduce viruses like Lily Symptomless Virus Cucumber Mosaic Virus, Lily Virus, Tulip Breaking Virus, etc.

Tulip Breaking Virus

Symptoms: This virus produces severe leaf mottling and distortion of the flowers, sometimes giving rise to notched petals and a general reduction in growth. Transmitted by the aphids *Myzus persicae, Macrosiphum euphorbiae* and *Aphis fabae*, diseased Darwin tulips commonly are the source of infection.

Lily Symptomless Virus

Symptoms: A general stunting is being reported.

Cucumber Mosaic Virus

Symptoms: Produces little effect other than growth reduction, but when both this and Lily Symptomless Virus occur together, characteristic elongated flecks develop on the leaves, at first yellowish but later grey or necrotic. The flowers may appear abnormal but this symptom is not invariable. The aphid *Aphis gossypii* transmits Lily Symptomless virus.

Management: Tissue cultured propagation has freed numerous taxa of these pathogens. Stunted and mottled plants should be collected and destroyed and bulbs from the diseased plants should not be used for propagation.

TUBEROSE

Other common names: Gul-shabu; *Polianthes tuberosa*; Family: Amaryllidaceae

Fungal Diseases

Foot and Tuber Rot (*Sclerotium rolfsii*)

Symptoms: The disease appears in patches and causes wilt and stem rot. In moist humid conditions, a characteristic fan-shaped mycelid strand of the fungus appears at the base of the infected plants, and in later stages brown mustard-like round sclerotia develop on the mycelial growth. This fungus initially attacks the roots and later spreads to the tubers and petioles, which induce rotting.

Management: Soil application of Thiram (0.2 per cent), Bavistin (0.5 per cent), Brassicol (0.1 per cent) or Zineb (0.3 per cent) three times at 20 days interval has been recommended to control the disease. Soil drenching with commercial Formalin (0.2 per cent) and mercuric chloride (0.1 per cent) was also found effective against this fungus.

Blossom Blight (*Fusarium equiseti*)

Symptoms: With the infection of this fungus, light brown lesions develop on petals, which soon darken and result in the drying of the tissue. The blighted blossom drops off from the plants. Tips of the florets become brown under humid conditions on which brown spore masses develop.

Management: Application of Benomyl (0.2 per cent) or thiophanate methyl (0.2 per cent) as a foliar spray reduces the disease.

Bacterial Diseases

Flower Bud Rot (*Erwinia* species)

Symptoms: The young flower buds are affected initially by this fungus coming in dry rotting of buds with brown scorched merotic discoloration of peduncles.

Management: The diseased plants are to be removed. Use of Streptocycline (0.01 per cent) was reported to be effective against the disease.

Alternaria Leaf Spot (*Alternaria polyanthi*)

Symptoms: It is characterised by faint concentric rings on the midrib and rarely on the margin of leaves. It is prevalent in the rainy season. Flower peduncle is also affected showing circular to oval spots. The leaves and peduncles become necrotic and dry up with the infection.

Management: Spray of Mancozeb (0.2 per cent) or Iprodione (0.2 per cent) at 10-day intervals helps in reducing the disease.

TULIP (*Tulipa* species and hybrids)

Family: Liliaceae

Fungal Diseases

Fire Disease [*Botrytis tulipae* (Lib.) Lind.]

Symptoms: The leaves prior to emergence get infected. They became stunted, malformed, distorted and fail to unroll. Under moist weather, masses of fungal spores (conidia) and sclerotia are formed on the outer surface of the infected tissues. Leaf spotting is of non-aggressive and aggressive type. Small necrotic areas are generally marked by dark green, water-soaked areas in non-aggressive lesion which do not increase in size and no conidia are produced on them. Aggressive lesions have whitish grey or brown with water-soaked margins that can enlarge and are able to produce more spores that may result in an epidemic form leading to significant reduction in bulb yield. Increase in lesion size depends entirely upon the production of spores and the duration of moisture during infection. Flowers may also be attacked. Fruit capsules are the most susceptible part and sclerotial bodies cover seeds. In the underground infection, bulbs are attacked first. Infection travels from mother to daughter bulb, sclerotia are formed in the outer husk of the affected bulbs. With this, the outermost fleshy scales develop small round to oval, sunken spots, the centre of which contains a yellow grey border. The stems produced by these bulbs are shorter and, in a few cultivars, a slight streak is apparent on flowers.

Management: Discard the infected bulbs. Removal of infected shoots just below ground level upon emergence and flower before anthesis is helpful in limiting the disease spread. Bulb dip prior to planting in Benomyl, Thiabendazole or 2 per cent Formalin for 30 minutes reduces the carry over of inoculum and subsequent production of fireheads. Iprodione and Vinclozolin provide significantly higher levels of disease control in the flowering period of the crop. Spraying with Benomyl (0.1 per cent) and Maneb (0.25 per cent) just after leaf development is recommended.

Fusarium Bulb Rot or Basal Rot (*Fusarium oxysporum* Schlenhtend ex Fr f.sp *tulipae* Apt.)

Symptoms: The first visible signs of infection on the bulbs are small, pale brown or grey sunken flecks on the outermost fleshy scales that become dark brown with age and enlarge in size. Rotting usually starts at or near the base of the bulb and results in exudation. Fungus produces white or pink mycelium and spores on bulbs in warm, damp sites. The infected tissues shrink, become firm and appear ring-shaped. Leaves in general do not wilt but remain upright and often develop a characteristic purple colour. Foliage produced on infected bulb senesces early. Flowers produced may be blasted or deformed.

Management: Delayed planting and early lifting avoid soil temperature that favours disease development and, therefore, provide natural resistance in bulbs. Diseased stock should be stored separately from the healthy stock in well-ventilated facilities that reduce production of ethylene. The storage temperature should be maintained below 15° C. Rotation for a minimum of 3 years should be practised. Dipping of bulbs immediately after lifting (within 48 hours) in Benzimidazole fungicides has provided excellent control. A satisfactory control is achieved by dipping bulbs for two hours in a 0.5 per cent Formalin solution. Planting when soil temperature is below 10° C reduces the infection. Benomyl and Thiophanate (0.05 per cent) provided very good control of the disease.

Antagonistic activities of *Minimodhura polyspora, Steptomyce* and *Trichoderma, Trichoderma viride* to *Fusarium oxysporum* f.sp. *narcissi* as pre-planting bulb treatment was studied by Beale and Pitts (1990) which revealed most effective results giving disease levels of 55.72 per cent and 56.88 per cent lower than the control with first two microorganisms, respectively, indicating reduction in disease. *Streptomyces* sp. alone can suppress infection up to 24 per cent to 18.7 per cent from 64 per cent in two consequent years of study thereafter after

Pythium Root Rot [*Pythium ultimum* (Trow) var. *ultimum*]

Symptoms: The attack in fields is visible by gaps and irregular growth of shoots, which may fail to emerge. Severe stunting of plants is apparent. Flowers and bulbs may wither before maturity and may remain small and lose their quality. The affected roots appear translucent with watery spots. The vascular tissues turn brown, roots are reduced to a fragile hollow cylinder containing the remnants of the conducting tissues and eventually get destroyed. Primary symptoms on the base of bulbs are soft grey spots often with brown borders that release a distinct odour.

Management: Soil sterilisation with steam or Formalin before planting gives good results. Bulbs should be planted in well-drained and sterilised soils. Beneficial effects of fluorescent *Pseudomonas* isolate E113 and *Trichoderma harzianum* against root rot fungi was achieved either by mixing the bacteria through the soil or by dipping the bulbs in bacterial suspensions immediately before planting.

Physiological Diseases

Sunscald

Drying and shrivelling of flower parts, especially along the upper edges has been found to be caused by sunlight under dry conditions.

Management: Shading the flowering plants in the greenhouse is effective.

Frost Injury

Among unfavourable conditions, frost has been reported as the cause of a disease. Oblong or oval spots develop along the veins. This injury is not common.

Flower Stalk and Stem Collapse 'Topple'

Certain varieties of tulips develop weak water-soaked spots, with an exudation of liquid either on the upper half of the flower stalk or on the lower parts, or even on the neck near the bulb. Flower stalks soon shrivel and collapse. Some stems may topple without first showing water-soaked spots. The plants otherwise appear to be in excellent condition. The disease is said to be caused by insufficient ripening of the bulb after wet, cool summer weather and by forcing of tulips in clay soil or at excessive temperatures.

Management: Avoid high temperatures and high humidity in forcing. See that bulbs used for forcing ripen thoroughly. Spreading out the bulbs in the forcing house for 2 to 3 weeks is recommended as a secondary ripening period. Do not force the bulbs too early, especially after wet, cold years. The varieties susceptible to this disease should be put in pots early to establish a strong root system. They may then be forced later. Some varieties are less susceptible and are to be preferred if this disease gives trouble.

Retarded Growth

If bulbs are stored or heeled-in under excessively warm conditions or are insufficiently ripened, the shoots are greatly retarded or are suppressed entirely.

Winter Injury

When bulbs are planted late in the fall in heavy clay soil or where drainage is not adequate, roots fail to grow and shoots are distorted and abnormal. Bulb decay follows.

Leaf-Flower Fusion

Some times there is an abnormal fusion of the leaves and flower petals. The cause of this condition has not been determined, but the use of hormone-type weed killers during the previous growing season is a possible cause.

PERENNIALS

BOUGAINVILLEA

Family: Nyctaginaceae

Fungal Diseases

Leaf Spots (*Phyllosticta bougainvilleae* DaCamara)

Small to large, irregular brownish-grey spots, delimited with brown margins appear on the leaves. The infected tissues fall off.

Alternaria Leaf Spot [*Alternaria alternata* (Kunze ex Fr.) Wiltshire]

The infection is noticed in the form of dark brown, circular to irregular spots covering large areas. The infection is also common on dead spines and inflorescence bracts. Most probably the fungus perpetuates on these parts. Affected portions give a blighted look.

Leaf Spot (*Colletotrichum gloeosporiodes* Penzig and *Colletotrichum dematium* f. *bougainvilleae* Area)

Symptoms: Large, irregular, brownish-yellow spots start appearing from the leaf margins and involve most of the leaf area. The infection is also noticed on the spines resulting in their death and drying. The twigs show dieback symptoms. It perpetuates on dead twigs and spines.

Management: Dead twigs, leaves and inflorescence should be removed and burnt. Bavistin (0.1 per cent) or Zineb (0.2 per cent) spray is recommended. It is most serious in cool humid weather. Spraying with Dithane Z-78 (0.2 5) or Bavistin (0.1 per cent) is recommended.

JASMINE (*Jasminum* species)

Fungal Diseases

Leaf Blight (*Cercospora jasminicola* and *Alternaria jasmini*)

Symptoms: This disease occurs in severe form on *Jasminum grandiflorum*. Reddish brown circular spots, 2 to 8 mm in diameter are produced on the upper surface of the leaves, spreading rapidly in the rainy season. Affected leaf margins show inward curling and become hard and brittle. In severe cases of infection, vegetative buds and young branches dry up. Since almost all the leaves are affected along with young branches, the plants look unhealthy with appearance of drying up from a distance. The disease appears in the months of May/June and the peak incidence occurs from August to November. Flower production is adversely affected and may cause 50 per cent loss in yield.

Varietal screening studies conducted at Tamil Nadu Agricultural University, Coimbatore, showed that *Jasminum arborescens, Jasminum paniculatum* and certain cultivars of *Jasminum sambac* remained free from leaf blight and *Jasminum grandiflorum* was highly susceptible to this disease. *Jasminum angustifolium, Jasminum auriculatum, Jasminum communis, Jasminum flexile, Jasminum humile, Jasminum pubescens* and *Jasminum regidum* and certain cultivars of *Jasminum sambac* showed only stray incidence of blight.

Management: Fungicidal trials conducted by using the highly susceptible cultivar of *Jasminum grandiflorum* revealed that 0.4 per cent Benlate, 0.2 per cent Dithane M-45, 0.1 per cent Bavistin and 1.0 per cent Bordeaux mixture were equally effective to control leaf blight. It was suggested to spray the fungicides at monthly intervals commencing from May and continuing up to pruning. The number of sprays may vary depending upon the weather conditions and severity of the disease. Spraying of Bordeaux or copper oxychloride to control the blight is also found effective.

Rust (*Uromyces hobsoni*)

Symptoms: This disease appears in July-August during monsoon rains, in severe form on *Jasminum auriculatum*. Other species, namely, *Jasminum sambac, Jasminum flexile, Jasminum communis* and *Jasminum angustifolium* are also susceptible to rust. The leaves show the presence of orange coloured aecial cups on both sides, but predominantly on the lower surface. Numerous blisters are produced in advanced stages of infection causing yellowing and crinkling of the leaves. The stems and branches are also infected, causing splitting of bark and subsequent death of the branches.

Management: The disease can be controlled by dusting sulphur 20-25 kg/ha. Spraying with Bordeaux or copper oxychloride was also recommended for the control of rust.

Wilt (*Fusarium solani* or *Sclerotium rolfsii*)

Symptoms: This disease is found in *Jasminum sambac*. The early symptom is yellowing of lower leaves which gradually spreads upwards, finally resulting in death of the plant. In the field, the disease occurs in patches and roots show black coloration. In the case of sclerotial wilt, in addition to the above symptoms, white mycelia are found generally girdling the roots and the sclerotia are found adhering to the roots of the wilted plants.

Management: Drenching the soil around the plants with 1 per cent Bordeaux mixture controls this disease.

Jasmines are subject to several other diseases, including crown-gall caused by the bacterium *Agrobacterium tumefaciens*, a blossom blight caused by a species of *Phoma* and root-rot caused by *Clitocybe tabescens*.

Viral Diseases

Mosaic

Symptoms: The diseased plants show stunted and yellowish green appearance with small leaves. Yellowish green to chlorotic flecks of 1-2 mm in diameter appear irregularly on the leaf and these streaks form a ring. The symptoms are more conspicuous on older leaves. The disease is graft-transmissible on *Jasminum simplicifolium, Jasminum sambac* and *Jasminum pubescens* was found to be highly susceptible to the disease while *Jasminum auriculatum* was relatively tolerant.

Management: Control of insect vectors will prevent the transmission of the disease.

Phyllody

Symptoms: This disease is suspected to be caused by mycoplasma. The affected plants produce malformed, reduced greenish flower-like structures instead of fragrant white flowers on panicles, which are highly congested and green in colour. The greenish corolla lobes become reduced and ovate in shape. Flower parts are transformed into leaf-like structures.

Management: The disease may be controlled by spraying with tetracycline hydrochloride (250 ppm). Cuttings from infected plants should never be used for planting.

ORCHIDS (Orchids, several genera and species)

Family: Orchidaceae

Fungal Diseases

Leaf Spots (*Gloeosporium, Colletotrichum, Phyllosticta* and *Phomopsis* sp.)

Symptoms: This disease is very common and can be found on almost all cultivated species of orchids. It is caused by species of *Gloeosporium, Colletotrichum, Phyllosticta* and *Phomopsis*. Within a few days of infection, sunken spots appear at any place on the leaves that later turn brown. Most leaf spot organisms do not kill the leaves but formation of spot make the leaves unsightly. Warm humid weather and lack of light encourage the spread of the disease.

Management: Spraying of Bordeaux mixture (6-6-50 or 4-4-50) will effectively control the disease. Blitox (0.3 per cent), or Difolotan (0.1 per cent) or Bavistin (0.1 per cent) may be sprayed as a remedial measure.

Pythium Black Rot [*Pythium ultimum* (Trow) var. *ultimum*]

Symptoms: This disease affects mostly seedlings in community pots. Affected plants turn black and leaves start falling. The pseudobulbs also start to rot within. Since in community pots the plants are close to each other, the disease spreads from plant to plant at a very rapid rate. Orchids such as *Cattleya* and *Epidendrum* species are found to be highly susceptible to damping-off.

Management: Withholding of watering for a few days and shifting the plants to a less humid part of the greenhouse helps to check the disease. Removal of infected plants and spraying with Mancozeb 2 g/l can effectively control the disease.

Flower Blight [*Botrytis cinerea* Pers. (ex) Fr.]

Symptoms: This disease is serious on orchids like *Cattleya, Dendrobium, Phalaenopsis, Oncidium* and *Vanda* under cool, damp and still conditions. It produces numerous, small, dark brown spots on petals, especially of older flowers. *Botrytis* sporulates heavily on dead and diseased flowers.

Management: Removal of the infected parts and proper sanitation is the most important management practice. Spray of following fungicides, namely, Benomyl, Bavistin (0.1 per cent) or Dithane M-45 (0.2 per cent) at regular intervals is recommended.

Brown Streak or Blight of Flowers

This disease affects a number of orchids and is caused by *Botrytis cinerea*. When infection is mild, spots appear on flowers. In severe cases, the entire flower may be covered with spore masses and a blighting effect is produced. Spray of Dithane M-45 or Bavistin (2 ml/l) was reported to be effective in controlling this disease.

Bacterial Disease

Bacterial Soft Rot

Symptoms: The disease is covered by *Erwinia carotovora* in *Cattleya* orchids. The disease starts at the upper end of the leaf as a small water-soaked and somewhat darker than normal green spot. Pseudobulbs turn soft and pulpy and become yellow in colour.

Management: Streptocycline (0.01 per cent) treatment was reported to be effective.

Viral Diseases

Blossom Brown Necrotic Streak

Symptoms: It attacks *Cattleya* species and produces brown spots, or whole flower break, yellow streaks on leaves. It is transmitted from one plant to another through the cutting knife.

Mosaic Flower Break of *Cattleya* and Other Orchids

Symptoms: It appears either as a mild colour break with flower variegation but no distortion, or as a severe colour break with flower distortion and variegation. Leaves are mottled and sometimes twisted. Transmission is by the green-peach aphid *Myzus persicae*.

Cymbidium Mosaic, Black Streak or *Cattleya* Leaf Necrosis

Symptoms: On *Cymbidium*, it first produced mosaic mottle then necrotic spots, streaks and rings on leaves. In *Cattleya*, sunken brown to black leaf patterns appear; sometimes rings, more often elongated streaks are formed on the older leaves. If leaves are killed prematurely, few small flowers of normal form and colour are formed.

Odontoglossum Ring Spot

Symptoms: Small necrotic spots and rings develop on the older leaves and light green to yellow areas develop on the younger leaves. The leaves may turn yellow and drop in 2 to 3 months. No flower symptoms develop. No vector has been identified.

Management: Using a hot or a disinfected knife while dividing the plants, sterilising or fumigating the potting medium, using disease-free planting material and spraying periodically with insecticides help prevent disease spread. Infected plants may be removed from the orchid house.

Tissue culture proved to be effective in rescuing valuable clonal orchids infected with virus. But it is not applicable for all orchids and all viruses. Meristem culture is one of the ways of getting rid of the virus in *Cymbidium* orchids.

Rhododendron (Rhododendron and other species)

Family: Eriaceae

Fungal Diseases

Botrytis Blotch (*Botrytis cinerea*)

Symptoms: Winter injury frequently leads to invasion by *Botrytis* as a secondary parasite. Under moist conditions slight injuries are aggravated by this grey mould, which continues to invade the healthy tissues of the leaves. The injury seldom calls for the application of a fungicide.

Bud and Twig Blight (*Briosia azaleae*)

Symptoms: This fungus disease is destructive to azaleas. In certain plantings it has been known to kill 90 per cent of the flowers. The disease may be described as a bud blast rather than a bud rot. Terminal flower buds are infected after which leaf buds and twigs are attacked. Diseased twigs do not flower the following season. The diseased buds remain attached for 2 or 3 years. The fungus gains entrance through the axils of the bud scales and penetrates deeply into the bark and pith. The conducting vessels of the twigs become plugged so that conduction of water is prevented. The fungus develops its spore stage on long, thin outgrowths capped by slimy balls, each of which contains thousands of spores. These outgrowths appear like fine bristles growing from the bud scales and bark of infected twigs. The spores are spread from plant to plant by insects (especially by bees during blossom time) and by the splashing of water.

Management: Prune and destroy infected branch tips. Spray with a copper fungicide, once a month, starting when the blooms fade (Pirone, 1978).

Grey Blight (*Pestalotia macrotricha*)

Symptoms: The fungus that causes this disease usually follows winter-killing, sunscald, or other injury, as a secondary parasite. The spot is at first white in the middle with a dark brown margin; later it becomes a blotch. Small bodies the size of pinpoints dot the surface; these are the fruiting bodies of the fungus.

Management: Plant rhododendrons where the blistering summer sun does not cause sunscald, and where the cold winds of winter do not freeze the leaves too severely (Pirone, 1988).

Dieback, *Botryosphaeria* Canker (*Botryosphaeria dothidea*)

Symptoms: Leaves are first attacked at the tips or around the margins. Spots are later formed that involve the entire leaf. Leaf stalks and twigs also become infected. The symptoms are much like those caused by *Phytophthora,* except that the surface of the parts infected with *Botryosphaeria* is much roughened by the protruding bodies of the fungus.

Management: Prune affected parts to sound tissue. Spray occasionally with a copper fungicide.

Crown Rot (*Phytophthora cryptogea*)

Symptoms: Collected *Rhododendron maximum, Rhododendron catawbiense* and *Rhododendron carolinianum* plants are susceptible to this disease. The fungus has been isolated from the main roots and basal portions of the stems, where it causes a brown discoloration of the wood. Infected branches wilt and die; severely infected plants become thin and leggy and eventually perish.

Management: Infected branches should be pruned out; severely infected plants should be removed from the planting, with the surrounding soil, and destroyed (Pirone, 1978).

Phytophthora Dieback (*Phytophthora cactorum*)

Symptoms: Terminal buds and leaves turn brown, roll up, and droop as though in a winter condition. The stem shrivels and a canker is formed which encircles the twig. All parts above the canker wilt and soon die.

Management: This fungus is the same as that which causes lilac twig blight; the two kinds of plants should, therefore, be widely separated. All diseased tips should be pruned out well below the infected parts. Bordeaux sprays applied as new leaves emerge and repeated two weeks later may help. Where losses have occurred in nurseries, drenching the soil with Dexon or Terrazole will kill the *Phytophthora* fungus present in the soil. Chemical treatments, however, will not cure already infected plants.

Damping-Off (*Pellicularia filamentosa*)

Symptoms: This fungus causes a stem rot at the soil level. It is common where young plants are crowded and over-watered. It occasionally kills larger plants.

Management: Plant seeds in milled sphagnum moss or drench the soil before planting with any systemic fungicide, for example, Bavistin/Benlate.

Azalea Petal Blight (*Ovulinia azaleae*)

Symptoms: Isolated cases have appeared elsewhere. Indian and Kurume azaleas are especially susceptible. In the first stage, pale, circular spots about the size of a pinhead appear on the undersides of the petals. In favourable climatic conditions, the spots enlarge rapidly and run together, appearing white on coloured flowers and brown on white flowers. The affected flowers become limp and are then covered with a delicate white bloom or a frost-like mat formed of spores of the fungus. These are easily carried away by wind and rain, also by seeds and insects, which spread the infection from flower to flower. After the flowers have been destroyed, the propagating bodies of the fungus, sclerotia develop. When young they are soft and blue-grey; afterwards they become hard and black. These bodies are very common on old flowers that have fallen to the ground, and here they over-winter.

Management: The most important factor in control is prompt and persistent picking of the infected flowers. In some plantings it has been found advisable to remove all the blossoms, including the uninfected ones. In larger plantings where it is impossible to do this, spraying the soil surface with Ferbam, before flowering, kills the ascospores, which are formed in the fruiting structures. Protection of the flowers can be provided with Benlate sprays applied at 5-day intervals during bloom. The first application should be made when the flowers start to show colour. This treatment should be effective for more than one year.

Azalea Gall (*Exobasidium vaccinii*)

Symptoms: The leaves become thickened or fleshy either wholly or in part and turn pale green or whitish. Sometimes a fleshy rosette of leaves appears at the tip end of the branch, or a gall which later becomes bladder-like as it decays and dries out. The flower parts, especially the petals of evergreen species such as *R. maximum*, become greatly thickened so that the whole bloom is turned into a hard, fleshy, waxy, irregular gall, the parts of which become covered with a whitish bloom. Even the seedpods become fleshy and gall-like. When found in nature, these fleshy galls are called 'pinkster apples,' and are eaten by those familiar with their qualities. Similar galls of a beautiful rose colour develop sometimes on cranberries; they are called 'rose apples.' The fungus that causes these abnormalities is one of the simple basidiomycetes, which forms basidia superficially instead of in fleshy fruit bodies. Azalea gall is largely composed of abnormal leaf tissue.

Management: In greenhouses and garden plantings, where the disease is important, young galls can be handpicked and destroyed. In larger plantings, just before the leaves unfurl, spray once with either Ferbam or with Zineb.

Leaf Spots (*Cercospora rhododendri, Colletotrichum* sp., *Corynespora cassiicola, Cylindrocladium scoparium, Coryneum rhododendri, Coryneum triseptatum, Crytostictics mariae, Diplodina eurhododendri, Gloeosporium rhododendri, Gloeosporium ferrugineum, Hendersonia concentrica, Laestadia rhodorae, Lophodermium melaleucum, Lophodermium rhododendri, Melasmia rhododendri, Monochaetia* sp., *Mycosphaerella* sp., *Phyllosticta maxima* and other species, *Physalospora rhododendri, Septoria rhododendri* and *Venturia rhododendri*).

A great many fungi cause leaf spots on azaleas and rhododendrons.

Management: Benlate, Ferbam, Maneb or Zineb sprays applied immediately after flowering and repeated twice at 10- to 14-day intervals, will control most leaf spot fungi. Include a few drops of a spreader-sticker like DuPont's SS or Triton B to help the spray adhere more firmly to the leaves.

Apply generous mulch over the soil, use windbreaks and control insects to keep foliage in sound condition and thus prevent the entrance of many leaf spot fungi, especially those that are weak parasites.

Leaf Scorch (*Septoria azaleae*)

Symptoms: Dark reddish-brown spots develop on azalea leaves, which fall prematurely. The spots are at first yellowish-red with brown centres and purplish margins. The fruiting bodies in the centre of the spots can be seen with a hand lens.

Management: Spraying with Bordeaux mixture, 2-2-50, or any readymade copper fungicide at two-week intervals from July 1 to August 15 will give adequate protection. Copper sprays may cause some injury to azaleas during periods of cool, wet weather.

Powdery Mildew (*Microsphaera alni*)

Symptoms: This disease appears as a white, powdery coating of the leaves.

Management: Spray with Acti-dione PM, Benlate, Karathane or wettable sulphur.

Rust (*Pucciniastrum vaccinii*)

Symptoms: Another disease common in nurseries and propagating houses is the rust on rhododendrons and azaleas. Sometimes seedlings are destroyed. The bright golden or brownish spores that break out from pustules on the undersides of the leaves make known the disease. These repeating spores carry the parasite from plant to plant rather rapidly, but the fungus is able to complete its life cycle only by attacking the hemlock. Other fungi may follow rust as secondary parasites on *Rhododendron*. Infection of *Rhododendron ponticum* is often followed by the leaf spot caused by *Pestalotia macrotricha*.

Management: The rust is carried over-winter on rhododendrons. Avoid planting azaleas and rhododendrons near hemlocks. Spray the ericaceous plants several times at 10-day intervals soon after the disease appearance with 2-3 g/litre of wettable sulphur.

Shoot Blight (*Monilinia azaleae*)

Symptoms: The spores of this fungus attack the leaves and young shoots when the plants are in full bloom. Later the capsules also are attacked; the pseudosclerotia develop in them and fill their cavities with solid masses of thick-walled cells.

Management: No controls have been developed for this disease.

Shoestring Root Rot (*Armillaria mellea*)

Symptoms: A wet decay of the cortex and bark accompanied by a distinct mushroom odour is the first symptom of the disease. Black strands of fungus tissue envelop the roots and enter the bark, and mushrooms also develop. White fans of mycelium are usually present under the loosened bark immediately above the soil level. The entire plant eventually succumbs.

Management: Plants in good condition are seldom attacked. The crown of the plant should be exposed to the air, so that the bark is well dried. Severely infected plants rarely recover.

Wilt (*Phytophthora cinnamomi*)

This is a seedling or nursery disease which affects plants up to 2 or 3 years old. The fungus is soil-borne; it enters the young roots and works its way up to the crown. The young leaves become yellowish and wilt. The fungus prefers cool situations and soil that is not sufficiently acid for good growth of the host; it is rare in soil adapted to good rhododendron growth. Over-watering of the plants causes death of the roots and thus affords entrance for the fungus. The fungus is more likely to attack plants recently transplanted, because of root injury in the operation.

Management: Temperature of the soil should be kept low by mulching, and the acidity of the soil should be increased by applying aluminium sulphate or sulphur. Heavily infested soil should be replaced with clean soil or be steam-pasteurised, or a soil drench of Dexon or Terrazole applied as directed by the manufacturer.

Non-Parasitic Problems

Salt Injury

Azaleas are sensitive to excessive amounts of salt in the soil.

Winter Injury

When strong cold March winds follow warm days, many rhododendrons may be killed. The damage may not appear until 2 or 3 weeks later. The leaves become brown, especially at the edges and near the tips. This is due to a loss of water by the leaves at a time when the water in the soil is still frozen or unavailable for other reasons.

Wilt (Black Walnut Injury)

If rhododendrons are planted near large black walnut trees, they may suddenly wilt and die. The greatest injury occurs along the roots of the walnuts and seems to be caused by toxins secreted by them.

Management: Remove black walnut trees growing near rhododendrons (Pirone, 1978).

ROSE (*Rosa* species and hybrids)

Family: Rosaceae

Fungal Diseases

Die-back *(Botryodiplodia theobromae, Colletotrichum gloeosporoides, Fusarium* sp. and *Phomopsis* sp.*)*

Symptoms: This is the most serious of the rose diseases in India. Earlier reports suggest that this disease was caused by fungus *Diplodia rosarum* Fr., but in recent years, several other fungi have been recorded to be present in the diseased twigs. These include *Botryodiplodia theobromae, Colletotrichum gloeosporoides, Fusarium* sp. and *Phomopsis* sp. (Shukla and Singh, 1991). As the term dieback implies, the disease causes the death of the plant from the top downwards. The disease starts from the pruned surface of the twigs. Initially it may be observed to the extent of a few centimetres below the pruned end, but in severe cases, the disease spreads further and kills the entire plant.

Outwardly, the symptoms appear as browning or blackening of the twig portion near the pruned end. In the advanced stage, the disease spreads from the branch twigs to the main stem and travels further down to the roots, killing the whole plant. The diseased twigs, when split open, reveal the infection to have travelled a few centimetres more than it appears, and the internal tissues show browning. Sometimes the flower stalks are also affected, resulting in direct loss of yields. The disease more frequently attacks the woody mature twigs of older plants than the younger herbaceous canes of younger plants. Some of the diseased twigs, when cut open, have been found to be tunnelled through the pith from the cut end downwards, showing the presence of Digger wasps (Grabrasp.). The twigs attacked by the wasp readily dry up and frequently lead to the dieback.

Management: Since the disease is usually associated with pruning, proper prophylactic measures following pruning would help prevent this disease to a large extent. Coating of a pruned end with a fungicidal paint containing 4 parts of copper carbonate, 4 parts of red lead and 5 parts of linseed oil, controls the disease effectively. Soil drenching with 2 g/l Bavistin, Benomyl or Demosan has also been reported to control the disease. M.P.K.V., Pune Centre, has recommended spraying with 0.2 per cent Captan or 0.2 per cent Mancozeb or 0.2 per cent copper oxychloride immediately after pruning and then twice at 10-day intervals to control the dieback of rose. In the Division of Floriculture and Landscaping, I.A.R.I., New Delhi, more than 800 varieties were screened against this disease (Sharma and Singh, 2001). None of the varieties was found free of this disease. Some level of tolerance was observed in the following varieties:

Tolerant/Moderately Resistant

Hybrid Teas

Angel Bells, Ariana, Abhisarika*, Arjun*, Bhim*, Catherine Denevue, Century Two,

Dr. B. P. Pal*, First Prize, Mercedes, Montezuma, Mrinalini*, Oklahoma, Priyadarshini*.

Floribundas

Arunima*, Delhi Princess, Else Poulsen, Fantasia, Gabriella, Folklore*, Jantar Mantar, Judy Garland, Lahar, Madhura, Neelambari*, Sadabahar, Suchitra*, Usha*.

(* Denotes names of I.A.R.I. released varieties).

Black Spot (*Diplocarpon rosae* Wolf)

Symptoms: Black spot disease has also been called leaf blotch, leaf spot, blotch, rose actinonema, rose leaf asteroma and star sooty mould. It is the most important disease of roses all over the world. Black spot is a minor problem in glasshouse roses, because greater care is taken to avoid syringing plants with water for spider mite control and because humidity is regulated very carefully. In outdoor roses, however, this disease is generally present, frequently epidemic, and a major problem. Characteristic black spot 2-12 mm in diameter develops on upper leaf surfaces. These leaf spots are circular or irregularly coalescent with characteristic feathery, radiate, fibrillose margins of sub-cuticular mycelial strands. Small black acervuli are often visible on the surface and may be distributed irregularly or in concentric circles. Conidia may be visible as white, slimy masses on the acervuli. Leaf tissue surrounding the spots turns yellow and chlorosis extends throughout the leaflet until abscission occurs. The pathogen actually presents only in the lesion itself; the yellow tissue is caused by pathogen metabolites. The yellow tissue exhibits high metabolic activity that is expressed by accumulation of total phenolics and ortho-dihydroxyphenols and amino acids as well as by high enzyme activity. Spots enlarge slowly taking several weeks to reach 12 mm in diameter. In resistant cultivars or under unfavourable environmental conditions, only tiny black flecks may form and leaves may not turn yellow or abscise.

Yellowing and abscission of leaflets are associated with ethylene. Leaves with black spots produce large quantities of ethylene; production decreases as leaves become yellow and ceases when leaves turn brown. Infected leaves contain less auxin than healthy ones. The pathogen degrades this abscission-retarding material, thereby hastening less abscission.

Raised, purple-red, irregular blotches develop on the immature wood of first year canes of susceptible cultivars. Spots later become blackened and blistered; they contain acervuli but lack fibrillose strands. Lesions are often small and rarely kill branches but are extremely important in the survival of the pathogen over the winter.

Petioles and stipules may have inconspicuous black spots similar to those found on leaves. Petioles may be girdled without abscising. Peduncles, fruit and sepals may have similar symptoms. Petals may have red flecks accompanied by moderate distortion. Acervuli frequently occur in the lesions. *Marssonina rosae* (Lib.) Lind (*Asteroma rosae, Actinonema rosae, Marsonia rosae*), are the imperfect stage of the black spot pathogen, the perfect stage of pathogen is *Diplocarpon rosae* Wolf. Epidemiological studies conducted at M.P.K.V., Pune, showed that black spot disease mostly occurs during June to October. The disease starts after the onset of the monsoon and reaches maximum intensity during September-October.

Management: Leaves should not be allowed to remain wet or at very high humidity for more than 7-12 hours. Removing leaves from the ground and pruning canopy that contains lesions will reduce over-wintering of the pathogen. Dense planting should be avoided to allow good air circulation through leaf canopy.

Due to its serious destructive nature, several workers in India have studied resistance to this disease under field conditions. Indian varieties showing field tolerance to this disease are Ganga, Gulzar, Hans, Jawahar, Kavita, Mrinalini, Noorjahan, Prima, Pusa Sonia, Pusa Sonora, Raktagandha, Sadabahar, Shabnam and Suryodaya. Varieties of exotic origin, namely, Babelune, Coronada, Ernest H. Morse, Forty Niner, Grand Opera, Lucy Cromphon, Sphinx and Tiara have been reported to be resistant to this disease. In the Division of Floriculture, I.A.R.I., New Delhi (1989-98) out of 166 hybrid teas screened for black spot, none of the varieties was resistant (Sharma and Singh, 2002). Sixty varieties, namely, Abhisarika, Alec's Red, American Heritage, Arjun, Atol, Avon, Belle of Punjab, Black Lady, Bonnie Nuit, Carina, Catherine Deneuve, Century Two, Chambe Di Kali, Charugandha, Chitralekha, Command Performance, Confidence, Dr. B. P. Pal, Dr. S. S. Bhatnagar, Doris Tysterman, Double Delight, Eiffel Tower, Flaming Sunset, Folklore, Fragrant Cloud, Galia, Ganga, Garden Party, Gladiator, Golden Giant, Golden Times, Gulzar, Happiness, Henkel Royal, Honour, Indian Princess, Jawahar, Jacquina, John F. Kennedy, Kankangini, Kardinal, Kiss of Fire, Lady-X, Lehar, Lalima , Medallion, Mechak, Mister Lincoln, Mother Teresa, Mount Shasta, Mrinalini, Orient Express, Peter Frankenfeld, Precious Platinum, Priyadarshini, Raja Mohan Roy, Royal Highness, Sonia Meilland, Surabhi and Surkhab were moderately resistant (I.A.R.I., New Delhi). None of the 30 Indian-bred rose cultivars that were screened were found resistant against black leaf spot disease. The cultivar Priyadarsini was moderately resistant (M.P.K.V., Pune). Of the 5 rootstocks, 'IIHR Thornless' and 'IARI Thornless' were tolerant to the disease.

Since the black spot fungus is located under the protecting cuticle, no fungicide can kill the fungus without destroying the leaves. For this reason, it is necessary to have the young leaves protected at all times against invasion by the fungus. The infected leaves, as soon as they are observed, should be clipped off and burnt. Preventive sprays of Ferbam (0.1 per cent) at fortnightly intervals, Benlate or Bayleton (0.1 per cent) applied just before the appearance of the infection helps in managing the disease. Under P.A.U., Ludhiana, conditions, Bavistin (0.1 per cent) was the most effective fungicide against leaf spot caused by *Diplocarpon* rose. Paste of some fungicides (1 per cent) in raw linseed oil or its spray (0.1 per cent) and spraying the crop with Mancozeb or 0.2 per cent Captafol or 0.2 per cent copper oxychloride six times at an interval of 10 days each were found effective (M.P.K.V., Pune). This Centre has also recommended six sprays of Captan 0.2 per cent or Bavistin 0.2 per cent at 8-day intervals (Pune) (Sharma and Bhattacharjee, 2002).

Powdery Mildew [*Sphaerotheca pannosa* var. *rosae* (Wallr.) Lev]

Symptoms: A major disease of the rose all over the world, it develops serious proportions in India in some seasons, particularly in areas with warm, humid weather with cool nights. The fungus *Sphaerotheca pannosa* var. *rosae* (Wallr) Lev. causes the disease. The disease affects all the aerial parts of the plant, though the leaves are found to be affected more. The younger leaves get curled, exposing the lower surface. Such leaves are likely to be more purplish than the normal leaves. Raised blisters like areas develop on these leaves, which become coated with the white powdery growth of the fungus. The young growing tips may get completely covered up with mildew. Badly infected flower buds fail to open. The petals are discoloured, dwarfed and finally die. Epidemiological studies conducted at M.P.K.V., Pune, observed that powdery mildew mostly occurs from November to March, having its maximum intensity till January-February.

Management: Protective sprays have mainly achieved control. Uses of systemic fungicides such as Benomyl and Triforine have been found to be effective.

Control measures in outdoor and in glasshouse roses differ somewhat. On outdoor roses, powdery mildew can be expected to occur when rainfall is low or absent, the temperature range is near optimal, and the humidity is high at night and low during the day. Should these conditions occur, protective sprays are necessary. The rapid production of susceptible shoots necessitates repeated application, and the timing of the application is extremely important. Pruning infected shoots at the end of the season and destroying these shoots in regions where they are severe will help prevent over-wintering of the fungus. Raking and destroying fallen leaves from around the bushes at the end of the season may also inhibit over-wintering.

On greenhouse roses, powdery mildew can be expected to occur when the temperature range is near optimal and humidity is high at night and low during the day. When these conditions exist, the occurrence of powdery mildew may be forecast three to six days before it appears. Thus, protective fungicidal sprays should be applied and repeated on a seven-day schedule as long as these conditions continue. Other preventive control measures include lowering the night humidity by fans and/or venting or by heating and venting.

Rosa multiflora was found tolerant to powdery mildew disease, while 'Edward' and Rosa indica var. odorata were susceptible to the disease.

Under the evaluation programme of rose germplasm against powdery mildew, the following lines were found to be resistant at I.A.R.I., New Delhi (Sharma and Singh, 2002) (Table 1)..

Deshpande (1980) reported nearly 40 varieties showing field tolerance to the disease. These include Africa Star, Baccara, Blue Moon, Careless Love, Crimson Glory, Eiffel Tower, King's Ransom, Kiss of Fire, Maria Callas, Michele Meilland, Mohini, Naavneet, Peace, Summer Holiday, Super Star and White Christmas. Under the All India Co-ordinated Floriculture Improvement Project trials, the Indian bred varieties Arjun, Arunima, Ganga, Himangiri, Kavita, Priyadarshini, Raktagandha, Sadabahar, Shabnam, Sindoor and Shola were found to show field tolerance to the disease (Anonymous, 1988).

In early years, applying wettable sulphur affected the chemical control of the disease. Foliar sprays of wettable sulphur, 1 kg in 500 litres of water at 5-7-day intervals controlled the disease (Pal, 1972). However, since roses are susceptible to sulphur injury, if sprayed on hot days having temperatures higher than 32° C, its use in the present days is limited. Sahni (1973) recommended Benlate and PP 675 (demethrimol) for the control of this disease. Kaul (1984) found fortnightly sprays of Saprol (triforine) at 0.15 per cent concentration to provide complete control of the disease. Dharam Vir and Sharma (1985) recommended sprays of synthetic fungicides like Bavistin, Benlate or methyl thiophonate at 1,000 ppm concentration at 3-4 week intervals or weekly sprays of 1,000 ppm

Karathane for effective control of this disease. Under the All India Floriculture Improvement Project Trials, Sulfex (0.2 per cent), Karathane (0.05 per cent) and Bavistin (0.1 per cent) were observed to be best for reducing the infection of powdery mildew. The centres of the AICRP on Floriculture have also recommended spray of Karach 0.2 per cent, Dinocap 0.05 per cent or Sulfex 0.2 per cent at an interval of 10 days.

Grey Mould or *Botrytis* Blight (*Botrytis cinerea* Pers. Ex. Fr.)

Symptoms: Severe blighting of buds and twigs during rainy season has been reported from the northern Indian hills. The petals of affected buds develop brownish patches, which soon engulf the entire surface and cause rotting. The discoloration extends below to the twigs killing them partially or wholly. Under foggy and high humid conditions, abundant whitish-grey fructification appears over the surface of the blighted portions, which results in quick spread and severity of the disease. The fungus causing the disease has been identified as the common grey mould fungus *Botrytis cinerea* Pers ex. Fr. On leaves, water-soaked lesions are formed which spread very fast. On flowers, infection develops in the form of water soaked lesions, which cause premature fading, discoloration and dropping. The infection is generally seen in the inner whorl of the flowers. In favourable conditions, the infection spreads very fast, coinciding with the period of senescence.

Management: Spray at regular intervals of Bavistin, Benomyl and Rovral has been recommended. Since the disease develops fast on wounded and senescent tissues it is advisable to remove crop debris and infected flowers.

In greenhouses and fields, try to avoid high moisture conditions and free water on leaves and flowers (Sharma, 2001).

Leaf Spot

This is caused by *Colletotrichum capsici* during the rainy season and is characterised in the form of circular blood red spots or rose leaves, with severity defoliation occurs.

Bud and Twig Blight

This is caused by *Phomopsis gulabia* yellowing and defoliation of leaves and light brown spots on the leaves is caused by *Curvularia lunata.*

Glomerella Leaf Blight

This is caused by *Glomerella cingulata* characterised by circular and irregular spots on leaves in rose. Spraying with Benlate (0.1 per cent) or Defolatan (0.1 per cent) was found effective in controlling the disease.

Botryodiplodia Leaf Blight

This is caused by *Botryodiplodia jaezevski* and characterised by circular and irregular leaf spots. The spots on leaves are amphigenous. Infected leaves become chlorotic and defoliate.

Rust (*Phragmidium mucronatum* Fr.)

Symptoms: The symptoms at early stages include chlorotic spots on the leaf surface. The orange coloured uredial spores are commonly seen. The infection results in defoliation of the plants.

Management: Sprays of chemicals like Saprol (0.2 per cent) and Bayleton (0.15 per cent) are effective in disease management.

TABLE 1
Sources of Resistance Against Powdery Mildew in Different Colours and Classes of Roses

Colour	Classes of Roses		
	Hybrid Teas	**Floribundas**	**Miniatures**
Red	Atol, Avon, Chitralekha, Fragrant Cloud, Gladiator, Gulzar, Happiness, Henker Royal, Jawani, Lalima, Mechak, Norma, Oklahoma, Papa Meilland, Precious Platinum, Raktagandha, Ranjana	Deepika, Happy Wanderer, Jantar Mantar, Mercedes, Nordia, Pandit Nehru, Shola, Sindoor	—
Pink	Belle of Punjab, Carina, Chambe de Kali, Catherine Deneuve, Eiffel Tower, First Prize, Homage, Jadis, Medallion, Nurjehan, Priyadarshini, Romantica, Sabine, Surabhi	Imperator, Kavita, Loree, Queen, Elizabeth, Sadabahar, Shrinagar	Tiki
Orange	Command Perfomance, Flaming Sunset, Folklore, Galia, Just Joey, Montezuma, Nayika, Orient Express, Raja Surendra Singh of Nalagarh, Sonia Meilland, Super Star, Sylvia	Angelique, Celestial Star, Flamengo, Suryakiran, Zambra, Zorina	—
Yellow	Black Lady, Cherry Brandy, Chitwan, Dutch Gold, Golden Splendour	Arthur Bell, Lehar, Playboy	—
White	Honor, Neelam, Prestine, Sweet Afton, White Christmas	Akito, Shabnam, Summer Snow	—
Bicolour	American Heritage, Anvil Sparks, Double Delight, Garden Party, Rose Gaujard, Surkhab	Banjaran, Chitchor, Judy Garland, Pink Parfeit, Prema, Sea Pearl	–

Bacterial Diseases

Crown Gall (*Agrobacterium tumefaciens*)

Symptoms: Cauliflower-like galls are produced most commonly at the crown region of the stem at the ground level. Galls are also produced on roots and on stems wounded due to harvesting, pruning and other cultural operations.

Management: This disease is a major problem to plant propagators. General control measures like clean nursery practices, use of disease-free planting material and removal of infected debris help prevent the disease. Commercial formulations consisting of *Agrobacterium rhizogenes* (K-84) are being successfully used in many countries to control crown gall.

Viral Diseases

Rose Mosaic Disease

Symptoms: The infection causes a range of symptoms including mosaic patterns, mottling growth abnormalities like reduced stem length, few flowers of poor quality. The infection also causes low survival, leading to transplanting problems and reduced rate of bud success.

Management: General control measures are advised. Clean nursery stock of mother plants. Rootstocks should be free from viruses.

Streak

Symptoms: Brown rings and brown vein banding appears in fully expanded leaves, and brownish or greenish rings in the canes. Necrotic areas develop in the vicinity of grafted buds, resulting in stem girdling and wilting of leaves. Grafting transmits this virus.

Management: Same as for mosaic.

WATER PLANTS

NELUMBO (Lotus)

Nelumbo nucifera; Family: Nymphaeaceae; Other common name: Kamal

Alternaria Leaf Spot (*Alternaria nelumbii*)

Symptoms: Small reddish-brown spots appear, usually bordered with light green. They increase in size. The leaves curl up and die from the margin inward. Another leaf spot that occasionally appears on this host is caused by *Cercospora nelumbonis.*

Management: Although no controls have been developed, it is likely that application of copper or dithiocarbamate sprays will control these leaf spots.

NYMPHEA (Water Lily)

Other common name: Kumudini; Family: Nymphaeaceae

Nymphaea Leaf Spot (*Ramularia nymphaerum*)

Symptoms: Spots on both leaf surfaces are at first rounded with concentric zoning but later enlarge irregularly and change from reddish to grey-brown or black and finally rupture in the centre. The mycelium is superficial and forms compact white spots on the lesions.

Management: Affected leaves may be removed. Fungicides should not be applied so as to save the water from contamination. Due precaution should be taken to avoid overcrowding of the plants in the pool.

REFERENCES

Alippe H. E. 'Gladiolus rot caused by *Sclerotinia draytoni* in Santa Fe province, Argentina'. *Rev. Fac. Agrpn. Univ. Nac. La Plata* 42: 131-141, 1966.

Anonymous. *Annual Report.* New Delhi, India: AICRP on Floriculture and Landscaping, Indian Agricultural Research Institute, 1988.

Bajaj, K. L., Arora, J. S. and Kaur, P. P. 'Biochemical difference in tolerant and susceptible varieties of gladiolus to Fusarium wilt'. *J. Res. Punjab Agric. Univ.* 26: 585-587, 1989.

Bakrdzhiv, N. 'Soft (wet) bacterial rot of hyacinth'. *Rast. Zashch.* 27: 23-24, 1979.

Bhattacharjee, S. K. and De, L. C. In: *Advanced Commercial Floriculture*. Jaipur, India: Aavishkar Publishers, Distributors, 2003.

Boer, J. H. De. 'A new method of practical heat treatment'. *Bloembollencultuur* 82: 53, 1972.

Bose, T. K. and Yadav, L. P. (Eds.). In: *Commercial Flowers*. Kolkata: Naya Prokash, 1989.

Bruhn. 'Untersuchungen uber die Fusarium-krankheit der gladiolen'. *Phytopathology* 25: 1-38, 1955.

Castillo, S. and Peterson, J. L. 'Cause and control of crown rot of New Guinea *Impatiens*'. *Plant Dis.* 74: 77-79, 1990.

Chandra, K. J., Negi, S. S., Raghava, S. P. S. and Sharma, T. V. R. S. 'Evaluation of gladiolus cultivars and hybrids for resistance to *Fusarium oxysporum* f. sp. *gladioli*'. *Indian J. Horticulture*. 42: 304-305, 1985.

Chen, L. Z., Gan, X. B., Song, J. Y. and Gu, W. 'A study on gladiolus root rot'. *J. Shanghai Agric. Coll.* 12: 240-246, 1994.

Crous, P. W. and Braun, U. 'Cercosporoid fungi from South Africa'. *Mycotaxon* 57: 233-321, 1996.

Darvas, J. M. 'The occurrence of *Stromatinia gladioli* in South Africa'. *Phytophylactica* 16: 255, 1984.

Deshpande, S. D. 'A note on the varietal reaction of hybrid tea roses to powdery mildew'. *Indian Journal of Mycology and Plant Pathology* 9(1): 120, 1980.

Dharam Vir and Sharma, R. K. *Indian Horticulture* 29: 14-15, 1985.

Dhiman, J. S. and Arora, J. S. 'Occurrence of leaf spot and flower blight of marigold (*Tagetes erecta* L.) in Punjab'. *J. Res. Punjab Agric. Univ.* 27: 231-236, 1990.

Drayton, F. L. 'The gladiolus dry rot caused by *Sclerotinia gladioli* (Massey) n. Comb'. *Phytopathology* 24: 397-404, 1934.

Duas, M. R., Des Look, M. T. and Lopes, M. C. '[Portuguese fungi XXVIII] Fungi Lusitanilae XXVII'. *Agron. Lusit.* 41: 77-92, 1981.

Dumock, A. W. and Tammen, J. 'Diseases'. In: *Easter Lilies: The Culture, Diseases, Insects and Economics of Easter Lilies.* (Eds.). Kiplinger, D. C. and Langehans, R. W. U.S.A.: The Ohio and New York State Flower Growers Associations, pp. 107-110, 1967.

Easthem, J. W. *Plant Pathology Branch Rep.* Br. Columb. Dep. Agric. 46: 9, 1931.

Eijk, J. P. van; Giessen, A. C. van der and Toxopeus, S. J. 'Perspective of breeding for resistance to 'yellow disease'. *Xanthomonas hyacinthi* (Wakker) Dowson, in hyacinths'. *Euphytica* 25: 131-138, 1976.

Ferrari, P. Iris diseases and pests'. *Inf. Agron.* 30: 42, 1974.

Georgieva, M. and Peikova, E. '*Fusarium and Penicillium* rots of stored gladiolus bulbs and their control'. *Zelench. I. Konservi* 8: 25-29, 1976.

Grigariu A. C. 'Mycological notes. II. Gladiolus rot caused by *Sclerotinia draytonii* Buddin & Wakef. in Greece'. *Phytopathol. Mediterr.* 15: 123-125, 1976.

Grimal'skaya S. L. 'Detection of resistance in daughter bulbs of gladiolus to *Fusarium* disease on artificially infested ground'. *Byulleten 'Vsesoyuznogo odrena Lenina I Ordena Druzhby Narodov Instituta Rastenievodstva Imeni NI. Vavilova* 78: 66-67, 1978.

Groen, N. P. A. 'Measurement before application is important. Nitrogen encourages *Fusarium* infection in gladioli'. *Bloembollencultuur* 101: 35, 1990.

Gupta, J. H. and Prasad, B. 'Reaction of rose varieties to dieback caused by *Diplodia rosarum*'. *Ind. J. Mycol. and Pl. Pathol.* 14(2): 194, 1984.

Harrison, D. E. 'Fusarium, yellow and corm rot of gladiolus'. *J. Agric. Viet.* 60: 377-387, 1962.

Hashimoto, T., Kimura, T. and Katsuki, S. 'Leaf spot disease of watercress (*Roippanasturtium-aquaticum* Hayek) caused by *Cercospora nasturtii* Passerini'. *Sci. Rep. Miyagi Agric. Coll.* 35: 77-79, 1987.

Holloman, A. and Young, R. A. 'Evaluation of fungicides for control of the leaf spot disease of gladiolus caused by *Botrytis gladiolorum* Timmermans'. *Plant Dis. Rep.* 38: 456-458, 1951.

Infantino, A. and Rumine, P. '*Fusarium oxysporum* f. sp. *gladioli* on montbretia cut flowers in Italy'. *Petria* 3: 65-68, 1993.

Kabashna, L. V. 'Soft rot of hyacinth in Ukraine'. *Mikrobiol. Zh.* (Kiev) 36: 733-737, 1974.

Kabashna, L. V., Korolova, I. B. and Palu, G. K. 'Sensitivity of causal agents of soft rots of hyacinth and calla to some antimicrobial substances'. *Mikrobiol. Zh.* (Kiev) 37: 492-495, 1975.

Katan, J. 'Solar pasteurisation of soil for disease control: Status and prospects'. *Plant Dis. Rep.* 64: 450-454, 1980.

Kaur, S., Arora, J. S. and Khanna, K. 'Fusarium wilt is a limiting factor in commercial cultivation of gladiolus'. *Indian Hortic.* 36: 21-22, 1989.

Kore, S. S., Nirmal, D. D. and Ansarwadekar, K. W. *South Indian Horticulture* 25: 167, 1977.

Lewis, J. A., Fravel, D. R., Lulmsden, R. D. and Shasha, B. S. 'Application of biocontrol fungi in granular formulation of pre-gelatinised starch-flour to control damping-off diseases caused by *Rhizoctonia solani*'. *Biol. Control* 5: 397-404, 1995.

Magie, R. O. 'Biological control of *Fusarium* disease of gladiolus'. *Acta Hortic.* 109: 415-420, 1980.

Magie, R. O. 'Soil fumigation in controlling gladiolus *Stromatinia* disease'. Proc. *Fla. State Hortic. Soc.* 70: 373-379, 1968.

Magie, R. O. and Poe, S. L. 'Disease and pest associates of gladioli'. In: *The World of Gladiolus.* (Ed.). Noreen Koenig. U.S.A.: N. Am. Gladiolus Council, pp. 155-181, 1972.

McCulloch, L. 'A leaf and corm disease of gladioli caused by *Bacterium marginatum*'. *J. Agric. Res.* 29: 159-177, 1924.

McCulloch, L. 'A vascular disease of gladiolus caused by *Fusarium*'. *Phytopathology* 34: 263-287, 1944.

McWhorter, F. P. 'Non-virus diseases of lilies (*Lilium* spp.)'. U.S.A.: *Lily Yb. N. Am. Lily Soc.* 16: 54-88, 1963.

Merlo, P. A., Rollan, M. C. and Ronco, B. L. 'Wilt of *Antirrhinum majus* L. caused by *Verticillium dahliae* Klebahn in the Republic of Argentina'. *Rev. Fac. Agron.* 61-62: 181-185, 1986.

Mirza, J. H. and Shakir, A. S. 'First report of fungal pathogens of gladiolus from Pakistan'. *Pak. J. Phytopathol.* 3(1-2): 74-76, 1991.

Nagler, M. 'Experiments on the chemical control of corm rot of gladiolus'. *Acta Bot. Hortic. Bucur.*, pp. 177-184, 1965.

Nene, Y. L. and Thapliyal, P. N. In: *Fungicides in Plant Disease Control.* New Delhi, India: Oxford & IBH Co. Pvt. Ltd., 1997.

Oh, B. Y., Shim, J. W., Jeong, Y. H., Park, I. S. and Lin, S. E. 'Development of smoking pellet fungicides for the control of major diseases of strawberry and lily under greenhouse conditions'. *Crop Prot.* 33: 37-38, 1991.

Paul, Y. S., Sharma, R. C. and Bhardwaj, L. N. 'Additions to fungi of Himachal Pradesh-IV. (Peronosporales)'. *Plant Dis. Res.* 6: 105, 1991.

Paulus *et al.* In: *Flower and Nursery Report.* U.S.A.: University of California, Summer and Spring (1-2), 1983.

Pirone P. P. *Diseases and Pests of Ornamental Plants.* Fifth edn. New York, U.S.A.: John Wiley and Sons, 1978.

Protsenko, E. P. 'Premature yellow of gladioli'. *Bull. Centr. Bot. Gdn. Moscow* 30: 78-84, 1958.

Sahni, M. L. 'Reaction of some rose cultivars to powdery mildew fungus'. *Indian Phytopath.* 40(4): 534-535, 1987.

Sellwood, J. E. and Lelliot, R. A. 'Internal flowering of hyacinth caused by *Erwinia thapontici*'. *Plant Pathol.* 27: 120-124, 1978.

Sezgin, E., Karcilioglu, A., Esentepe, M. and Onan, E. 'Investigations on the determination of susceptibility of some gladiolus cultivars against Fusarium corm rot'. *J. Turk. Phytopathol.* 12: 73-75, 1983.

Sharma, P. 'Recent developments in disease management of greenhouse crops'. *Intensive Agriculture* Jan.-Feb., 2001.

Sharma, P. and Bhattacharjee, S. K. *Plant Protection in Ornamental Crops.* AICRP on Floriculture, I.C.A.R. Tech. Bull., 19, 2002.

Sharma, P. and Raghawa, S. P. S. In: *Disease Management of Ornamental Crops.* New Delhi, India: Division of Floriculture and Landscaping, I.A.R.I., 2000.

Sharma, P. and Singh, A. P. 'Evaluation of rose cultivars against dieback'. *Indian Phytopathology* 54(3): 388-389, 2001.

Sharma, P. and Singh, A. P. 'Multiple disease resistance against foliar diseases of roses'. *Indian Phytopathology* 55(2):169-172, 2002.

Shukla, P. and Singh, B. *The Indian Rose Annual* 9: 146-148, 1991.

Smith, C. O. 'Leaf and scape spot of the Amaryllidaceae in California'. *Phytopathology* 19: 1138, 1929.

Srivastava, D. N. 'Dieback of roses'. *Indian Horticulture* 5: 24-25, 1961.

Tarabeih, A. M., Al-Zarari, A. J. and Michael, S. H. 'Leaf scorch of Amaryllis, new to Iraq'. *Phytopathol. Mediterr.* 19: 150-152, 1980.

Valaskova, E. 'Combined effect of heat and fungicides in the treatment of gladiolus corms against *Fusarium* disease'. *Vyzkunueho a Slechtitelskeho Ustav Oktrasueho Zahraduictvi V Pru honicich* 7: 265-282, 1977.

Vlasova, V. I. and Shltan, N. 'Means for increasing resistance of plants to *Fusarium* wilt'. *Nauchn. Trudy Stavrool Sk.* 37: 127-133, 1974.

Wiedemann, G. J., TeBeest, D. O. and Cartwright, R. D. 'Host specificity of *Colletotrichum gloeosporiodes* f.sp. *aeschynomene and C. truncatum* in the leguminosae'. *Phytopathology* 78: 986-990, 1988.

Woltz, S. S. and Jones, J. P. 'Micronutrient effects on the *in vitro* growth and pathogenicity of *Fusarium oxysporum* f. sp. *lycopersici*'. *Phytopathology* 58: 336-338, 1968.

Woltz, S. S., Magie, R. O., Switkin, C., Nelson, P. E. and Toussoun, T. A. 'Gladiolus disease response to pre-storage corm inoculation with *Fusarium* sp'. *Plant Dis. Rep.* 62: 134-137, 1978.

Wu, W. S. and Cjou, J. K. 'Chemical and biological control of *Alternaria carthami* on zinnia'. *Seed Sci. Technol.* 23: 193-200, 1995.

Wu, W. S. and Yang, Y. H. Alternaria blight, a seed transmitted disease of zinnia in Taiwan'. *Plant Pathol. Bull.* 1: 115-123, 1992.

Yadahalli, K. B., Srikant, K., Anahosur, K. H. and Kulkarni, S. '*In vitro* and *in vivo* evaluation of fungicides against leaf spot of zinnia caused by *Cercospora zinnia* Ell. & Mart'. *Karnataka J. Agric. Sci.* 7: 363-365, 1994.

—oo(O)oo—

11

NEMATODE DISEASES OF ORNAMENTALS AND THEIR MANAGEMENT

M. NAGESH AND T. JANAKIRAM

INTRODUCTION

NEMATODES are important soil-borne pests of most crop rhizospheres, including vegetables, fruits, ornamentals, medicinal, aromatic, plantation, spices and condiment crops besides several other field crops. The importance of plant parasitic nematodes can no longer be underrated, although the damage they cause is rarely visible to the naked eye. Because of their short life cycle and very high rates of reproduction, nematodes are able to build up such high population densities rapidly, especially in close cropping (for example, carnation, gerbera, tuberose, gladiolus) or in permanent crops and monocultures (for example, jasmine, crossandra, rose) when there is no possibility of crop rotation, that no longer can tolerable levels of damage must be expected. The microscopic size of nematodes (length 0.5-1.5 mm; diameter 30-50 mmm) and the fact that they inhabit the soil and damage the root means that a superficial inspection of the crop usually does not immediately disclose the adverse effects they have on the growth of the plants.

Monoculture, the common feature of the ornamental crops, provides a congenial environment for one or the other nematodes, the population of which, if unchecked, may touch an alarmingly high level and may even wipe out a whole crop. Apart from this, even a low nematode population may render planting stocks like tuberose and gladioli bulb, unmarketable. The infested planting material may also act as a carrier of some nematode genera/species of quarantine significance to newer areas. In the U.S.A., losses in ornamentals due to nematodes were estimated to the tune of $ 60 million annually (Hague, 1972). Throughout the world in ornamental crops, plant nematodes are responsible for 11.1 per cent losses (Sasser and Freckman, 1987). However, in India, there is no precise information about the losses in ornamental crops alone.

HISTORICAL

Ornamentals, like other crops, harbour a multitude of nematode fauna globally (Table 1), of which some nematodes based on their intrinsic potentials have caused tremendous economic losses.

In India, almost the first report on nematode problems on ornamentals dates back to 1964 when Prasad and Dasgupta reported the association of some nematodes on rose. Subsequently, several reports on the association/occurrence of plant parasitic nematodes on ornamental crops were made from different parts of the country (Table 2). Nematodes like *Ditylenchus dipsaci* have been reported to be responsible for completely wiping out the narcissus industry of Britain (Hague, 1972). *Meloidogyne* sp. on gladioli, *Aphelenchoides ritzemabosi, Belonolaimus longicaudatus* and *Trichodorus* sp. on chrysanthemum in the U.S.A., *Xiphinema americanum* on roses in the U.S.A. and *Xiphinema diversicaudatum* in Western Europe have been listed as the major pests resulting in considerable growth reduction (Hague, 1972).

The literature collected is presented under two heads, namely, global and Indian scenarios, separately, for the convenience of the reader.

Global Scenario (Table 1)

Indian Scenario (Table 2)

ECONOMIC IMPORTANCE

The avoidable yield losses due to plant parasitic nematodes in some of the ornamental crops have been presented in Table 3. However, in India, there is no precise information about the losses in ornamental crops alone.

Though farmers, knowingly or unknowingly, practice crop rotation and cropping systems that suppress nematode populations naturally, nematodes like root-knot, reniform, lesion, etc. multiply rapidly and cause decline in plant vigour and yield due to their polyphagous feeding and wide host range. In tropical and subtropical climates, build-up of nematode populations is more rapid. Plant parasitic nematodes, besides causing direct yield losses by debilitating the plant health, also cause loss to quality and market value. The quality of planting material, including rhizomes, tubers, transplants, saplings and cuttings (namely, tuberose, gladiolus, carnation, gerbera, crossandra, jasmine, etc.), raised in nematode infested fields/soil is not only poor but also such plants are pre-disposed to other soil-borne pathogens.

SYMPTOMS

General Symptoms Due to Plant Parasitic Nematodes

In the Roots: The various nematode species in search for food inflict visible damage mainly in the form of lesions (necrotising of cortical tissue) or swellings, cessation of growth and death of root tips or deformations of the root system due to root knot. All these impairments of the feeder roots inhibit their growth and cause them to die. As well as the purely mechanical destruction or pathological changes to the cells directly parasitised by nematodes, the physiology of the root tissue is also altered in such a way as to generally reduce the functional efficiency of the root. Serious concomitant effects are marked disturbances of water and nutrient uptake by the active root system.

The parts of roots attacked by nematodes offer avenues of entrance to soil-borne microorganisms (fungi and bacteria). These secondary pathogens invade the wounds caused by nematodes or can also infect the weakened plant more easily at different points. Interactions, mainly with soil-borne fungi such as *Fusarium* sp., *Rhizoctonia, Verticillium* sp. and *Pythium* sp., greatly intensify the damage already caused by the two groups of pathogens (fungi and bacteria). The necroses spread rapidly to the vascular system of the central cylinder of the roots, directly impeding, and at an advanced stage, completely abolishing the supply of water and nutrients to the plant. Complete decay of the affected area of the root is the result.

TABLE 1
Plant Parasitic Nematodes Associated with Ornamental Crops in the World

Ornamental Crops	Nematode Associated
Flowering Plants	
Rose	*M. hapla*
	Pratylenchus vulnus
	Helicotylenchus labiata
	H. dihystera
	Hoplolaimus galeatus
	Xiphinema diversicaudatum
	Helicotylenchus nannus
	Pratylenchus pratensis
	Hemicycliophora typica
	Longidorus macrosoma
Chrysanthemum	*Aphelenchoides ritzemabosi*
	Pratylenchus penetrans
	Helicotylenchus
	Meloidogyne
Carnation	*Rotylenchulus reniformis*
	M. arenaria
	M. hapla
	M. javanica
	M. incognita
	Ditylenchus dipasaci
	Heterodera trifolii
	Pratylenchus dianthus
	Criconemella curvata
	Pratylenchus
	Tylenchus,
	Pratylenchus
	Macroposthonia cuvata
	Pratylenchus curvitatus
Dahlia	*Aphelenchoides ritzemabosi*
	Ditylenchus diestructor
Crossandra	*Longidorus africanus*
	Meloidogyne incogita
Jasmine	*M. incognita*
	Aphelenchoides besseyi
Hyacinth and Tulip	*Ditylenchus*

Continued...

...Continued

Ornamental Crops	Nematode Associated
Lily	*Pratylenchus penetrans*
Phlox	*Xiphinema vuittenema*
Gladiolus	*Scutellonema brachyurum* *Helicotylenchus rotundicaudatus*
Asparagus sprangeri	*Helicotylenchus cavensii*
Cineraria hybrida	*Helicotylenchus* sp.
Begonia	*M. incognita*
Zinnia elegans *Salvia, Aster*	*Aphelenchoides ritzemabosi*
Hibisicus rosa sinesis	*Helicotylenchus varicaudatus*
Orchids	*Aphelenchoides ritzembosi*
Succulent Leaf Plants	
Cacti	*Cactodera cacti* *M. incognita* *Heterodera cacti*
Croton	*H. cacti*
Crenatha and Oppenheimiana	*Helicotylenchus dihystera*
Momordica	*Pratylenchus coffeae*
Hoya carnosa *Notocactus schuman*	*Scutellonema brachyurus*
Palm	*Helicotylenchus dihystera*
Other Ornamentals	
Pratylenchus coffeae *Ficus elastica* *Hippestrum vittarum* *Peperomia* *Zebrina pendala* *Abenda arandifolara* *Cornus florida* *Hydrangera*	*Meloidogyne hapla*

Continued...

...Continued

Ornamental Crops	Nematode Associated
Photima *Spiraea* *Viburnum cartesii*	
Ferns	
Asplenium nidus	*Aphelenchoides ritzemabosi*
Aucuba japonica *Buxus microphylla* (Japanese boxwood)	*M. arenaria*
Shrubs and Trees	
Camellia japonica	*Tylenchorhynchus claytoni*
Gardenia jasminoides	*M. arenaria*
Illex corunata (Chinese holly) var *.rotunda* *Illex crenata compacta*	*Meloidogyne arenaria*
Juniperus coniferta	*Criconemelola xenoplex*
J. excelsa	*Pratylenchus vulnus*
J. horizontalis	*P. vulnus*
Nandina *Rhododendron indicum* *R. obstusum*	*T. claytoni*

Source: Singh and Sharma, 1998.

In the Shoots: The symptoms of damage are much the same for all crops. High population densities of nematodes may cause the following symptoms due to water stress and reduced nutrient uptake: wilted, chlorotic and later necrotic foliage (nutrient deficiency) leading to death of entire branches; stunting of shoots, leaves and fruits, deficient fruiting, retarded fruit ripening and, in bananas, reduced standing power due to the weakened root system.

Irregular growth and different rates of fruit ripening within the same plantation may create difficulties for harvesting. Even if no direct damage to the aerial parts of the plants is yet apparent, substantial yield losses are often still to be expected. Moreover, a high infestation level in the crop often means increased production of costs for replanting, additional fertilisation and irrigation of the diseased plants as well as control measures to combat subsequent infection by fungal pathogens.

Specific Symptoms

The specific symptoms of damage in different ornamental crops as reported are listed in Table 4.

TABLE 2
Plant Parasitic Nematodes Associated with Ornamental Crops in India

Nematode	Crop	Place	Reference
Bitylenchus vulgaris *Hoplolaimus galeatus* *Xiphinema diversicaudatum* *Helicotylenchus nannus* *Pratylenchus pratensis* *Hemicyclophora typica* *M. javanica*	*Rosa indica*	Delhi	Prasad & Dasgupta, 1964
M. arenaria, *M. javanica*	*Chrysanthemum*	Karnataka	Chandwani & Reddy, 1967
Tylenchorhychus zeae	*Clerodendran splendens*	Rajasthan	Sethi & Swarup, 1968
T. chonai	*Bougainvillea* sp.	Delhi	Sethi & Swarup, 1968
Pratylenchus delattrei	*Crossandra*	Tamil Nadu	Srinivasan, 1974
M. incognita	*Rosa indica*	Aligarh	Alam *et al,* 1973b
Hemicyclophora labiata	*Rosa indica*	Tamil Nadu	Muthukrishnan *et al,* 1975
M. incognita *M. javanica*	*Polianthes tuberosa*	Tamil Nadu	Jayaraman *et al,* 1975
Meloidogyne incognita	*Crossandra*	Tamil Nadu Karnataka Andhra Pradesh	Rajendran *et al,* 1976
Aphelenchoides ritzemabosi	*Chrysanthemun* *Zinnia elegans, ,* *Savia* *Aster* *Dahlia*	Himachal Pradesh	Gill & Sharma, 1976
Longidorus africanus	*Crossandra*	Tamil Nadu	Muthukrishnan *et al,* 1977
M. javanica	*Chrysanthemum*		Sen & Dasgupta, 1977
M. incognita	*Jasminum sambac,* *J. flexile*	Tamil Nadu	Rajaraman & Rajendran, 1979

Continued...

...Continued

Nematode	Crop	Place	Reference
M. incognita	Gladiolus	Karnataka	Parvatha Reddy *et al*, 1979
M. indica *R. reniformis* *Helicotylenchus crenatus* *Tylenchorhynchus dubius* *Hoplolaimus indicus*	*Chrysanthemum* sp.	Karnataka	Krishnappa *et al*, 1980
M. incognita *M. javanica* *M. arenaria*	*Polianthes tuberosa*	Tamil Nadu	Sundarababu & Vadivelu, 1988b
Radopholus similis	*Jasminum pubescens*	Karnataka	Khan and Parvatha Reddy, 1989
Tylenchulus semipenetrans	*J. sambac*	Haryana	Bajaj, 1989
M. incognita *M. javanica* *Rotylenchulus reniformis*	*Polianthes tuberosa*	Karnataka	Khan and Parvatha Reddy, 1991
Pratylenchus coffeae	*Chrysanthemum* sp.	Karnataka	Khan and Parvatha Reddy, 1991
Apratylenchoides homoglans	*Chrysanthemum* sp.	Karnataka	Siddiqui *et al*, 1991
M. incognita	*Carnation, Gerbera*	Karnataka	Nagesh & Parvatha Reddy, 1996
Helicotylenchus *Tylenchorhynchus* *Xiphinema basiri*	*Rosa indica*	Rajasthan	Trivedi, 1998
M. incognita	*Chrysanthemum* sp.	Karnataka	Nagesh *et al*, 2001

Meloidogyne incognita is one of the serious limiting factors in commercial cultivation of carnation and gerbera under polyhouse conditions. Most of the highly fetching exotic cultivars of carnation and gerbera from Europe have shown 40 to 60 per cent mortality in polyhouse beds due to root-knot nematode infection in and around Bangalore (Nagesh and Parvatha Reddy, 1996).

INTERACTION WITH OTHER MICROORGANISMS

Bacteria

Incidence of gladiolus scab caused by *Pseudomonas marginata* was found to increase in the population of *Meloidogyne javanica* (Elgoorani *et al*, 1976).

TABLE 3
Avoidable Yield Losses in Ornamental Crops Due to Plant Parasitic Nematodes Recorded in India

Crop	Nematode(s)	Yield Loss (%)	Reference(s)
Crossandra (*Crossandra undulaefolia*)	*M. incognita*	21.64	Khan & Parvatha Reddy, 1994
Tuberose (*Polianthes tuberosa*)	*M. incognita*	13.78	Khan & Parvatha Reddy, 1994
Patchouli (*Pogostemon patchouli*)	*M. incognita*	47.00	Prasad & Reddy, 1984
Carnation (*Dianthus caryophyllus*)	*M. incognita*	26.60	Nagesh & Parvatha Reddy, 2000a
Gebera (*Gerbera chinensis*)	*M. incognita*	31.10	Nagesh & Parvatha Reddy, 2000b

Fungi

The nematode interacts with soil-borne fungi such as *Fusarium* sp. and *Rhizoctonia solani* in causing wilt/root-rot complex in several ornamentals. Edward *et al* (1969) reported *Pseudomonas chrysanthus* associated with root-rot of chrysanthemum. The presence of *Pseudomonas coffeae* has been found to aggravate the disease severity of *Pythium aphanidermatum* and *Rhizoctonia solani* in chrysanthemum (Hassan, 1988). *Meloidogyne incognita* interacts synergically with wilt fungi *Fusarium oxysporum* and *Fusarium solani* and results in premature death of Crossandra plants (Khan and Parvatha Reddy, 1992). Nagesh *et al* (1999) observed Fusarium wilt in Gladiolus followed by lesions and rotting of corms due to *Fusarium oxysporum* and *Meloidogyne incognita* complex.

Histopathology and Histochemistry

Pratylenchus coffeae infects and remains confined to the cortical cells in chrysanthemum. Parenchymatous cells are completely destroyed due to nematode feeding (Rashid and Khan, 1975).

MANAGEMENT METHODS

The nematode control and management practices studied by various workers in the world are listed in Table 5.

The nematode control and management practices studied by various workers in the country are listed in Table 6.

Hot-Water Treatment of Planting Material

Hot water treatment is widely used for the killing of nematodes within plant tissues before planting and has proved useful with nematode-infested bulbs, tubers, seeds and roots of plants such as tuberose and gladiolus (Table 7).

Disinfestation of Nematode-Infested Re-Circulated Water: An exposure time of at least 1 hour to 20 g ozone was required to eliminate *Radopholus similis* and prevent reproduction on

TABLE 4
Symptoms of Damage Caused by Plant Parasitic Nematodes

Nematode	Symptom	Reference
Rose		
Pratylenchus vulnus	Stunting and chlorosis, affected roots show patchy discoloration in cortical region, which coalesce and lead to root rotting.	Schindler, 1956; Sher, 1957
Xiphinema diversicaudatum	Unproductive flower and increased susceptibility to other diseases. Affected roots show crinkled terminals, poor secondary roots.	Schindler, 1957
Hoplolaimus galeatus	Stunting in rose	Prasad & Dasgupta, 1964
X. diversicaudatum	Stunting in rose	Prasad & Dasgupta, 1964
Helicotylenchus nannus	Stunting in rose	Prasad & Dasgupta, 1964
Hemicycliophora typica	Stunting in rose	Prasad & Dasgupta, 1964
Hemicycliophora labiata	Infects 2 rose varieties Edward & Red rose and causes debilitation	Muthukrishnan *et al*, 1975
P. zeae	Poor growth of rose plants showing chlorotic, symptoms, poor and stunted growth with necrotic lesions on roots.	Sunderababu & Vadivelu, 1988a
Chrysanthemum		
Aphelenchoides ritzembosi	Necrosis, stunting, distortion distortion & formation of high rough feeding areas with die-back symptoms	Southey, 1952
A. ritzembosi	Angular spots in leaves reduced number of leaves with shrunken appearance, shed untimely production of undersize flowers	Gill and Sharma, 1976 Mcleod 1981

Continued...

...*Continued*

Nematode	Symptom	Reference
Pratylenchus penetrans	Root lesions causing decreased root weight and shoot growth	Netherlands, 1986
P. coffeae	Heavy root damage which subsequently leads to poor growth. Stunting of plants with premature yellowing, drying of leaves and reduced size of flowers, common above ground symptoms	Rashid and Khan, 1975
Root-knot nematode	Yellowing and severe chlorosis of foliage of the affected plant, root galling and necrosis	Nagesh *et al,* 2001
Crossandra		
P. delattrei	Stunting, chlorosis of leaves and wilting. The leaves show mottled appearance which turn to brown and become pinkish eventually. Later the leaf tips dry and entire leaves turn yellow and completely defoliate. The roots exhibit brown to black lesions. Heavily infested plants do not produce tertiary spike thereby flower yield is reduced	Srinivasan & Muthukrishnan, 1975
Longidorus africanus	Distortion, forking and swelling of root. Hyperphasia in root parenchyma pushing root meristem laterally and disarrayed leading to stunting of plants; branching, swelling and darkening of root tips.	Muthukrishnan *et al,* 1977

Continued...

...Continued

Nematode	Symptom	Reference
M. incognita	Stunting and reduced plant growth	Rajendran *et al*, 1976
Jasmine		
M. incognita	Pale coloured leaves, dieback. Root exhibited small swelling and enlarged rootlet	Rajaram and Rajendran, 1979
Aphelendoides besseyi	Stem galls	Stoke, 1979
M. incognita	Galls or knots on roots	Nagesh and Parvata Reddy, 1998
Radopholus similis	Lesions and necrosis in cortical cells, initially patches which coalesce and cause root decay	Nagesh and Parvata Reddy, 1998
Carnation and Gerbera		
M. incognita	Drying of leaves from tip which spreads to whole plant, gradual drying and loss of plant	Nagesh & Parvatha Reddy, 1998
Dahlia		
Aphelenchoides ritzembosi	Brown spots on leaves	Gill & Sharma, 1976 Lamberti *et al*, 1987
Gladiolus		
Meloidogyne sp.		
Trichodorus sp.		
Pseudomonas marginata	Scab	
Polianthes tuberosa		
M. incognita *M. javanica*	Affected plants exhibit stunting, yellowing and drying up of leaves and rotting of bulbs	Jayaraman *et al*, 1975
Gladiolus		
M. incognita	Heavy galling on roots results in yellowing of leaves which subsequently lead to stunted growth	Parvatha Reddy *et al*, 1979

Revised after Singh and Sharma, 1998.

TABLE 5
Nematode Control/Management Practices in Different Ornamental Crops Studied Worldover

No.	Nematode	Practice	Reference
I.	**Physiotherapeutic and Cultural Method**		
1.	*Aphelenchoides ritzemabosi* in *Heliconia* sp.	Hot water treatment at 48° C for 1 hr or at 50° C for 30 min	Criley, 1988
2.	*A. ritzemabosi* in chrysanthemum	Hot water treatment at 46° C for 5 min.	Staniland, 1950
3.	Nematodes in gladiolus	Hot water treatment at 57.8° C for 30 min	
4.	*A. fragariae* in lily bulbs	Hot water treatment at 36° C for 6 hr or at 41° C for 1 hr.	
5.	*Ditylenchus dipsaci* in tulip bulbs	Hot water treatment at 43.3° C for 3 hr	Staniland, 1937
6.	*Meloidogyne incognita* in jasmine	Raise cuttings or planting material in nematode-free or treated nursery beds	Nagesh and Parvatha Reddy, 1998
II.	**Chemical Method**		
1.	*Aphelenchoides ritzemabosi*	Parathion (25 WP) 2-4 kg a.i./100 gallon of water	Sher, 1957
2.	*M. hapla* in rose	Nemacur @ 0.1 % as bare root dip treatment for 30 min	Dale, 1973
3.	*Hemicycliophora* sp. *Xiphinema basiri* in rose	Oxamyl 25 @ 10 ppm a.i. in potted plant soil or 500 ppm a.i. as aerial spray	Muthukrishnan *et al*, 1975; Rajendran and Muthukrishnan, 1980
4.	*Aphelenchoides ritzemabosi* on *Zinnia elegans*	Metacid and Ekalux	Gill and Walia, 1980
5.	*M. incognita* in Gladiolus	Plant fumigation with vorlex	
6.	*Aphelenchoides ritzemabosi* in chrysanthemum	Soil fumigation with methyl bromide	
7.	*Pratylenchus penetrans* in chrysanthemum	Soil application with Aldicarb @ 500 q/ha	

Continued...

...Continued

No.	Nematode	Practice	Reference
8.	*Ditylenchus dipsaci* in tulip bulbs	Hot water treatment and spray of Thionazin 2 l / 100 gallon water in flowers	Winfield, 1972, 1973
9.	*M. incognita* in jasmine	a. Treat nursery beds with 25 g Carbofuran 3G or phorate 10G/sq m. b. Treat nursery beds with 2 kg/sq m neem or pongamia cakes. c. Treat nursery beds with 1 kg/sq m neem or pongamia cakes along with mycorrhizal spores /sq m. d. Treat soil with carbofuran 3G or phorate 10 G @ 2.5 kg ai/ha before planting in main field	Nagesh and Parvatha Reddy, 1998

Source: Singh and Sharma, 1998.

Anthurium andreanum in laboratory tests. Activated hydrogen peroxide was effective from a concentration of 400 ppm during at least 24 hours. With a new formulation, only 200 ppm for 24 hours is required for the complete elimination of *Radopholus similis* (Runia and Amsing, 1996).

Cultural Methods

Inter-cropping of Crossandra with enemy plants like marigold significantly reduced the root knot and lesion nematode population both in soil and roots (Khan and Parvatha Reddy, 1994). Application of farmyard manure has been found to reduce the lesion nematode population in Crossandra.

Chemical Methods

Seedling Bare-Root Dip Treatment: Dipping of gladiolus corms in Thionazin or Fensulfothion solution (0.5 g a.i. per litre) gave reduced root-knot nematode infestation. Root dip treatment for 8 hours in 0.05 to 0.1 per cent solution of Aldicarb, Carbofuran, Fensulfothion, Fenamiphos or Phorate protected Crossandra seedlings from *Pratylenchus delattrei* for 30 days after transplanting (Vadivelu and Muthukrishnan, 1979).

Soil Treatment in the Main Field: Carbofuran, Phorate, Fenamiphos and Disulfoton at 1 g per plant effectively controlled *Pratylenchus delattrei* and increased flower yield of Crossandra (Vadivelu and Muthukrishnan, 1979). Carbofuran at 2.5 kg a.i. per hectare was found to be effective against root-knot and lesion nematodes on Crossandra (Srinivasan and Muthukrishnan, 1976).

Aldicarb and Phorate, each at 1.5 kg a.i. per hectare were effective in controlling *Aphelenchoides ritzemabosi* on chrysanthemum (Gill, 1981). For the control of *Meloidogyne incognita*

TABLE 6

Nematode Control/Management Practices in Different Ornamental Crops Studied in India

No.	Nematode	Practice	Reference
I.	**Physiotherapeutic and Cultural Method**		
	Meloidogyne incognita in jasmine	Raise cuttings or planting material in nematode-free or treated nursery beds	Nagesh and Parvatha Reddy, 1998
II.	**Chemical Method**		
	Hemicycliophora sp, *Xiphinema basiri* in rose	Oxamyl 25 @ 10 ppm a.i. in potted plant soil or 500 ppm a.i. as aerial spray	Muthukrishnan *et al*, 1975 Rajendran and Muthukrishnan, 1980
	Aphelenchoides ritzemabosi on *Zinnia elegans*	Metacid and Ekalux	Gill & Walia, 1980; Gill, 1981
	M. incognita in Gladiolus	Plant fumigation with Vorlex	
	Aphelenchoides ritzemabosi in chrysanthemum	Soil fumigation with methyl bromide	
	Pratylenchus penetrans in chrysanthemum	Soil application with Aldicarb @ 500 q/ha	
	M. incognita in jasmine	a. Treat nursery beds with 25 g Carbofuran 3G or phorate 10 G/sq m. b. Treat nursery beds with 2 kg/sq m neem or pongamia cakes. c. Treat nursery beds with 1 kg/sq m neem or pongamia cakes along with mycorrhizal spores/sq m. d. Treat soil with Carbofuran 3 G or Phorate 10 G @ 2.5 kg ai/ha before planting in main field	Nagesh and Parvatha Reddy, 1998

on tuberose, Aldicarb, Fensulfothion and Phorate were effective and increased flower yield.

In field trials carried out in Hawaii, the insecticide-nematicides Aldicarb and Oxamyl were effective against burrowing nematode, *Radopholus similis* on Ozaki anthuriums (*Anthurium*

TABLE 7
Time and Temperature Recommendations for Hot-Water Treatment for Denematising Planting Stock

Nematode	Planting Stock	Time Min.	Temp.° C
Meloidogyne sp.	Tuberose bulbs	60	49.0
	Caladium tubers	30	50.0
	Gladiolus corms	30	57.8

andreanum). Fenamiphos controlled *Radopholus similis.* Plants treated with Aldicarb, Fenamiphos and Oxamyl had increased flower production and flower stem length. Spathe area was increased by Aldicarb and Fenamiphos. Increase in flower production, flower stem length and spathe area was mainly attributed to control of the burrowing nematode (Hara *et al,* 1988).

Foliar Application: Foliar application of Methyl Parathion, Chlorpyriphos and Quinalphos each at 0.5 per cent were effective in the control of *Aphelenchoides ritzemabosi* infecting zinnia and chrysanthemum (Gill, 1981; Gill and Walia, 1980).

HOST RESISTANCE

Screening of Germplasm and Development of Resistant Varieties: Research on crop resistance to economically important nematodes in ornamentals is in its infancy. So far very little work was carried out in India in this direction except for identification of resistance sources in certain ornamental crops (Table 8). There is a need to develop a strong database on available sources of resistance, genetics and nature of resistance-breeding programmes to evolve commercial germplasm with resistance/tolerance to important nematodes.

Biochemical Mechanism of Nematode Resistance: Attempts were made to record the sequential biochemical changes in roots of *Callistephus chinensis* lines resistant and susceptible to *Meloidogyne incognita* Chitwood race 1 (Nagesh *et al,* 1999). There were significantly higher contents of total phenols and total proteins, and higher activity of PPO in nematode infested roots of resistant cultivars in the first two weeks of infection, which declined in the third and fourth weeks of infection, compared to that of its healthy roots and healthy and infected roots of susceptible cultivar. Phenols, PPO and PAL increased as the infection advanced, especially in the resistant line. The initial increase in phenols and PPO activity in resistant lines would have prevented the nematode penetration and development. With ageing of the roots, (3rd and 4th week after infection) the phenols and PPO activity would have naturally declined. However, with ageing of the roots, the PAL

TABLE 8
Sources of Nematode Resistance in Ornamental Crops

Crop	Nematode	Resistant Cultivars	Reference
China aster	*M. incognita*	Shashank-Resistant Poornima-Mod. Resistant	Nagesh *et al*, 1995
Tuberose	*M. incognita*	Sringar-Resistant Suvasini-Tolerant	Nagesh *et al*, 1995

TABLE 9
Reports of Opportunistic Fungi Effective Against Plant Parasitic Nematodes in India

Antagonists	Nematodes	Crop	Reference
Paecilomyces marguandi	*M. incognita*	Crossandra	Khan and Parvatha Reddy, 1994
P. lilacinus *Verticillium lecanii*	*M. incognita*	Crossandra	Nagesh & Parvatha Reddy, 1995
P. lilacinus *T. harzianum* *T. viride*	*M. incognita*	Carnation, Gerbera	Nagesh & Parvatha Reddy, 1996
P. lilacinus	*M. incognita*	Crossandra	Nagesh & Parvatha Reddy, 1997a
P. lilacinus	*M. incognita*	Tuberose	Nagesh *et al,* 1977
P. lilacinus *Trichoderma harzianum* *T. viride*	*M. incognita*	Gladiolus	Nagesh *et al,* 1999
Glomus fasciculatum *G. intraradices* *G. mosseae* *Aucalospora laevis*	*M. incognita*	Crossandra	Nagesh *et al*, 1999

activity could be high in order to bio-synthesise isoflavanoids that would increasingly activate cellular lignification, thus preventing nematodes from establishing and developing.

BIOLOGICAL METHODS

Several studies were carried out on the feasibility of biological control of plant parasitic nematodes associated with ornamental crops. However, the studies mainly focussed on antagonistic fungi (Table 9). The parasitic fungus, *Paecilomyces marguandi* grown on paddy seeds at 2 g per kg soil gave effective control of root-knot nematode infecting Crossandra (Khan and Parvatha Reddy, 1994).

Application of farmyard manure has been found to reduce the lesion nematode population in Crossandra. The parasitic fungus, *Paecilomyces marguandi,* grown on paddy seeds at 2 g per kg soil gave effective control of root-knot nematodes infecting Crossandra (Khan and Parvatha Reddy, 1994).

INTEGRATED NEMATODE MANAGEMENT METHODS

Crossandra

Integrated management of root-knot nematode infecting Crossandra was achieved by rational combination of a biocontrol agent (*Trichoderma harzianum* or *Verticillium chlamydosporium*) with a nematicide (Aldicarb) or oil cake (neem cake) (Khan and Parvatha Reddy, 1994). Combinations of *Verticillium lecanii* and *Paecilomyces lilacinus* (2×10^4 spores/ml each) with 5 per cent neem leaf extract resulted in significantly higher plant growth parameters and Crossandra flower yield (Nagesh and Parvatha Reddy, 1995). Root galling and nematode multiplication factor were least and the parasitisation of eggs and egg masses was highest in *Verticillium lecanii* + 5 per cent neem leaf extract. Integration of neem, karanj and castor cakes with VAM *Glomus mosseae* significantly enhanced plant growth parameters and flower yield of Crossandra, root colonisation and sporulation of VAM. The above treatments also reduced root-knot nematode multiplication and root galling (Nagesh and Parvatha Reddy, 1997a).

Four VAM fungi, namely, *Aucalospora laevis*, *Glomus fasciculatum*, *Glomus intraradices* and *Glomus mosseae* in combination with neem cake were evaluated for their comparative efficacy to colonise Crossandra roots and suppress *Meloidogyne incognita* infection. *Glomus fasciculatum* and *Glomus mosseae* colonised Crossandra roots better compared to *Glomus intraradices* and *Aucalospora laevis* with comparatively lower nematode infection and multiplication, thus promoting better root health and plant growth in former species. Although all the 4 VAM fungi in combination with neem cake gave better control of nematodes over Carbofuran treatment, two species, namely, *Glomus fasciculatum* and *Glomus mosseae* were found to be more efficient (Nagesh *et al*, 1999).

Soil solarisation followed by application of antagonistic fungi not only reduced root-knot nematodes but also enhanced seed germination, plant establishment and plant growth parameters significantly compared to that of the plants that received only the antagonistic fungi, solarisation or control. Solarisation not only reduced nematode populations in soil by itself, but also reduced the competition of other biotic factors and aided in better establishment of introduced antagonists, and their parasitisation (Nagesh and Parvatha Reddy, 2000b).

Tuberose

Integration of vesicular arbuscular mycorrhizae (VAM) such as *Glomus mosseae* or *Glomus fasciculatum* with neem cake or/and Aldicarb gave effective management of root-knot nematodes infecting tuberose (Khan and Parvatha Reddy, 1994). Tuberose bulb treatment for 20 minutes in neem leaf extract (5 per cent) mixed with spores of *Paecilomyces lilacinus* (2×10^4 spores/ml) was effective for the management of *Meloidogyne incognita* and increased flower yield and egg and egg mass parasitisation by *Paecilomyces lilacinus* (Nagesh *et al*, 1996).

Carnation and Gerbera

Integration of *Paecilomyces lilacinus/Trichoderma harzianum* at 0.5 litre per m^2 (aqueous spore suspension containing 2×10^4 spores per ml) with neem cake at 0.5 kg per m^2 or Fenamiphos at 2 g a.i. per m^2 increased plant growth parameters and flower yield of Carnation and Gerbera. The above treatments also increased root-knot egg parasitisation by the parasitic fungi (Nagesh and Parvatha Reddy, 1996).

Pre-plant treatment of carnation and gerbera beds with Dazomet followed by the application of neem cake (1 kg/m^2, 15 days later) along with *Paecilomyces lilacinus* or *Verticillium chlamydosporium* significantly reduced populations of *Meloidogyne incognita*, mortality of plants and suppressed the nematode infection for nearly 2 years. On long term basis substratum management with pre-plant treatment of Dazomet, followed by application of oil cakes plus antagonistic fungi was more effective against *Meloidogyne incognita* than post-plant treatment with Carbofuran, Carbosulfan and Chlorpyriphos in carnation and gerbera polyhouses (Nagesh and Parvatha Reddy, 2001).

Jasminum

Application of Phorate at 4 g a.i. per plant during May and September months and incorporation of 20 kg FYM increased flower yield of jasmine by 50 per cent and reduced the root-knot nematode population by 70 per cent (Sundarababu, 1992).

Rose

Oxamyl 25 at the rate of 10 ppm a.i. in potted plant soil or 500 ppm a.i. as aerial spray could effectively control *Hemicycliophora* sp., *Xiphinema basiri* in rose (Muthukrishnan *et al*, 1975; Rajendran and Muthukrishnan, 1980). Application of phorate at 4 g a.i. per plant during May and September, pruning and incorporation of 20 kg FYM per plant increased the yield of jasmine by 50 per cent and reduced the root-knot nematode population by 70 per cent (Sundarababu, 1992).

Gladiolus

Gladiolus plants treated with combinations of *Paecilomyces lilacinus* + *Trichoderma harzianum* + neem cake or *Paecilomyces lilacinus* + *Trichoderma viride* + neem cake before planting the corms not only controlled root-knot nematode infection (*Meloidogyne incognita)* but also *Fusarium* wilt till the harvest of flower spikes. The corms and cormels obtained from these combinations were free from *Fusarium* fungus, while 20-30 per cent cormels from the plants that were treated with fungi alone (Nagesh *et al,* 1999).

Chrysanthemum

Integrated use *Paecilomyces lilacinus* formulations (talc or pesta granules) at the rate of 4-6 kg/acre with neem cake (150 kg/acre) increased fungal antagonism against *Meloidogyne incognita* in chrysanthemum and increased flower yield by 23-28 per cent under farmers' field conditions (Nagesh *et al,* 2001). Negative correlation between increased PL dosage and RKI and NMR was recorded.

QUARANTINE

Cadet *et al* (1991) devised an easy method to intercept and eliminate nematodes from infested planting material of ornamentals. Rooted plants of *Alpinia* and *Anthurium* were placed in a mist chamber for extraction of *Meloidogyne arenaria, Paecilomyces incognita* and *Radopholus similis*. Rooted shoots of *Alpinia purpurata* were able to produce about 20,000 *Meloidogyne arenaria* juveniles during the first 2 months of storage and to avoid re-infestation entire removal of roots or use of flower shoots were recommended for planting. Most *Anthurium ferrierense* infested with *Meloidogyne incognita* could be freed of nematodes by mist watering for 2-3 weeks before planting. *Anthurium andreanum* could not be freed of *Radopholus similis* because multiplication occurred in roots and stalk.

Methods of inspection of consignments and identification of particular intercepted pests are detailed by Karnkowski (1999). The results of inspections of ornamental plant consignments (cuttings, bulbs, rhizomes, pot plants and cut flowers), imported to Poland in 1993-98, were presented. Leaf and bud nematode (*Aphelenchoides fragariae*) was the most numerous among nematodes found in planting stock, pot plants and cut flowers imported mainly from the Netherlands. The interception of *Ditylenchus destructor* in bulbs of *Erythronium denscanis* was also noted. This pest was not subject to compulsory control on bulbs of *Erythronium denscanis*, as the plant was not normally a host of *Ditylenchus destructor*.

CONCLUSIONS

The nature and magnitude of important nematode diseases of ornamental crops as reviewed in this chapter indicate that nematode problems can no longer be ignored and need immediate and greater research emphasis. The need to phase out the use of synthetic molecules to control nematodes in these crops was nevertheless realised and felt, now more than before, due to pesticide residue problems. Integrated management with an emphasis on cultural methods, tolerant hosts and biolocal agents is now being practised and popularised in a wider perspective. The aspects that need emphasis in future to attain the sustainable nematode management in horticultural crops are as follows:

- There is considerable lacuna in our knowledge about nematodes associated with ornamental crops. Intensive and systematic surveys of ornamental crops should be conducted with the dual objectives of determining the incidence, prevalence and severity of such diseases and the geographical distribution of the nematode involved. This would go a long way in developing an advisory diagnostic service for the farmers.
- Adequate emphasis should be given to studies on the biology and host-parasite relationship of major nematode pests, which may lead to formulation of control methods on sound basis.
- Studies have to be carried out on biotypes and on intra-specific variation in nematodes already known to be of economic importance in ornamental production in India.
- Techniques for precise determination of damage thresholds of populations and assessment of crop losses have to be standardised.
- Work on disease complexes involving nematodes should be intensified with adequate collaboration between nematologists and plant pathologists.
- Data has to be generated on crop rotation as a method of nematode management. The effect of deep summer ploughing, field sanitation and other cultural practices, hot water treatment of planting material, are the areas which need more attention in the coming years.
- Regulatory and quarantine measures against introduction of exotic nematode pests need to be strengthened.
- There is a great need to develop varieties that are resistant or tolerant to nematodes. Varietal screening and subsequent breeding programmes should be intensified. Fundamental investigations in relation to biochemical and physiological basis of resistance may also be taken up.
- Development of integrated nematode management models for ornamental crops should be taken up. Research on the use of tolerant/resistant cultivars, parasitic fungi and bacteria, organic amendments/plant products and vesicular arbuscular mycorrhizae should be evaluated and, if acceptable to the farmers, should be popularised through demonstration plots.
- Development of horticultural cropping systems-based nematode management components.
- Development of database for region/agroclimate/crop specific nematode management packages.
- Extensive development of diagnostic and advisory service centres through electronic media for the farmers.

- Search for novel and potent agromolecules of biological origin for nematode control as curative measures.
- Standardisation of protocols for mass production, application methods, dosages, integration of components, etc. for biological control agents.
- Creation of increased awareness among farmers regarding nematode problems on ornamental crops through extension services, literature, audio-visuals in local languages, demonstrations in farmer's fields and training programmes; strengthening of various extension programmes in nematology under lab-to-land and training and visit systems have to be explored.

REFERENCES

Alam, M. M., Khan, A. M. and Saxena, S. K. 'Some new host records of root-knot nematode, *Meloidogyne incognita* (Kofoid & White, 1919) Chitwood, 1949'. *Indian J. Nematol.* 3: 159-160, 1973.

Bajaj, H. K. In: *Phyto and Soil Inhabitating Nemic Fauna of Haryana.* Hisar, India: Deptt. Of Nematology, Haryana Agricultural University, 1989.

Cadet, P., Queneherve, P. and Topart, P. 'Nematode infestation transmitted by planting material of *Alpinia* and *Anthurium*'. In: Proc. Ninth Annual Conference, *Barbados Society of Technologists in Agriculture*, Rockley Resort, Barbados, November 13 and 15, 1991 and Supplement, pp. 116-121, 1991.

Edward, J. C., Misra, S. L., Rai, S. B. and Peter, E. 'Association of *Pratylenchus chrysanthus* n. sp. with chrysanthemum root rot' *Allahabad Fmr.* 43: 175-179, 1969.

Elgoorani, M. A., Abo-Eldahar, M. E. and Menair, F. F. 'Interaction between root-knot and *Pseudomonas marginata* on gladiolus corms'. *Phytopathology* 64: 271-272, 1976.

Gill, J. S. 'Fighting the foliar nematode of chrysanthemum'. *Indian Horticulture* 25: 21-22, 1981.

Gill, J. S. and Sharma, N. K. 'Additional hosts of the foliar nematode, *Aphelenchoides ritzemabosii* from India'. *Indian Journal of Nematology* 6: 169-171, 1976.

Gill, J. S. and Walia, R. K. 'Control of chrysanthemum foliar nematode, *Aphelenchoides ritzemabosi* on *Zinnia elegans*'. *Indian Journal of Nematology* 10: 86-88, 1980.

Hague, N. G. M. 'Nematode disease of flower bulbs, glass-house crops and ornamentals'. In: *Economic Nematology.* (Eds.). Webster, J. M. London, U.K.: Academic Press Inc., pp. 409-434, 1972.

Hara, A. H., Nishijima, W. T. and Sato, D. M. 'Impact on Anthurium production of controlling an orchid thrips (Thysanoptera: Thripidae), and Anthurium white fly (Homoptera: Aleyrodidae), and a burrowing nematode (Tylenchida: Tylenchidae) with certain insecticide-nematicides'. *Journal of Economic Entomology* 81: 582-585, 1988.

Hassan, A. 'Interaction between *Pratylenchus coffeae* and *Pythium aphanidermatum* and/or *Rhizoctonia solani* on chrysanthemum'. *Zeist Phytopathology* 123: 227-232, 1988.

Jayaraman, V., Rajendran, G. and Muthukrishnan, T. S. 'Occurrence of root-knot nematodes in *Polianthes tuberosa* L. in Tamil Nadu'. *Indian Journal of Nematology* 5: 101-102, 1975.

Karnkowski, W. 'Quarantine insects, mites and nematodes intercepted in consignments of ornamental plants imported to Poland in 1993-1998'. *Progress in Plant Protection* 39: 312-320, 1999.

Khan, R. M. and Parvatha Reddy, P. '*Jasminum pubescens*—hitherto unrecorded host of *Radopholus similis* from India'. *International Nematology Network Newsletter* 6: 19-20, 1989.

Khan, R. M. and Parvatha Reddy, P. 'Nematode problems of ornamental crops and their management'. In *Floriculture—Technology, Trades and Trends.* (Eds.). Prakash, J. and Bhandary, K. R. New Delhi, India: Oxford & IBH Publishing Co. Pvt. Ltd., pp. 468-472, 1994.

Khan, R. M. and Parvatha Reddy, P. 'Nematode pests of ornamental crops'. In: *Problems and Progress in Economic Phytonematology.* (Eds.). Bhatti, D. S. and Walia, R. K. New Delhi, India: CBS Publishers and Distributors Pvt. Ltd., pp. 205-207, 1991.

Khan, R. M. and Parvatha Reddy, P. 'Nematode pests of ornamental crops'. In: *Nematode Pests of Crops.* (Eds.). Bhatti, D. S. and Walia, R. K. New Delhi, India: CBS Publishers & Distributors, pp. 250-257, 1992.

Krishnappa, K., Krishnaprasad, K. S. and Reddy, B. M. R. 'A preliminary note on the plant parasitic nematodes associated with ornamental plants at Hebbal campus of U.A.S., Bangalore'. *Current Research* 9: 101-103, 1980.

Lamberti, F., Divito, N. M., Scrano, Chinappaen, M. and Bucha, J. 'Root-knot nematode *Meloidogyne* sp. in Mauritius and Rodrigues'. F.A.O. Plant Protection Bulletin *Veg. Ornamental Fruit Plant* 35(2): 43-50, 1987.

McLeod, R. W. 'Nematodes, a costly pest of cut flower'. *Agril. Gazette of New South Wales* 92(3): 41-42, 1981.

Muthukrishnan, T. S., Chandrasekhar, J., Lakshmanan, P. L. and Chinarajan, A. M. 'Occurrence of *Hemicycliophora labiata* Colbran, 1960 on rose in Tamil Nadu'. *South Indian Hort.* 23: 75-76, 1975.

Muthukrishnan, T. S., Vadivelu, S. and Rajendran, G. '*Longidorus africanus* and its pathogenic effect on *Crossandra undulaefolia*'. *Indian Journal of Nematology* 7: 69-72, 1977.

Nagesh, M. and Parvatha Reddy, P. 'Comparative efficacy of *Paecilomyces lilacinus* and *Verticillium lecanii* in combination with botanicals against *Meloidogyne incognita* infecting *Crossandra undulaefolia*'. *Journal of Biological Control* 9(2): 113-115, 1995.

Nagesh, M. and Parvatha Reddy, P. 'Crop loss estimation in carnation and gerbera due to the root-knot nematode, *Meloidogyne incognita,* (Kofoid & White) Chitwood'. *Pest Management in Horticultural Ecosystems* 6: 158-159, 2000a.

Nagesh, M. and Parvatha Reddy, P. 'Management of *Meloidogyne incognita* in *Crossandra undulaefolia* by integration of soil solarisation with antagonistic fungi, *Paecilomyces lilacinus* and *Trichoderma viride*'. Bangalore, India: *Emerging Trends in Plant Disease Management*, 7-8 December, 2000., IPS Zonal Meeting, U.A.S., G.K.V.K., 2000b.

Nagesh, M. and Parvatha Reddy, P. 'Management of *Meloidogyne incognita* on carnation and gerbera in commercial polyhouses'. *Second International Crop Science Congress*, New Delhi, 1996b.

Nagesh, M. and Parvatha Reddy, P. 'Management of *Meloidogyne incognita* on *Crossandra undulaefolia* using vesicular arbuscular mycorrhiza, *Glomus mosseae*, and oil cakes'. *Mycorrhiza News* 9(1): 12-14, 1997a.

Nagesh, M. and Parvatha Reddy, P. 'Nematode management in ornamental crops'. In: *Progressive Floriculture*. (Eds.). Yadav, I. S. and Choudhary, M. L. Bangalore, India: The House of Sarpan (Media), 1997b.

Nagesh, M., Hussaini, S. S. and Gopinatha, K. V. 'Management of root-knot nematode, *Meloidogyne incognita* (Kofoid & White) Chitw. in chrysanthemum using formulations of *Paecilomyces lilacinus* in combination with neem cake'. Proc. 2nd National Symposium *Integrated Pest Management in Horticultural Crops. New Molecules, Biopesticides & Environment. IPM in Horticultural Crops: Emerging Trends in the New Millennium.* (Eds.). Verghese, A. and Parvatha Reddy, P. AAPMHE, I.I.H.R., Bangalore, pp. 149-150, 2001.

Nagesh, M., Janakiram, T., Rao, T. M. and Parvatha Reddy, P. 'Evaluation of China Aster varieties/pure lines for resistance to *Meloidogyne incognita* (Kofoid & White, 1919) Chitwood, 1949'. *Pest Management in Horticultural Ecosystems* 1: 101-104, 1995.

Nagesh, M., Meenakshi, S., Murthy, N. and Parvatha Reddy, P. 'Screening of some tuberose cultivars/ hybrids for resistance against *Meloidogyne incognita* race-1'. *Indian Perfumer* 39: 138-140, 1995.

Nagesh, M., Parvatha Reddy, P. and Ramachandran, N. 'Integrated management of *Meloidogyne incognita* and *Fusarium oxysporum* f. sp. *gladioli* in gladiolus using antagonistic fungi and neem cake'. Proc. *3rd International Symposium of Afro-Asian Nematologists.* SBI, Coimbatore, India, pp. 263-266, 1999.

Nagesh, M., Parvatha Reddy, P. and Rao, M. S. 'Comparative efficacy of VAM fungi in combination with neem cake against *Meloidogyne incognita* on *Crossandra undulaefolia*'. *Mycorrhiza News* 11(3): 11-13, 1999.

Nagesh, M., Parvatha Reddy, P. and Rao, M. S. 'Integrated management of *Meloidogyne incognita* on tuberose using *Paecilomyces lilacinus* in combination with plant extracts'. *Nematologia Mediterranea* 25: 3-7, 1996.

Nagesh, M., Parvatha Reddy, P., Janakiram, T. and Rao, T. M. 'Sequential biochemical changes in roots of *Callistephus chinensis* lines resistant and susceptible to *Meloidogyne incognita* Chitwood race 1'. *Nematologia Mediterrania* 27: 39-42, 1999.

Ohkawa, K. and Saigusa, T. 'Resistance of rose root stocks to *Meloidogyne hapla, Pratylenchus penetrans* and *P. vulnus*'. *Hort. Sci.* 16: 559-560, 1981.

Parvatha Reddy, P., Singh, D. B. and Rao, V. R. 'The root-knot nematode on Gladiolus from India'. *Current Science* 48: 82, 1979.

Prasad, S. K. and Dasgupta, D. R. 'Plant parasitic nematodes associated with roses in Delhi'. *Indian J. Entomol.* 26: 120-121, 1964.

Rajaraman, B. and Rajendran, G. 'Report of *Meloidogyne incognita* (Kofoid and White, 1919) Chitwood, 1949 in *Jasminum sambac* and *J. flexila*'. *South Indian Hort.* 27: 70, 1979.

Rajendran, B. and Rajendran, G. 'Report of *Meloidogyne incognita* in *Jasminum flexile*'. *South Indian Horticulture* 27: 70, 1979.

Rajendran, G., Vadivelu, S. and Muthukrishnan, T. S. 'Pathogenicity of *Meloidogyne incognita* on Crossandra'. *Indian'. Journal of Nematology* 6: 115-116, 1976.

Rashid, A. and Khan, A. M. 'Studies on the lesion nematode, *Pratylenchus coffeae*, associated with chrysanthemum'. *Indian Phytopathology* 28: 22-28, 1975.

Runia, W. Th. and Amsing, J. J. 'Disinfestation of nematode-infested re-circulation water by ozone and activated hydrogen peroxide'. In Proc. *9th International Congress on Soil-Less Culture.* St. Helier, Jersey, Channel Islands, 12-19 April, 1996. International Society for Soil-Less Culture, pp. 381-393, 1996.

Sasser, J. N. and Freckman, D. W. 'A world perspective on Nematology: The role of the society'. In: *Vistas on Nematology.* (Eds.). Veech, J. A. and Dickson, D. W. Hyattsville, MD, U.S.A.: Society of Nematologists Inc., pp. 7-14, 1987.

Schindler, A. F. 'Nematode associated with rose in survey of commercial grain house'. *Pl. Dis. Reptr.* 40(1): 277-280, 1956.

Schindler, A. F. 'Parasitism and pathogenicity of *Xiphinema diversicaudatum*-ecto parasitic nematode'. *Nematologica* 2: 25-31, 1957.

Sen, K. and Dasgupta, M. K. 'Additional hosts of the root-knot nematode, *Meloidogyne* spp. from India'. *Indian Journal of Nematology* 7: 74, 1977.

Sethi and Swarup. 'Plant parasitic nematodes of Western India. I. The genus *Tylenchorhynchus*'. *Nematologica* 19: 77-78, 1968.

Sher, S. A. 'A disease of rose caused by root lesion, *Pratylenchus vulnus*'. *Phytopath* 47: 703-706, 1957.

Singh, R. V. and Sharma, H. K. 'Nematode problems and their management in ornamental crops'. In: *Nematode Diseases in Plants*. (Ed.). Trivedi, P. C. New Delhi, India: CBS Publishers and Distributors, pp. 168-176, 1998.

Srinivasan, A. Studies on root lesion nematode, *Pratylenchus dellattrei*. M.Sc. Thesis. Tamil Nadu Agri. Univ., Coimbatore, India, 1974.

Srinivasan, A. and Muthukrishnan, T. S. 'A note on nematode fungal complex in Crossandra (*Crossandra undulaefolia*) in Coimbatore'. *Current Science* 44: 743-744, 1975.

Srinivasan, A. and Muthukrishnan, T. S. 'Control of root lesion nematode *Pratylenchus delattrei* Luc 1958 on crossandra'. *Indian Journal of Nematology* 6: 111-114,. 1976.

Stoke, D. E. 'Some plant symptoms associated with *Aphelenchoides* species in Florida'. *Nematological Circular,* Div. of Plant Industry, Florida, Deptt. of Agril. and Consumer Services, No 89, 1979.

Sundarababu, R. 'Nematodes in jasmine'. *Tamil Nadu Agricultural University Newsletter* 21(11): 2, 1992.

Sundarababu, R. and Vadivelu, S. 'Studies on the nematodes associated with rose'. *Indian Journal of Nematology* 18: 139, 1988a.

Sundarababu, R. and Vadivelu, S. 'Pathogenicity of *Meloidogyne* species to tuberose (*Polyanthes tuberosa* L.)'. *Indian Journal of Nematology* 18: 146-148, 1988b.

Trivedi, P. C. 'Plant nematology in Rajasthan: Aspects and Prospects'. In: *Phytonematology in India*. (Ed.). Trivedi, Pravin Chandra. New Delhi, India: CBS Publishers & Distributors, 1998.

Vadivelu, S. and Muthukrishnan, T. S. 'Control of root lesion nematode, *Pratylenchus delattrei* on Crossandra'. *Indian Journal of Nematology* 9: 86, 1979.

Winfield, A. L. 'Chemical control of *Ditylenchus dipsaci* in flower bulbs and onions'. *Ann. Appl. Biol.* 75: 454-460, 1973.

Winfield, A. L. 'Observation on the control of the stem nematode *Ditylenchus dipsaci* of tulips with hot water or thionazin treatment'. *J. Hort. Science* 47: 357- 364, 1972.

—oo(O)oo—

12

WEED MANAGEMENT IN ORNAMENTAL CROPS

B. K. DHADUK AND N. Y. WAKCHAURE

INTRODUCTION

PLANTS are man's most valuable assets without which he could not have survived on earth. But we cannot afford to let them grow indiscriminately all over the land and water bodies, else a place for our activities and us will be difficult to find. Of more than 300,000 species of plants known in the world, hardly 3.000 are of economic value to us. When one grows any of these economic species of plants, invariably a variety of volunteer vegetation comes up simultaneously, which is competitive and undesirable. Besides, several plants are unwanted because either they are responsible for human health hazards or they hamper the activities of man in some other ways.

The history of weeds is a history of man's efforts to collect and, later, cultivate food. Man discriminated between useful and unwanted plants even when he was a primitive about a million years ago. He then strove to remove shrubs and tall grasses with simple stone tools to reach the edible nuts and fruits. This, incidentally, allowed more light to reach the useful plants for photosynthesis and reduced plant competition. The crop cultivation about 10,000 years ago comprised mainly two operations, namely, (1) planting of certain food plants, and (2) pulling out associated volunteer vegetation, the weeds. In other words, farming started with a fight with weeds, and this fight continues even today. However during all those years man has slowly but surely improved upon his capacity to deal with the weeds. It is truly said, "agriculture is a controversy with weeds".

LOSSES CAUSED BY WEEDS

(i). Weed Menace in Agriculture

Directly, weeds reduce the crop yield and, indirectly, they elevate farm production costs through energy spent in controlling them. Further, the negative value of weeds is depicted during harvesting, marketing, storage and dockage of the weedy crop. Add to this the increased pests and diseases in different crops in the succeeding years and the menace to our irrigation system. Recent estimates show that weeds cause an annual loss of about Rs. 1,980 crore to Indian agriculture, which is more than the combined losses caused by insect pests and disease.

(ii). Reduction in Crop Yields and Production Efficiency

Reduction in yields due to uncontrolled growth of weeds ranges from 34.3 to 89.8 per cent. Although the farmers adopt some kind of weeding on their fields, it still leaves us with a conservative estimate of at least 10 per cent reduction in crop yield.

In weedy fields, farm operations like application of fertiliser, insecticides and irrigation become cumbersome. When spiny weeds invade the crop field, it makes harvesting difficult. Some weeds, like bind weed, bind the crop plants together so well that their harvesting becomes troublesome. Add to this the cost of separating weed seeds or fruits from the grain and other farm produce. Growing within and outside the fields, weeds frequently provide shelter to insect pests and disease organisms of crops and act as alternate hosts to these, both during the crop season and off-season.

(iii). Erosion of Crop Quality

Weeds reduce the quality of farm produce in many ways. Contamination of food grain with weed seeds fetches low prices. The weedy grains produce flour with bad odour. In warehouses, the weed seeds and weed fragments continue respiration and thus cause the grain to heat and rot. In tea, the presence of Loranthus (*Dendrophthoe falcata*) leaves impair its quality. In cotton, the dry weed fragments adhere to the lint and hinder the spinning process. In dryland agriculture, weeds cause severe moisture stress and force the food grains to shrivel. Vegetables, fruits and flowers are discoloured and mis-shaped in the presence of weeds.

(iv). Weeds and Aesthetics

In many advanced countries, lowering of the aesthetic values is the primary objection to weeds. Their presence around living and working places and in gardens makes the surroundings dull and insipid.

WHAT ARE WEEDS?

Agricultural weeds may be defined as that part of the non-crop vegetation that reduces the quantity or quality of existing crops to such an extent that any net benefit of the weed plants themselves cannot compensate for these unfavourable effects on crops. In simple words, a weed is "a plant out of place", "a plant interfering with the intended use of the land", "and a plant with negative value". However, it should also be kept in mind that not all non-crop vegetation in the field is harmful to crop production.

WEEDINESS DEFINED

Weediness is defined as "the state or condition of a field, flower bed, lawn, and so forth in which there is an abundance of weeds" (Winburne, 1962). Thus, weediness connotes a condition. It also connotes numbers. To suggest that a crop is weedy implies that it has many weeds—enough to reduce crop yield.

WEED MANAGEMENT DEFINED

While weed control aims at only putting down the weeds already present by some kind of physical or chemical energy, weed management is a systemic approach in which weed prevention and weed control have companion roles. It includes whole land use planning done in advance to minimise the very invasion of weeds and give crop plants a competitive advantage over the weeds.

The following are the major principles of integrated weed management:

- Modification of crop habitat with the objective to place the crop in competitive advantage over the weeds.
- Measures should be directed to reduce the survival mechanism of weeds in the soil.
- Crop management practices such as crop rotation, summer fallowing, suitable time and method of planting should be utilised to discourage the establishment of colonies of perennial and parasitic weeds.
- Any individual element of weed management should be, as far as possible, eco-friendly.
- The plan of weed management should be flexible so as to accommodate innovations and practical experiences of the progressive farmers.

Thus, weed management has both immediate and long-term objectives.

DIFFERENT METHODS OF WEED MANAGEMENT

Weed management is a combined approach that includes both weed prevention and weed control.

Weed Prevention

This includes all the measures which prevent the entry and establishment of new weeds in an area. Some major aspects of weed prevention are:

1. Use weed free crop seeds: The dispersal of weed seed can be prevented either by production of weed free crop seeds at the government farm or by cleaning the crop seeds before storage as well as at the time of sowing.

2. Keep non-crop areas clean: Weed control efforts should be extended to non-crop areas like fences, farm boundaries, irrigation ditches, bunds, etc. which are often neglected by farmers.

3. Avoid contamination of manure pits: Farmyard manure serves as a source of weeds to cropland. Hence, contamination of the manure with weed seeds should be prevented.

4. Prevent mechanical movement of weeds: Movement of livestock, farm machinery, nursery stock, etc. from weed-infested areas to clean areas without cleaning should be prevented.

5. Keep vigilance: One should inspect the farm areas periodically for weed seedlings even if different control measures were used. As the seedlings are noticed, they should be uprooted immediately and the soil in that spot should be treated with a suitable sterilant.

6. Quarantine measures: Legal measures are necessary to check inter-State or inter-country movement of noxious weeds.

Weed Control

Weed control is the process of limiting weed infestation so that crops could be grown profitably and other activities of man conducted efficiently.

The major elements are:

- Crop husbandry control of weeds.
- Physical control of weeds.
- Herbicidal control of weeds.

- Biological control of weeds.
- Crop breeding for weed control.
- Non-living mulches.
- Burning and flaming.

CROP HUSBANDRY CONTROL OF WEEDS

These are referred to as Ecological Control Methods. They weaken the weed growth by crop competition and improve the efficacy of other methods of weed control. Crop husbandry methods include: selective crop stimulation, proper crop stand and early seedling vigour, proper planting method, proper planting time, crop rotation, clean seedbeds, summer fallowing, minimum tillage, lowering area under bunds, competitive cropping, etc.

PHYSICAL CONTROL OF WEEDS

Physical methods of weed control utilise manual energy, animal power or fuel to dig out weeds. They include pre-plant tillage, post plant tillage, row cultivation, mowing, cutting, dredging and chaining and soil solarisation. The implements may vary from simple hand tools to complex tractor-drawn machinery. Physical control is safe to the crop, environment and user. However, it is a labour intensive method. Also, timely operation of these methods is very important. The stage of weed growth and soil condition affects the success of these methods. However, it may increase weed problem as it cuts the rhizomes and roots of the perennial weeds, which give rise to new plants. It also buries the weed seed under soil which may serve as further source of infestation.

HERBICIDAL CONTROL OF WEEDS

Herbicides are chemicals capable of killing or inhibiting the growth of plants. They have saved farmers repeated inter-cultivation and have helped in obtaining satisfactory weed control where physical methods often fail. Today, we have over 150 herbicides in common use. These chemicals vary in molecular structures, response to environment, selectivity, mobility within plants and fate in soils.

Benefits of Herbicides

Besides less labour intensity and less money requirement, herbicides still have many advantages. Some of these are as follows:

- Physical weeding may be unfeasible during monsoon season. Under such a condition, herbicides can be very useful. Also, during early crop growth when many fields need simultaneous weeding, herbicides can be of great help even in labour rich countries.
- By physical methods, weeds are removed after they have offered considerable competition to the crops. But herbicides can control the weeds at a very early stage of crop growth, as soon as the weeds emerge from the soil. Thus, timely weed control is possible with herbicides.
- Some weeds survive by mimicry. They resemble crop plants and escape physical weeding, for example, wild oat (*Avena* sp.) in wheat. Herbicides are useful in controlling such weeds.
- Herbicidal control does not require strict row spacing as is the case with the physical method, where the crop rows have to be sufficiently wide to accommodate weeding implements.

- Many modern herbicides get translocated deep in the underground system of weeds and bring about longer lasting control of perennial weeds and bushes.
- Spiny weeds can be very conveniently controlled by herbicides.
- Herbicides are safe on erodable lands as they do not cause soil or water erosion.
- Herbicides do not fragment the vegetative propagules of weeds. They kill the weeds *in situ* without permitting their dissemination.
- Herbicide sprays easily reach the weeds growing in obstructed situations.

Some other benefits of using herbicides are fewer labour problems, greater possibility of farm mechanisation, easy crop harvesting and lower cost of farm produce.

Limitations of Herbicides

Important limitations in the use of herbicides are as follows:

- In herbicidal control, there is no automatic signal to stop a farmer who may be applying the chemical inaccurately till he sees the results in the crop sprayed.
- Herbicides may interact with the environment to produce unintended results. Herbicides drift, wash-off and run-off may cause considerable damage to the neighbouring crops.
- In case of diversity in farming, a variety of herbicides are needed to control weeds in different fields, while for physical control, the farmer has to possess only one or two kinds of implements.
- Herbicidal control requires considerable skill. The user must be able to identify the weeds and possess considerable knowledge about the herbicides and their proper usage.
- In herbicide treated soils, the selection of the replacement crop has to be based on its tolerance to the herbicides.
- Military use of herbicides is the greatest misfortune of their discovery. In Vietnam, 2,4-D and 2,4,5-T were used for defoliating forests and crops. In future, chemical warfare with residual herbicides may be more devastating.

BIOLOGICAL CONTROL OF WEEDS

The biological control of weeds involves the use of living organisms, such as insects, snails, mites, herbivorous fish, fungi and competitive plants to limit their infestation. An important aspect of biological weed control is that it is applicable to the control of only one major weed species that has spread widely. The operation of biological control of weeds is mainly a responsibility of technicians; the farmer has little to contribute to it.

CROP BREEDING FOR WEED CONTROL

Hardly any attempt has been made to breed crop plants tolerant to competition from weeds. Recently, attention of a group of plant breeders has been drawn towards breeding millet varieties tolerant to *Striga* sp.

NON-LIVING MULCHES

Mulching stunts or kills the weeds by cutting light to them. Straw, hay, dry sugarcane leaves, farmyard manure, rice hull, saw dust and bark dust are natural, partial mulch materials, which can at best stunt the weed growth. Perennial weeds grow through them and natural mulches tend to harbour insect pests and disease organisms. Synthetic mulches namely black paper or polythene films provide stronger mechanical barriers to all kinds of weeds. But as they are costly, they are used only with high value vegetables and ornamentals.

BURNING AND FLAMING

Burning is the cheapest method of eliminating mature, unwanted vegetation from non-crop situations in dry season. Weed seeds that have already shattered on the ground before flaming will not be killed by it. Burning is a potential source of fire hazard while flaming is a momentary exposure of green weeds to as high as 1000° C temperature from flame-throwers.

GENERAL CONSIDERATIONS IN WEED MANAGEMENT

The following are the pre-requisites of a successful weed management programme:

- One must gain knowledge of the ontogeny of weeds under attack before choosing a system for their control.
- The nature of the weed problem must be surveyed in the target area.
- Weed control measures must be planned for the whole farm and not just against weeds in a field (Table 1).

Large number of weeds affect the cultivated crops; a few important weed species are mentioned in Table 1. Research work carried out on different ornamental crops for controlling weeds are enumerated below crop-wise:

GLADIOLUS

Weed control is an important practice in commercial production of gladiolus. In India, it is generally done manually, and if not properly done, damages the plants. Results of experiments on chemical weed control of gladiolus have revealed that several chemicals are very effective in eradicating weed flora without affecting the growth and flowering of gladiolus. These herbicides are applied either pre-plant or pre-or post- emergence of weeds.

The kinds of chemicals, their concentration, time and method of application vary according to soil types and the prevalent weed population. Only one chemical cannot be used universally.

Under a study, 16 species of weeds were found in gladiolus fields, among which the dominant weeds were *Digitaria sanguinalis* and *Portulaca oleracea*. The most effective herbicides for controlling grass weeds were Alachlor, Simazine, Napropamide, Linuron and Pendimethalin. These herbicides gave excellent weed control with very slight injury to flowers. However, in vinyl house cultures, the incidence of abnormal plants increased and slight injury to flowers was observed at high rates of Simazine, Linuron and Oxadiazon + Pendimethalin; no injury was observed with Napropamide or Alachlor + Pendimethalin application (Kwon-Oh Do *et al*, 1996).

Field studies were undertaken at Nauni to evaluate different weed management practices including hand weeding, mulching and 13 combinations of two pre-emergence herbicides (Atrasine and Pendimethalin), three post-emergence herbicides (Paraquat, Glyphosate and Isoproturon) with or without hand weeding, in gladiolus cv. Psittacinus hybrid. Out of a total of 25 weed species observed, *Galinsoga parviflora, Cyperus rotundus, Echinochloa colonum*, *Echinochloa crus-galli* and *Digitaria sanguinalis* were the prominent ones. All treatments were effective in reducing the total weed population during earlier stages; however, Atrazine (1-2 kg/ha) was more effective than Pendimethalin (1-2 kg/ha) during the first period; only mulching (3 inches of local grasses)

TABLE 1
Some Important Weed Species in Monsoonal South Asia

Botanical Name	Common Name	Family
Celosia argentea		Amaranthaceae
Chenopodium album		Chenopodiaceae
Cynodon dactylon	Bermuda grass	Poaceae
Cyperus rotundus	Purple nutsedge	Cyperaceae
Digitaria sanguinalis	Crab grass	Poaceae
Echinochloa crus-galli	Barnyard grass	Poaceae
Eichhornia crassipes	Water hyacinth	Pontederiaceae
Imperata cylindrica		Poaceae
Parthenium hysterophorus		Asteraceae
Phalaris minor	Canary grass	Poaceae
Saccharum spontaneium	Kans, Sarkanda	Poaceae
Stiga asiatica	Witchweed	Scrophulariaceae
Sorghum halepense		Poaceae
Convolvulus arvensis		
Typha latifolia		Poaceae

proved effective against monocotyledonous weeds as well as total weeds. Monthly hand weeding (6 times), mulching and treatment with Paraquat or Glyphosate produced lower dry weight of weeds during later stages of crop growth (Chahal *et al*, 1994).

Matazachlor at 1.5 kg/ha, Pendimethalin 1.32 kg and Pendimethalin 1.32 kg + Metamitron 2.1 kg were well tolerated when applied after sowing gladiolus cv. Eurovision cormels (<4 cm) and corms (8-10 cm). Linuron at 1.5 kg ± Napropamide or Fluazifop-P 1.5 kg and Cyanazine alone at 1.5 kg caused leaf tip chlorosis and reduced the weight of cormels, but corms were not affected (Mynett and Jagusz, 1990).

Almost 100 per cent control of the persistent weed *Cyperus esculentus* was observed in bulb fields when plots were covered with black plastic sheeting at least 0.1 mm thick. Practical advice is to use herbicides and plastic covers on infested land. In order to avoid spreading the small weed tubers when crops of gladiolus cormlets are lifted, it is advised that the smallest grade material is discarded (Koster and Meer, 1989).

Oryzalin, Prodiamine and Pendimethalin at 2.2 and 4.5 kg/ha pre-emergence did not adversely affect vigour or corm yield of gladiolus cv. Promenade, April, Ajax or Priscilla. Metolachlor at 4.5 kg/ha applied one day after planting seriously reduced germination, vigour and corm yield (Bing *et al*, 1988).

Over a two-year period, 11.2 kg/ha of Chlorthal and 0.7 litre/ha of Trifluralin each gave better control of *Convolvulus arvensis, Tribulus terrestris, Amaranthus retroflexus, Setaria glauca, Portulaca oleracea, Polygonum convolvulus, Chenopodium vulvaria, Sonchus oleraceus, Cynodon dactylon, Cepsella burea-pastoris* and *Papaver rhoeas* in gladiolus cv. Oscar than did 8.9 kg/ha of Lenacil (Talia *et al*, 1986).

Applications of 0.5, 0.75 and 1.0 kg Oxyfluorfen, 0.75 kg Oxadiazon and 1.6, 2.4 and 3.2 kg Metoxuron/ha were made pre- and post-emergence to gladioli cv. Peter Pearce and Trader Horn. Main infesting weed species were *Amaranthus, Malva, Heliotropium, Convolvulus* sp. and *Sonchus oleraceus.* Oxyfluorfen and Oxadiazon were safe at all rates tested, while Metoxuron was safe only at the lowest rate. *Convolvulus* was controlled better by pre- than post-emergence application of Oxyfluorfen and Oxadiazon (Brosh *et al*, 1985).

Pre- and post-emergence applications of 1.5 lb Alachlor + 2 lb CIPC (Chlorpropham), 2 lb Napropamide, Oryzalin, Pronamide (Propyzamide) and 4 lb Thiobencarb/acre were evaluated for control of *Digitaria ciliaris* and *Amaranthus hybridus* in gladiolus cv. Manatee White. Control of *Digitaria ciliaris* was 90 per cent with 2 and 4 applications of all herbicides, while this level of control of *Amaranthus hybridus* was only achieved with Alachlor + CIPC, Napropamide and Oryzalin (Gilreath, 1984).

CHINA ASTER

The effect of pre-emergence application of Diuron (0.25-1.25 kg/ha), Simazine (1.0-1.5 kg), Metolachlor (1.0-1.5 kg/ha), Alachlor (1.0-1.5 kg) and Pendimethalin (0.75-1.25 kg/ha) on weeds of *Callistephus chinensis* and on plant height, number of branches, flower yield per plant and flower yield per hectare at harvest, was investigated in field studies at Bangalore, Karnataka, during 1988-89. All herbicide applications reduced the number of weeds, with the exception of sedges. Higher numbers of branches per plant were observed after treatment with Diuron (1.25 kg) and least with the unweeded control. *Callistephus chinensis* tolerated all treatments except Pendimethalin (1.25 kg). Diuron gave highest flower yield per hectare and maximum number of flowers per plant (Basavaraju *et al*, 1992).

Chloramben and Napropamide granules and Diphenamid, Napropamide and Oryzalin sprays gave excellent grass control in aster. Napropamide and Oryzalin also controlled broad-leafed weeds, as did Chlorprophan granules (Harasmaki and Kuhns, 1983).

In field and laboratory trials, Alachor, Chlorprophan, Oryzalin, Simazine and Trifluralin were evaluated for weed control in direct sown and transplanted China Aster cv. Powder Puff and Kurenai, which were grown for seed. China Aster plants could not tolerate 0.25 kg/ha Simazine, while 1.125 kg Alachlor stunted growth. A single application of Oryzalin (3.75 kg at 4 days after transplanting) provided excellent weed control and 10-fold increase in seed yield to 568 kg/ha. Trifluralin, applied at 2 kg pre-sowing, was the only herbicide not to significantly reduce China Aster emergence. Oryzalin (applied at 4.5 kg 10 days after sowing) provided the best aster seedling survival (83.2 per cent). Trifluralin and Oryzalin resulted in 597 and 23,577 weed plants m^2 compared to 843-1,560 in untreated plots (Phetpradap and Hampton, 1991).

Single application of Treflan (trifuralin), Kerb 50w (propyzamide), Ramrod (propachlor), Tenoran (chloroxuron), Pyramin (chloridazon) or Venzar (lenacil) to *Callistephus chinensis* grown for seed did not reduce germination (Duczmal and Bartosewicz, 1984).

Oxadiazon applied at 2-4 kg/ha pre-emergence was safe on transplanted *Aster chinensis*. It provided 100 per cent control of all weeds in summer (Lamont and O'Connell, 1986).

CARNATION

The efficacy and selectivity of Pendimethalin (132 g/da), Metolachlor (200 g/da), Oxadiazon (100 g/da) and Chloroxuron (450 g/da) for controlling annual weeds in *Dianthus caryophyllus* (cv. Regina and Niky) were investigated during 1995-96 at Sofia, Bulgaria. Pendimethalin and Metolachlor gave good control against annual grass weeds (Poaceae), but inhibited growth. Oxadiazon and Chloroxuron controlled annual dicotyledonous weeds and did not inhibit the growth of the crop (Atanassova and Ivanova, 1998).

The most important weeds in glasshouse grown carnations were *Amaranthus albus, Capsella bursa-pastoris, Chenopodium album, Ambrosia elatior, Galinsoga parviflora* and *Digitaria sanguinalis*. The best weed control was obtained with Tenoran (chloroxuron) at 300 g + Adol (lenacil) at 200 g/100 m^2, applied after the establishment of rooted cuttings. Where there was a high proportion of *Digitaria sanguinalis,* the addition of Dual (Metolachlor) at 300 ml/1000 m^2 to the mixture enhanced the weed control (Molnar, 1982).

Studies conducted in China during 1993-95 showed that weeds in carnations could be effectively controlled using a soil treatment of Pendimethalin at 450 g a.i./ha.. There were no adverse affects of this treatment on the growth of carnations (Shen *et al*, 1996).

TUBEROSE

Weeds become a serious problem in the cultivation of tuberose as the crop demands heavy manure and irrigation. If the weeds are not removed in time, growth and productivity will be markedly reduced. Manual weeding is effective but is laborious and increases the cost of cultivation. Hoeing between plants is also necessary to loosen the soil and destroy the weeds.

Pal and Das (1990) conducted field trials with 3-4 kg/ha Saturn (Thiobencarb), 3-4 litres/ha Gramoxone (paraquat dichloride) and Basagran (bentazone) and 2-3 litres Glycel (glyphosate) applied 7 days prior to planting or 3 times (at intervals of 40 days) after planting tuberose cv. Single. They reported that Gramoxone reduced weed biomass from weedy control values of 55-57.5 g DW/0.25 m^2 to 23-27.5 g, whereas all other treatments resulted in 37-52 g weeds. The following tuberose growth parameters were increased by herbicide treatments: height (from 27-27.5 to 27.5-41 cm), the number of leaves/plant (from 6.5-7 to 6.7-13), the number of bulbs/plant (from 12 to 12-19), the weight of bulbs (from 100-105 to 102-180 g), the duration of flowering (from 14-15 to 14-19 days), the length of flower spikes (from 72-73 to 73-94 cm), the number of flower spikes/plot (from 90-95 to 100-180) and the flower yield/plot (from 1-1.2 to 1.3-4.2 kg). Gramoxone at 4 litres resulted in the largest increments in each case, whereas 3 kg Saturn and 2 litres Glycel generally had the least effect.

Murthy and Gowda (1993) evaluated the effects of Alachlor, Metolachlor and 2,4-D at 1.0-2.0 kg/ha and Diuron, Pendimethalin, Atrazine and Butachlor at 0.75-1.25 kg for the control of weeds in tuberose cv. Double. They reported that all herbicide treatments reduced weed DM at harvest and increased crop flower yields from untreated control values of 67.8 g/0.25 m^2 and 38,890 plants/ha, respectively, to 11.7-26.0 g and 67,780-113,330 plants, respectively. Metolachlor at 2.0 kg and Pendimethalin at 1.25 kg resulted in the lowest weed DM at harvest and the greatest flower yields, respectively.

Treatments with Prodiamine or Oryzalin did not hinder the crop quality. The treated plants were as vigorous as plants in hand weeded plots (Gilreath, 1995).

Ciritto *et al* (1980) studied the herbicidal effects of Alipur (Chlorbufan), Treflan (Trifluralin), Prevenol (chlorprofam) and TOK E 25 (nitrofen) on tuberose cv. Perla, grown for cut flowers. They reported that Treflan at 2 kg/ha and emulsified in 20 litres of water controlled most weeds except *Cyperus longus*. The other herbicides were less effective.

ROSE

Weeds pose a very serious problem in rose cultivation as heavy manure and irrigation requirements of this crop create conditions conducive for the rank growth of numerous monocot and dicot weeds. The weeds not only consume water and nutrients from the soil but also act as hosts for a number of diseases and pests. Manual weed control is effective if done frequently. But this procedure is very expensive, laborious and time consuming and, if not properly done, damages the root system. Chemical weed control is comparatively economical, convenient and efficient in eradicating weeds by one or two applications. Certain chemicals have proved very effective in controlling both monocot and dicot weeds. These chemicals are applied either as pre-plant or pre-emergence of weeds after the roses are planted or post-emergence to established weeds.

Sudhakara and Narayan Gowda (2002) tested Pendimethalin at 1.0 kg a.i./ha, Oxyfluorfen at 0.1 kg a.i./ha, Atrazine at 1.0 kg a.i./ha and Glyphosate at 1.0 kg a.i./ha in rose at Bangalore. All the herbicides were applied 3 days after punning. Among 4 herbicides, glyphosate gave excellent control of weeds and, in turn, gave more flower yield. Similar results were obtained with Oxyfluorfen. However, hand weeding was found superior to herbicides. Unweeded control lowered 60 per cent flower yield.

On sandy loam soil, Cyanazine at 1.5 kg/ha, Diuron at 1.0 kg/ha and Methabenzthiazuron at 2.5 kg/ha were applied with Simazine at 1.0 kg/ha for weed control in rose root stocks (*Rosa cariifolia* var. Froebelii and *Rosa canina*) before budding and in budded stocks. None of the treatments injured the roses. Applied to rootstock in the year of planting, Cyanazine gave the most effective weed control; Diunon also performed well but failed to control *Polygonum persicaria.* Applied in the spring following budding, all treatments gave excellent weed control, probably due to the effect of Simazine treatment before budding and to the hardening of the ground in winter.

It is also established that Metha-benthiazuron plus Diuron gave good weed control in 10 polyantha rose cultivars. The weed cover in untreated plots depended on the growth habit of the cultivar. No symptoms of phytotoxicity were observed. Annual weeds were well controlled by Simazine at 2.5-3.5 kg/ha, Alachlor 3.3 kg/ha, Propyzamide at 1.5 kg/ha and Nitrofen at 2.0-3.0 kg/ha applied pre-emergence of weeds on bare soil. On mulched soil, the herbicide activity was much reduced and mixtures of the above gave better results. For perennial weed control, only Glyphosate at 2.16 or 4.32 kg/ha gave better results (Prasad and Kumar, 1998).

The effect of herbicides Bladex 500 SC (Cyanazine), Acer 25 EC (Flurochloridone), Arelon 50 (isoproturon), Butisan 500 SC (Metazachlor), Ronstar 25 EC (Oxadiazon) and Goal 240 EC (Oxyfluorfen) on rose rootstock *Rosa canina* 'Schmid's Ideal' was investigated in field studies conducted in Poland. Herbicides were applied in spring after rootstock planting but before bud emergence. Increase in root neck diameter in herbicide treated plots was equal to those of rootstock grown in control (hand weeded) plots. It was concluded that all herbicides were suitable for use with *Rosa canina* (Gabarkiewicz, 1997).

Botta and Prando (1988) compared weed control treatments in roses. These were (a) 0.75 litres Fomesafen/ha + 1.5 litres Fluazifop-butyl, (b) 1.0 litres Fomesafen + 2.0 litres Fluazifop-butyl, (c) 2.0 litres Fluazifop-butyl + 1.5 litres Bentazone, and (d) 2.0 litres Fluazifop-butyl. Predominant broad-leafed weeds were *Rumex* sp., *Plantago* sp., *Capsella bursa-pastoris* and *Artemisia vulgaris*. The main monocotyledonous were *Digitaria* sp. and *Setaria* sp. Treatment (d) confirmed the selectivity of Fluazifop-butyl for grass-weed control. The mixtures (a) and (b) gave good weed control without being phytotoxic to the roses.

Himme *et al* (1988) got excellent control of *Poa annua* by Propyzamide at 1.25-1.5 kg/ha in grafted roses in the Netherlands.

In other applications, 0.75 kg Oxyfluorfen/ha as soil drench and 5 kg Oxyfluorfen/ha as 2 per cent granules, 0.75 kg Oxadiazon/ha and 0.5 kg Simazine/ha were applied as soil drenches with a hand sprinkler at the base of rose bushes cv. Mercedes, prior to irrigation.

There were no undesirable effects of any treatment on rose growth or yield. *Convolvulus* sp. was reduced by up to 80 per cent by Oxyfluorfen or Oxadiazon and all annual weeds were eliminated (Brosh *et al*, 1985).

In a trial with *Rosa laxa* (*Rosa cariifolia* cv. Froebelii) rootstocks comprising several herbicides alone or together, Simazine + Propyzamide (625 g + 1,000 g/ha) gave the best control of most weeds. In a trial with several scion cultivars on *Rosa canina* cv. Pfander and *Rosa rubiginosa* (eglanteria) rootstocks, comparing the effects of Propyzamide at 1.5 or 2.0 kg/ha, both rates were effective against several weeds, but the higher rate had the better overall effect. Rose growth was not adversely affected (M, 1983).

Strelets *et al* (1982) carried out trials in plantation of 2 roses (*Rosa canina* cv. Vitaminnyi VNIVI and Vorontsovskii I). Of the herbicides tested, Caragard (terbumeton) at 3 kg/ha applied in early spring gave the best weed control without adverse effects on rose growth and cropping.

Terbacil at 2 kg/ha and Caragard-combi (terbuthylazine + terbumeton) at 15 l/ha gave the best weed control in a 5-year-old essential oil rose plantation on meadow soil. But it reduced the essential oil content of the flowers and yield/ha (Georgieva and Kazakova, 1982).

Ustinex Z (methabenthiazuron + Diuron) gave good weed control in 10 polyantha rose cultivars. No symptoms of phytotoxicity were observed. In a 10 year trial with 15 hybrid Tea Rose cultivars, no phytoxicity was observed (Kolbe, 1981).

Simazine (1,000 g/ha) and Cyanazine (1,500 g/ha) used alone gave reasonable control of the 4 weed species (*Poa annua, Stellaria media, Urtica urens* and *Polygonum persicaria*) when they were tested on bushes of rose cv. Orange Morsdag, Meinastur and Junior Miss (M, G. D. 1981).

CHRYSANTHEMUM

Post planting care concentrates on checking the weed growth and keeping proper moisture condition of the soil to promote sustained growth. Frequent shallow cultivation should be done to conserve the moisture and control the weeds. The role of weedicides for controlling the weeds in chrysanthemum was studied by various workers.

Two litres Oxyfluorfen/ha applied pre-emergence provided 100 per cent control of all weeds in summer and was safe on *Chrysanthemum morifolium* (Lamont and O'Connell, 1986).

The most important weeds in the glasshouse ornamentals were *Amaranthus albus, Cepsella bursa-pastoris, Chenopodium album, Ambrosia elatior, Galinsoga parviflora* and *Digitaria sanguinalis*. In chrysanthemum, acceptable weed control was obtained with Tenoran (Chloroxuron) at 300 g + Dachthal (chlorthal dimethyl) 900 g/1000 m^2, applied 1 to 2 days before planting out the rooted cuttings (Molnar, 1982).

On a silty soil, 1 to 2 kg granular Oxadiazon/ha caused no injury to any of the 6 chrysanthemum cultivars. The lowest rate did not prevent re-growth of *Galinsoga parviflora* but weed control with the higher rates was satisfactory (Deuber and Aguiar, 1982).

A covering of black plastic or green woven cloth prevented weed growth in the soil under pot grown chrysanthemums. Herbicides applied before planting the pots also controlled weeds effectively. Under pot treatment with 4 lb Oxadiazone (e.c. formulation) followed by the same rate of a w.p. formulation over the top gave good control as did 3 lb Metolachlor, 0.5 lb Oxyfluorfen granules, 4 lb Oxadiazon granules and 4 lb Oryzalin (w.p. formulation)/ac; all applied under the pots (Bing, 1983).

MARIGOLD

In marigold production, control of weeds is an important operation. If the weeds are not removed in time, a great loss would occur in terms of growth and productivity of marigold. In India, 3-4 manual weedings are required for the entire period.

Chemical weed control in marigold has also been recommended by various workers. Single applications of Treflan (Trifluralin), Kerb 50 W (propyzamide), Ramrod (Propachlor), Tenoran (Chloroxuron), Pyramin (Chloridazon) or Venzar (Lenacil) did not reduce germination in *Tagetes erecta* grown for seed (Duczmal and Bartosewicz, 1984).

Chloramben and Napropamide granules and Diphenamid, Napropamide and Oryzalin sprays gave excellent grass control; Napropamide and Oryzalin also controlled broad-leafed weeds as did Chlorprofam granules. Marigold (*Tagetes patula*) grew and flowered well when treated with Diphenamid, Chloramben and Chlorpropham (Haramaki and Kuhns, 1983).

Alachlor at 4 lb, Chloramben 4 lb and Oxadiazon 4 lb/acre were very effective in controlling seedling weeds when applied after transplanting in French marigold (*Tagetes patula*) (Bing, 1981).

JASMINE

Weeds cause enormous losses to jasmine crop and increase the cost of cultivation. Manual weed control is effective if done properly, but this is very expensive. Chemical weed control is comparatively economical, convenient and efficient in eradicating weeds by one or two applications. Treatment with Oryzalin controlled weeds for 10 weeks on silt loam soil. Good vegetative growth and flowering of jasmine was also recorded with the application of this chemical (Haramaki and Kuhns, 1983). Mulching also reduced the weed population.

GERBERA

The effects of 4-8 kg Devrinol 50 WP (Napropamide), 1-2 kg Adol 80 WP (Lenacil), 6-8 kg Tenoran 50 WP (Chloroxuron), or mixtures of 4 litres Dual (Metolachlor) + 1 or 2 kg Adol or + 6-8 kg Tenoran/ha on weed control in gerbera were studied. Mixtures with Dual improved the control of *Digitaria sanguinalis* but it did not improve control of other weeds like *Oxalis europaea*, *Stellaria media*, *Galinsoga parviflora*, *Amaranthus* sp., *Chenopodium* sp., *Portulaca oleracea* or *Urtica urens* after 2 weeks. *Digitaria sanguinalis* control remained excellent after 6 weeks. Mixtures of Dual with Tenoran were less phytotoxic to gerbera than those with Adol (Molnar, 1986).

ANTHURIUM

In trials to evaluate herbicides for use in *Anthurium andreanum* cv. Ozaki Red, 2.2-9.2 kg Oxyfluorfen/ha + 1.1 or 4.5 kg Oryzalin, 3.3 or 6.6 kg Oxadiazon and woven plastic ground cover provided excellent weed control. Spray applications of Oxadiazon were phytotoxic to *Anthurium* leaves and emerging flowers, but the granular material, used more than 268 days after planting, was not. Diuron at 1.2 kg/ha provided commercially acceptable control for approximately 2.5 months. Diuron increased the number of side shoots relative to other treatments, resulting in vigorous, bushy plants (DeFrank *et al*, 1989).

GAILLARDIA

Tolerance of transplanted blanket flower (*Gaillardia aristata*) to pre-emergence herbicides was evaluated in container trials. Herbicides were applied at the maximum use rate and twice the maximum use rate. Dithiopyr, Pendimethalin and Prodiamine provided excellent control of spotted spurge (*Euphorbia maculata*) and yellow wood sorrel (*Oxalis stricta*) with little injury to the herbaceous perennial. DCPA (chlorthal dimethyl), Oxadiazon and Metolachlor were tolerated by Gaillardia but these chemicals provided lower control of one or both weed species (Derr, 1994).

Chopped newspaper mulch at 2.3 and 3.6 kg/m^2 has shown good potential in controlling the weeds in Gaillardia × grandiflora cv. Goblin (Pellett and Heleba, 1995).

REFERENCES

Aldrich, R. J. and Kremer, R. J. In: *Principles in Weed Management.* New Delhi, India: Panima Publishing Corporation, 1998.

Atanassova, B. and Ivanova, I. 'Influence of some soil herbicide applied in mini carnation (*Dianthus caryophyllus* F. Spray) stands'. *Rasteniev'dni Nauki* 31(1): 41-44, 1998.

Basavaraju, C., Gowda, J. V. N. and Muniyappa, T. V. 'Effect of pre-emergent herbicides on yield of China Aster [*Callistephus chinensis* (L.) Ness)]'. *Curr. Res.* Univ. of Agric. Sci., Bangalore 21(3): 50-51, 1992.

Bhattacharjee, S. K. and De, L. C. In: *Advanced Commercial Floriculture.* Jaipur, India: Aavishkar Publishers, Distributors, 2003.

Bing, A. 'Weed control in annual flowering plants'. Proc. *North Eastern Weed Science Soc.* 35: 262-266, 1981.

Bing, A. 'Weed control in field grown pot chrysanthemum'. Proc. *37th Ann. Meeting North Eastern Weed Science Soc.*, pp. 362-366, 1983.

Bing, A., Senesac, A. F. and Macksel, M. 'Pre-emergence weed control n gladiolus cormel production.' Proc. *42nd Annual Meeting of the North Eastern Weed Science Soc.*, pp. 137-140, 1988.

Bose, T. K. and Yadav, L. P. In: *Commercial Flowers.* Kolkata, India: Naya Prokash, 1989.

Botta, G. and Prando, G. 'Trials on post-emergence weed control herbicides on rose bushes'. *Informatore Agrario* 44(47): 43-44, 1988.

Brosh, S., Hadar, E. and Zilberstein, J. 'Weed control in roses grown in greenhouses'. *Phytoparasitica* 13: 3-4, 1985.

Brosh, S., Zilberstein, J. and Boshwitz, Y. 'Weed control in gladioli for flower production'. *Phytoparasitica* 13(3-4): 245-246, 1985.

Chahal, D. S., Sehgal, O. P. and Singh, K. K. 'Effect of chemical and agronomic treatments on population and growth of weeds in gladiolus field'. *Annals of Biology* Ludhiana 10(2): 245-249, 1994.

DeFrank, J., Easton-Smith, V. A., Higaki, T. and Imamura, J. S. 'Response of Anthurium and weeds to four pre-emergence herbicides'. *HortScience* 24(6): 1044, 1989.

Derr, J. F. 'Weed control in container-grown herbaceous perennials'. *HortScience* 29(2): 95-97, 1994.

Deuber, R. and Aguiar, J. C. 'Use of Oxadiazon in chrysanthemum (*Chrysanthemum* sp.) and rose (*Rosa* sp.) crops'. Abstracts of the *XIV Brazilian Congress on Herbicides and Herbaceous Weeds* (SBHED) and the *VI Congr. Latin American Weed Association* (ALAM), Campinas, pp. 157-158, 1982.

Duczmal, K. W. and Bartosewicz, A. 'The effect of herbicides on germination of different ornamental plants'. *Biultyn Instytutu Hodowli i Aklimatizacji Roslin* 153: 263-265, 1984.

Gabarkiewicz, R. 'The effect of herbicides on rose root stock *rosa canina* 'Schmid's Ideal'. *Zeszyty Naukowe Instytutu Sadownictwa i Kwiaciarstwa w Skierniewicach* 4: 183-190, 1997.

Georgieva, S. and Kazakova, K. Studies on the effect of some herbicides on the carbohydrate metabolism of the Kazanl'k rose'. *Rasteniev'dni Nauki* 19(5): 69-74, 1982.

Gilreath, J. P. 'Chemical weed control in flowering gladiolus'. Proc. *Florida State Horticultural Soc.* 97: 297-299, 1984.

Gilreath, J. P. 'Pre-emergence weed control in tuberose'. Proc. *Florida State Horticultural Soc.* 108: 53-57, 1995.

Haramaki, C. and Kuhn, L. J. 'Chemical weed control in bedding plants'. Proc. *37th Annual Meeting North Eastern Weed Science Soc.*, pp. 357-361, 1983.

Haramaki, C. and Kuhns, L. J. Proc. *37th Annual Meeting North Eastern Weed Science Soc.*, pp. 335-338, 1983.

Himme, M. van; Bulcke, R. and Van, Himme, M. 'Control of triazine-resistant *Poa annua* in nurseries'. *Mededelingen van de Faculteit Landbouwwetenschappen Rijksuniversiteit Gent.* 53(3b): 1279-1288, 1988.

Kolbe, W. 'Long term weed control trial with Ustinex Z in roses (1972-1981)'. *Pflanzenschutz Nachrichten Bayer* 34(3): 302-313, 1981.

Koster, A. T. J. and Meer, L. Jv. D. 'Darkness does for the weed. Black plastic controls tiger nut'. *Bloembollencultuur* 100(9): 34-35, 1989.

Kwon, OhDo; Kim, HongJae; Kim, Joungkeun; Han, KyuPyong; Kwon, Do; Kim, H. J., Kim, J. K. and Han, K. P. 'Herbicide evaluation for pre-emergence weed control under various cropping patterns in gladiolus'. *RDA Journal of Agricultural Sciences, Crop Protection* 38(1): 495-500, 1996.

Lamont, G. P. and O'Connell, M. A. 'An evaluation of pre-emergent herbicides in field grown flowers'. *Plant Prot. Quarterly* 1(3): 95-100, 1986.

M, G. D. 'A trial of materials for rose rootstocks. A trial of materials for rose bushes. Comparison of rootstocks'. *Verbondsnieuws voor de Belgische Sierteelt* 27(18): 907-909, 1983.

M. G. D. 'Weed control in rose bushes'. *Verbondsnieuws Voor de Belgische Sierteelt* 25(12): 562, 1981.

Molnar, M. 'Chemical weed control experiments in Gerbera'. *Novenyvedelem* 22(2): 86-88, 1986.

Molnar, M. 'Chemical weed control in glasshouse ornamentals: carnations, chrysanthemums, freesias'. *Novenyvedelem* 18(10): 412-413, 1982.

Murthy, G. M. A. and Gowda, J. V. N. 'Evaluation of herbicides for weed control in tuberose (*Polianthes tuberosa* Linn.) cv. Double'. *Crop Research,* Hisar 6(1): 176-178, 1993.

Mynett, M. and Jagusz, M. 'Effect of several herbicides on gladiolus plants cultivated from corms and cormels'. *Acta Horticulturae* 266: 557-560, 1990.

Pal, A. K. and Das, S. N. 'Effect of weedicides on growth and flowering of tuberose'. *South Indian Horticulture* 38(3): 143-149, 1990.

Pellet, N. E. and Heleba, D. A. 'Chopped newspaper for weed control in nursery crops'. *J. Env. Hort.* 13(2): 77-81, 1995.

Phetpradap, L. and Hampton, J. G. 'Herbicide evaluation in China Aster grown for seed'. Proc. *44th New Zealand Weed and Pest Control Conf.*, pp. 300-303, 1991.

Prasad, S. and Kumar, U. *Commercial Floriculture.* Bikaner, India: Agro Botanica, 1998.

Shen, Jian Ying; Gu, ZhenFang; Jiang, XueJun; Chen, CuiFang; Shen, SiJun; Shen, J. V., Gu, Z. F., Jiang, X. J., Chen, C. F. and Shen, S. J. 'A study on the occurrence and control of weeds in cut carnation production shed'. *Journal of Shanghai Agricultural College* 14(2): 95-100, 1996.

Strelets, V. K., Bukina, N. V., Kochetkov, V. P. and Tsybul-Ko, N. S. 'Herbicides in young dog rose plantations'. *Khimiko Farmatsevticheskii Zhurnal* 16(11): 85-89, 1982,

Styrbjorn, Alstrom. *Fundamentals of Weed Management.* Jodhpur, India: Vigyan Prakashan, 1996.

Sudhakara, K. A. and Narayana Gowda, J. V. 'Screening herbicides in roses'. Abstracts of Nat. Symp. *Indian Floriculture in the New Millennium,* 2002.

Talia, M. C., Traversa, E. and Cocozza Talia, M. 'Weed control for increasing the size of gladiolus cormels'. *Colture Protette* 15(11): 51-55, 1986.

Windburn, J. N. *A Dictionary of Agricultural and Allied Terminology.* E. Lansing, U.S.A.: Michigan State University Press, 1962.

—oo(O)oo—

13

PHYTOSANITARY REQUIREMENTS FOR IMPORT/EXPORT OF FLORICULTURE PRODUCTS/PLANTING MATERIAL

O. R. REDDY

INTRODUCTION

LIBERALISATION of world trade in agriculture with advent of WTO has thrown up challenges and opportunities for the developing countries for expansion of trade in agriculture exports. Compliance with international standards on food safety, animal and plant health has gained utmost importance to enable access to international markets. Quarantine, both of exports and imports, will be given utmost importance so that Indian agriculture is protected from the ingress of exotic pests and diseases and also to facilitate safe exports. Thus, there is an imperative need for developing and strengthening quarantine facilities and scientific capacities and capabilities in meeting the stringent sanitary and phytosanitary requirements of our most developed country trading partners.

The import/export of floriculture products/planting material, which has boomed consequent to the New Policy on Seed Development brought out by the Ministry of Agriculture in 1988, has reached a plateau due to improper management of pest/disease problems, poor quality and high cost of production.

This paper discusses the provisions of the WTO-SPS Agreement and revised text of the International Plant Protection Convention. Further, the phytosanitary requirements in import/export of floriculture products/planting material in order to safeguard the interest of the floriculture industry are also discussed and strategies suggested to promote quality exports.

IMPLICATIONS OF WTO-SPS AGREEMENT

The sanitary and phytosanitary measures outlined under the WTO-SPS Agreement are aimed at protecting animal, human and plant life and health. The SPS agreement sets out the rights and responsibilities of WTO members that wish to take action to restrict the imports in order to protect their animal, human or plant life or health. However, the restrictions so placed should be the minimum necessary to achieve the objects and, at the same time, be least restrictive of the trade. The SPS measures should be based on scientific principles and not maintained without sufficient scientific evidence and, further, these measures are not to be applied as arbitrary, discriminatory, unjustified, unnecessary or disguised barriers to trade. Under the SPS Agreement, members are allowed to adopt stricter measures to conform to higher levels of protection than achieved by measures based on international standards; however, these are required to be scientifically justified and based on risk assessment. Further, the members are allowed to adopt provisional SPS measures as a precautionary approach, in case where the scientific evidence is insufficient but probable risks exist. The SPS agreement allows the members to accept the sanitary and phytosanitary measures of other members as equivalent, even if different, where the same provide appropriate levels of protection. The members are required to notify each other of SPS measures in the course of preparation. Also the control, inspection and approval procedures should be undertaken in no less favourable manner for imported products than for like domestic products.

Consequent to adoption of the SPS Agreement, there is a growing concern among the developing countries that certain SPS measures may impede the export of agriculture produce from developing countries, as the latter are not well equipped to address various issues related to the SPS Agreement due to lack of information on measures that affect exports; lack of infrastructure facilities and scientific support for inspection, testing, diagnosis, and treatment; lack of risk assessment/ surveillance capabilities; and lack of preparedness to associate with setting up of international standards and participation in dispute settlement.

The key phytosanitary issues involved in the implementation of provisions of the WTO-SPS Agreement are as under:

- Harmonisation of phytosanitary measures;
- Risk assessment;
- Appropriate level of protection;
- Adoption of equivalent measures;
- Maintenance of pest-free areas;
- Identification of areas of low pest prevalence; and
- Information on pest lists and pest distribution.

Particularly, the developing countries are facing the following challenges in the wake of implementation of the WTO-SPS Agreement and other International agreements:

- Review and updating of SPS legislation and regulations;
- Existence of national standards on SPS measures;
- Notification and transparency;
- Scientific justification of high level protection;
- Non-discrimination and national treatment;
- Risk assessment and appropriate level of protection;
- Recognition of equivalence of specific SPS measures; and
- Contact points and information exchange.

THE NEW REVISED TEXT OF IPPC AND INTERNATIONAL STANDARDS

The International Plant Protection Convention (IPPC) was established in 1951 under the aegis of the Food & Agriculture Organisation (FAO) of the United Nations to facilitate international co-operation in controlling plant pests and to prevent their international spread. The IPPC role in relation to trade, particularly the establishment and recognition of international standards, was strengthened in the New Revised Text of the Convention approved in 1997. The SPS Agreement recognises the IPPC as the relevant international organisation responsible for the establishment of international standards on phytosanitary measures and encourages WTO Members to base their phytosanitary measures on IPPC standards, guidelines and recommendations to facilitate global harmonisation of phytosanitary measures. Recognising the position of the IPPC in relation to the SPS Agreement, FAO established the IPPC Secretariat in 1992 followed by setting up standard procedures and a Committee of Experts on Phytosanitary Measures, which has now been replaced by the Interim Standards Committee. Amendments to the convention resulted in the establishment of the Commission on Phytosanitary Measures with an objective to review the status of plant protection in the world, provide directions to the working programme of the IPPC Secretariat and approve standards. An Interim Commission on Phytosanitary Measures is currently open to all FAO members to perform these functions until the new revised text of the IPPC comes into force through acceptance of the amendments by two-thirds of the contracting parties to the Convention. The Convention is a legally binding agreement, but standards developed and adopted by the IPPC as well as those of Codex and OIE are not legally binding. The SPS Committee may invite the IPPC or its subsidiary bodies to examine specific matters with respect to phytosanitary standards, guidelines and recommendations and the Committee may recommend areas where IPPC standards are elaborated. The WTO may establish panels to review phytosanitary-based trade disputes at the request of members, which may seek advice from, or consult with, the IPPC Secretariat in order to obtain technical information or identify technical experts in the resolution of phytosanitary-based trade disputes. The IPPC also has a dispute settlement provision pertaining to the implementation of the Convention, in particular where phytosanitary measures are challenged as unjustified barriers to trade.

QUARANTINE LEGISLATION AND REGULATIONS

The plant quarantine regulations are operative in India through The Destructive Insects & Pests (DIP) Act, 1914 and an amendment issued to Section 3 vide The Destructive Insects & Pests (Amendment and Validation) Act, 1992 (No. 12 of 1992) published in Part II-Section 1 of the Gazette of India dated 31st March, 1992, empowering the Central Government to levy and collect fees at prescribed rates by a notification issued under Section 3. The DIP Act confers the powers to Customs to operate the notifications issued under Section 3 of the Act as if the same were issued under Section 11(k) of the Customs Act. However, the PQ authorities were not vested with policing powers under the DIP Act. A new draft Plant Quarantine Bill is under scrutiny of the Ministry of Agriculture to replace the existing DIP Act. The New PQ Bill provides for constitution of a 'Plant Quarantine Authority of India' vested with independent executive and financial powers for effective implementation of plant quarantine measures and to give effect to the international agreements.

The import of seeds/plants/plant materials both for sowing/planting/propagation and consumption; and soil/peat-moss, etc. is regulated by 'The Plants, Fruits & Seeds (Regulation of Import into India) Order, 1989' [Notification S. O. 867 (E) dated 26th October, 1989] and amendments issued thereunder.

PRE-ENTRY QUARANTINE REQUIREMENTS

Both the permits issued by an authorised officer notified vide amended Notification S. O. 844 (E) dated 30th August, 2001 (Appendix-I) and the phytosanitary certificate issued at the country of origin are essentially required for import of seeds/plants/plant material for sowing/planting/ propagation as well as consumption of plant material; provided that cut flowers, garlands, bouquets, fruits and vegetables weighing less than two kilograms imported for personal consumption may be allowed to be imported without a phytosanitary certificate or an import permit. Also no permit is required for import of flower seeds and tissue cultures of plants [Notification S. O. 768 (E) dated 11th October 1993) and mushroom spawn cultures [Notification S. O. 523 (E) dated 16th July 1993). A special permit is issued by the PPA as required for import of soil/peat-moss as per PFS Order, 1989.

NOTIFIED POINTS OF ENTRY

All consignments of plants/plant material can be permitted import only through the entry points notified vide Notification S. O. 263 (E) dated 27th March, 1990. However, consignments of seeds/plants/plant material meant for propagation can only be imported through Amritsar, Kolkata, Chennai, Mumbai and New Delhi.

PROHIBITED/RESTRICTED PLANT SPECIES

The list of prohibited/restricted plant species covered under Schedule-II of The Plants Fruits & Seeds (Regulation of Import into India) Order, 1989 (PFS Order) is reproduced in Appendix-IIA. Besides this, a list of plant species covered under the New Seed Policy, for which the PPA prescribed additional declarations and special conditions, is reproduced in Appendix-IIB. The importer should ensure that no prohibited plant species are imported except those by authorised institutes and that every consignment of plant species imported shall be accompanied by a phytosanitary certificate issued by an authorised officer at the country of origin of consignment, containing additional declarations for freedom from specified pests. Also the importer should ensure that the special conditions prescribed thereunder are complied with.

GUIDELINES FOR ISSUANCE OF PERMIT

(i). Import of Seeds/Plants/Plant Materials for Sowing/Planting/Propagation

The importer shall apply for a permit for import of seeds/plants/plant materials for sowing/ planting in advance to an authorised officer of the concerned area/port of jurisdiction vide Notification S. O. 844 (E) dated 30th August, 2001 (Appendix-I) in form 'B' of PFS Order, 1989, in duplicate along with a demand draft for Rs. 300/- [vide S. O. 467 (E) dated 16th May, 2000] drawn in favour of 'The Pay & Accounts Officer, Department of Agriculture & Co-operation' of the concerned area of jurisdiction. A copy of catalogue/invoice shall accompany the application and a registration certificate issued by the NSC or Director of Horticulture/Agriculture as the case may be. In addition, a certificate of approval of PEQ facility issued by DIA, and an undertaking by the importer to grow the plants/ plant material under post-entry quarantine is required in respect of seeds/plants/plant materials that are subject to post-entry quarantine.

The application thus received shall be registered and the permit is issued within 3-4 days in form 'D' of PFS Order, 1989, in five copies, in case the applicant satisfies all the necessary conditions. The issue of the permit may be refused or withheld by the competent officer after giving reasonable notice to the applicant and reasons for withholding or refusal shall be recorded in writing.

The import permit issued shall be valid for a period of six months from the date of its issue and validity may be extended on request by the importer for a further period of six months provided the reasons for seeking extension are substantiated. Two copies of the permit shall be forwarded along with an orange and green colour tag specified in form 'E' of the PFS Order, 1989, to the importer with an advice to send one copy of the permit along with the above tag to the exporter in order to facilitate the incorporation of permit number on the phytosanitary certificate issued at the country of origin. The orange/green colour tag facilitates the identification of the consignment at the time of import at the prescribed point of entry. The permit issued is not transferable and no permit shall be issued for landed consignments.

(ii). Import of Consumption Plant Material

An application for issuance of permit for import of consumption plant material shall be made in duplicate in form 'A' of PFS Order, 1989, to an authorised officer of the concerned port/area of jurisdiction notified vide Notification S. O. 844 (E) dated 30th August, 2001 (Appendix-I) along with a bank draft/pay order for Rs. 150 [vide Notification S. O. 467 (E) dated 16th May, 2000] drawn in favour of 'Pay and Accounts Officer, Department of Agriculture & Co-operation' of concerned area of jurisdiction, sufficiently in advance of despatch of consignment. The application duly completed in all respects shall be registered and permit is issued in form 'C' of PFS Order and in five copies. Two copies of the permit are issued to the importer with an advice to forward one copy of the permit to the exporter so as to facilitate incorporation of import permit number on the phytosanitary certificate issued at the country of origin. The permit issued is valid for a period of six months from the date of its issue and further extended for another six months provided substantial reasons are given for the same. No permit shall be issued for landed consignments and the permit is not transferable.

(iii). Import of Soil/Peat-Moss, etc. and Biocontrol Agents, etc.

The application for import of soil, earth, clay and similar material for any microbiological, soil mechanics or mineralogical investigations and peat/sphagnum moss for horticultural purposes shall be made to the PPA at least one month in advance in form 'F' of PFS Order, 1989. The PPA or any other authorised officers, after scrutiny of application, if satisfied of the purpose for which such consignment is being imported, shall issue a special permit in form 'G' of PFS Order, 1989. Separate application is required to be filed with PPA for authorising import of biocontrol agents. The application for import of germplasm and genetically modified organisms shall be made to the Director, National Bureau of Plant Genetic Resources, New Delhi.

IMPORT INSPECTION AND PQ CLEARANCE

The importer or his agent shall file an application in the prescribed format in duplicate in respect of each consignment immediately upon its arrival at the prescribed point of entry or in advance in case of perishable consignments along with copies of the following documents: permit (importer's copy), if applicable; phytosanitary certificate (original) issued at the country of origin or PSC-Re-export format, in case of re-exported consignment along with attested copies of PSC issued from the country of origin; bill of entry duly endorsed by the Customs; shipping/airway bill, as the case may be; invoice and packing list and fumigation certificate, if any, to the officer-in-charge of the plant quarantine station located at the prescribed point of entry or area of jurisdiction. However, no such application is required to be filed in the case of plants and plant material imported through passenger baggage, couriers and post parcels. On receipt of the application, the PQ officer shall scrutinise the application and, if found complete in all respects, shall register the application and assess the inspection fees. The importer or his agent shall pay the inspection/PEQ inspection fees as the case may be by a bank draft drawn in favour of the 'Pay & Accounts Officer, Department of Agriculture & Co-operation' of the concerned area of jurisdiction at the rate prescribed under Schedule-

III of the PFS Order, 1989, as amended by Notification No. S. O. 574 (E) dated 14th June 2000 (reproduced in Appendix-III).

The importer or his agent shall arrange for inspection/sampling of the consignment on the scheduled date and time at the prescribed place by the nominated plant quarantine officer as per the quarantine order issued. Sampling of seed for propagation shall be in accordance with the International Seed Testing Association (ISTA) Rules, 1976. Sampling of cereals, pulses, oil seeds and others for consumption shall be as per the Bureau of Indian Standards (IS: 2814/1978 and IS: 3714/1978). The importer or his agent shall always associate with the inspecting officer while undertaking inspection. The PQ officer deputed for inspection will draw appropriate samples for detailed laboratory testing. In the event live insect infestation is noticed, the importer or his agent shall arrange for fumigation of consignments exceeding 1,000 m^3, by an approved pest control operator at his own cost under the supervision of the PQ officer.

A release order is issued to Customs, if a consignment on inspection is found to be free from infestation/infection. However, in case of consignments found infested/infected the same shall be permitted for clearance only after fumigation/treatment and after re-inspection. A detention order is issued in respect of consignments that are found infested/infected with a quarantine pest or imported in contravention of PQ regulations, for arranging deportation of consignment, failing which the same shall be destroyed at the cost of the importer.

The period of quarantine clearance for seeds for sowing is 30-35 days with the exception of fleshy seeds or seeds having shorter viability or *in situ* germination, which are cleared within 48 hours. Perishable plant materials such as cuttings/saplings/bud wood, etc. of plants are cleared within 24-72 hours. However, plant materials for consumption are cleared within a day or two, except those requiring fumigation which are cleared after 3 days and tissue cultures and mushroom spawn cultures are cleared within 6 hours.

POST-ENTRY QUARANTINE INSPECTION AND CLEARANCE

In accordance with the provisions of Chapter IV of PFS Order, 1989, the importer shall be required to establish the post-entry quarantine facilities such as an isolated field/nursery/glass house/screen house/polyhouse, etc., that are duly certified by the Designated Inspection Authorities (DIAs) notified vide Notification S. O. No. 261 (E) dated 27th March, 1990, in accordance with guidelines prescribed by the PPA for importing plants and plant materials that require post-entry quarantine. The importer shall establish these facilities sufficiently in advance so that the same may be ready for use at the time of arrival of the consignment but not later. A certificate issued by the DIA shall be submitted to the permit issuing authority at the time of applying for import permit or to the officer-in-charge of the PQ station at the time of arrival of the consignment at the prescribed point of entry. After the initial inspection of the consignment, if satisfied, the PQ officer shall accord quarantine clearance with a condition to grow the imported consignment under post-entry quarantine and the DIA will be intimated accordingly. The importer shall inform in advance the concerned DIA about the time of planting of imported plant material in an approved facility and permit complete access to the post-entry quarantine facility for the inspection of the plants and also abide by DIA's instructions concerning plants growing under post-entry quarantine. The importer shall provide necessary facilities to the DIA for undertaking inspection at the time of planting and at such intervals as may be considered necessary by him in accordance with guidelines issued by the PPA for the detection of pests and diseases, if any, and general observance of conditions governing the approval of post-entry quarantine. The DIA may permit the release of plants grown under post-entry quarantine, if they are found free from pests for the period specified in the permit. Where the plants are found affected by quarantine pests, the DIA shall order the destruction or return to the country of origin of the affected consignment of plants, of the whole or part of it.

If the pest is not of quarantine importance, the DIA may advise necessary control measures and release the plant material if satisfied that the pest is controlled, or issue an order for destruction. The importer shall destroy the affected plant material when ordered in a prescribed manner under the supervision of DIA. The DIA may submit a copy of PEQ inspection report to the PPA under intimation to the officer-in-charge of PQ station through which consignment was imported.

(i). Requirements of Open Field Quarantine Facility

1. The farm/nursery site shall be fenced all round up to a height of 1.25 metres with a single entry point with a lockable gate. It shall be isolated from similar cropped areas by at least one km and be weed-free.
2. The nursery or cropped area shall be bordered with 3-4 rows of densely populated crops such as daincha, sesbania, sun hemp, subabul, etc. to serve as an insect barrier.
3. An isolation distance of 100 metres shall be adopted to segregate different varieties of the same crop or different sources of the same crop species or, alternatively, each population is isolated by a polythene barrier up to a minimum height of 1.25 metres.
4. A suitable signboard must be displayed at the gate indicating restricted area and entry to the area requires the approval of the owner of PEQ facility.
5. The farm/nursery shall have the facilities for soil pasteurisation or solarisation or fumigation, incineration, spraying and watering facilities.
6. A buffer zone of not less than 3 metres in width must be maintained in vegetation-free condition around the inside perimeter of the approved area by regular weeding.

(ii). Requirements of Closed Quarantine Facility (Glass/Screen/Polyhouse)

1. The glass/screen/polyhouse facility shall have double door entry with entrance porch with inner door fitted with automatic spring door closure and outer sliding door provided with external lock.
2. The vents/openings of glass/screen/polyhouse facility shall be screened with insect-proof mesh of stainless steel/phosphor bronze, 40 meshes to a linear inch and woven of 30-gauge wire. If nylon mesh is used, it should be of good quality and evenly woven.
3. The drainage must be through a gully trap with a provision for rat proofing in case of concrete flooring.
4. The facility shall have a small isolated cubicle for growing indicator plants for virus detection and a head house for storage of pots, green house tools/ equipment and storage bins for storing sterile soil or potting mixture, soil sterilisation or solarisation/ incineration, spraying equipment/watering facilities, etc.

5. The entrance shall be provided with a disinfectant pad for sterilisation of feet and a basin for disinfecting hands.
6. The facility shall have a minimum floor space area of 10 sq. m. However, larger houses shall have insect-proof screened partitions for isolation of different kinds of plant species/varieties.
7. Suitable yellow sticky traps shall be hung at various intervals at crop canopy level to monitor insect population.

(iii). Requirements for Tissue Culture Facility

1. The tissue culture facility shall be accredited by/registered with the appropriate authority and meet the required standards for aseptic multiplication of plant tissue cultures.
2. The tissue culture facility shall have isolated growth room for holding imported tissue cultures and inspection room for observation of contaminants.
3. The facility shall have an isolated poly/screen house facility for hardening tissue cultures. However, where import of tissue cultures is made for *in vitro* propagation, such facility is not required.
4. The tissue culture laboratory shall have diagnostic facilities for virus-free testing.

EXPORT INSPECTION AND PHYTOSANITARY CERTIFICATION

Export inspection and phytosanitary certification of plants and plant products is carried out in accordance with Article IV of the International Plant Protection Convention (IPPC) to meet the legal obligations of the member countries. The phytosanitary certificates are issued in the formats set out under Article V of the Convention in consistence with current PQ regulations of the importing country. Such certificates are issued after careful inspection and treatment of plants and plant products by technically qualified and duly authorised officers at the country of export and shall include additional declarations as may be required by the importing country and also particulars of treatment, if any, given.

The exporter or his agent shall submit an application in the prescribed format in duplicate to the officer-in-charge of the PQ station at the designated port through which he intends to export or to the concerned inspecting and certifying authority notified vide Notification No. 8-97/91-PP.I, dated 26th November, 1993, issued by the Ministry of Agriculture, sufficiently in advance or at least 2-3 days prior to the shipment of consignment. The application shall be accompanied by a copy of invoice, packing list, shipping/airway bill, letter of credit or trade agreement or purchase order, export license (if applicable) and fumigation certificate, if any. Besides these, a copy of the permit issued by the importing country in case of export of seeds/propagating plant material, and wildlife clearance certificate if the export is covered under the Convention on International Trade in Endangered Species of Wild Flora and Fauna (CITES).

On receipt of the application, the PQ officer shall scrutinise the application and if found complete in all respects shall register the application and assess the inspection fee. The applications made in respect of export/import-prohibited consignments are withheld or refused for issuance of PSC and also applications made in respect of preserved or pickled or frozen plant products. The exporter or his agent shall pay the inspection fee at the prescribed rates vide letter No. 16-10/58-PPS dated 21st September 1960, issued by Ministry of Agriculture.

On specific request from the exporter or his agent, the consignment may be inspected at places outside PQ station on payment of Rs. 10/- per visit towards outside inspection charges within municipal limits of town or corporation limits of city as the case may be. Further the exporter or his agent shall meet the travelling and dearness allowances of the PQ officer and staff deputed for inspection outside port city or town. The above charges are paid in advance or immediately upon completion of the tour but before issuance of PSC. The inspection fee shall not be refunded in the case of cancelled or rejected application. The exporter shall abide by the terms and conditions stipulated by the inspecting and certifying authority.

The exporter or his agent shall present the consignment at the office of PQ station or arrange for inspection at his premises or present the containers at his premises or any other approved place on the scheduled date and time of inspection as per the quarantine order issued. The exporter or his agent shall provide necessary transport, labour and other facilities for opening, sampling, re-packing, sealing, etc. Sampling of seed for propagation shall be in accordance with the International Seed Testing Association (ISTA) Rules, 1976. Sampling of cereals, pulses, oil seeds and others for consumption shall be as per the Bureau of Indian Standards (IS: 2814/1978 and IS: 3714/1978). The exporter or his agent shall always associate with the inspecting officer while undertaking inspection. After thorough inspection, the PQ officer shall draw an appropriate sample for detailed laboratory testing.

In the event live insect infestation is noticed, the exporter or his agent shall arrange for fumigation of the consignment or container at his premises or any other approved place by an approved pest control operator under the supervision of the PQ officer. The exporter or his agent shall submit an undertaking for this purpose in the prescribed format along with payment of supervision charges of Rs. 25 per container. The exporter or his agent shall provide necessary transport/labour facilities if the fumigation is carried out in the fumigation chambers at the PQ station and pay fumigation or disinfestation or disinfection charges as prescribed above. The exporter or his agent shall pay storage charges at the rate of Rs. 10 per cu. m. space per day or part thereof, if the consignment is not immediately removed after degassing and re-inspection. The consignment shall be re-inspected after degassing to ensure freedom from live infestation.

Phytosanitary Certificates are issued in duplicate, that is, original for the exporter and duplicate copy for office record, if the consignment on inspection is found to be free from infestation/ infection. On specific request from the exporter or his agent, the PSC is re-issued after cancelling the earlier original certificate to facilitate the incorporation of corrections/amendments, subject to the production of revised shipping documents. Such re-issuance of PSC for incorporating amendments/ corrections, is done within 7-10 days from the date of issue of the original certificate and thereafter no such requests will be considered.

The issue of PSC will be rejected if the commodity on inspection is found to be a prohibited one or found affected by quarantine pests of concern to the importing country or the commodity is prohibited for export or could not be fumigated to render it pest-free or packed with hay, straw, wood shavings, etc. or contaminated with soil or noxious weed seeds or processed food containing additives and preservatives. The reasons for rejection shall immediately be communicated in writing to the exporter or his agent under intimation to Customs/port authorities. A minimum of 8-10 days or more is required for certification of seed consignments. Perishable commodities such as nursery plants, tissue cultures, fresh fruits, cut flowers, etc. are certified within a maximum period of 24 hours and consignments that require fumigation are certified within 3 days.

APPENDIX-I

List of Permit Issuing Authorities for Import of Seeds, Plants and Plant Products and Other Articles

[As per Clause 2(a) and Clause 3(ii) of the principal order issued vide Notification S. O. 867 (E) dated 27th October 1989; S. O. 262 (E) dated 27th March 1990; and subsequently amended by Notifications S. O. 707 (E) dated 6th October 1997; S. O. 207 (E) dated 16th March 1998; S. O. 712 (E) dated 1st August 2000; and finally substituted by S. O. 844 (E) dated 30th August 2001].

No.	Issuing Authority	Jurisdiction	Authorised to Issue Permits for
1.	Plant Protection Adviser to the Government of India, Ministry of Agriculture, Directorate of Plant Protection, Quarantine & Storage, NH-IV, Faridabad, Haryana – 121 001	All notified points of entry	All kinds of plants/plant material and other items as: insects, microbial cultures, bio-control agents, etc.
2.	Director, National Bureau of Plant Genetic Resources, PUSA Campus, New Delhi	New Delhi	All kinds of import of plant germ plasm for public public/ private sector institutions in the country
3.	Joint Director (In-charge), Plant Quarantine Division, Directorate of Plant, Protection, Quarantine and Storage, NH-IV, Faridabad, Haryana	All notified points of entry	All kinds of plants/plant materials
4.	Deputy Director (PP/Ent.), National Plant Quarantine Station, Rangpuri, . New Delhi 121 037	i. New Delhi airport ii. All notified points entry in Northern Zone in the States of Delhi, Haryana, Himachal Pradesh, J&K, Rajasthan, Uttar Pradesh and Uttaranchal	Import of all kinds of plants/ plant materials for sowing, planting, propagation and consumption

Continued...

...Continued

No.	Issuing Authority	Jurisdiction	Authorised to Issue Permits for
5.	Deputy Director (PP/Ent.), Regional Plant Quarantine Station, Ajnala Road, Near Air force Station, Raja Sansi Airport, Amritsar	i. Amritsar airport ii. All notified points of entry bordering Pakistan in the State of Punjab, UT Chandigarh	Import of all kinds of plants/ plant materials for sowing, planting, propagation and consumption
6.	Deputy Director (PP/Ent.), Regional Plant Quarantine Station, GST Road, Near Trident Hotel, Meenambakkam, Chennai 600 027	i. Chennai airport/ seaport ii. All notified points of entry in Southern Zone in the States of Andhra Pradesh, Karnataka, Kerala, Tamil Nadu, Andaman and Nicobar Islands, Lakshwadeep and Pondicherry	Import of all kinds of plants/ plant materials for sowing, planting, propagation and consumption
7.	Deputy Director (PP./Ent.), Regional Plant Quarantine Station, Haji Bunder Road, Sewri, Mumbai 400 014	i. Mumbai airport/ seaport ii. All notified points of entry in Western Zone in the States of Goa, Gujarat, Madhya Pradesh, Chattisgarh, Maharashtra and UT Dadra & Nagar Haveli, Daman & Diu	Import of all kinds of plants/ plant materials for sowing, planting, propagation and consumption
8.	Deputy Director (PP/Ent.), Regional Plant Quarantine Station, F-B Block Sector III, Salt Lake City, Kolkata 700 024	i. Kolkata airport/ seaport ii. All notified points of entry in Eastern Zone in the States of Arunachal, Pradesh, Assam, Bihar, Jharkhand, Meghalaya, Manipur, Nagaland,	Import of all kinds of plants/ plant materials for sowing, planting, propagation and consumption

Continued..

...Continued

No.	Issuing Authority	Jurisdiction	Authorised to Issue Permits for
		Orissa, Sikkim, Tripura, West Bengal and Mizoram	
9.	Plant Protection Officer (E/PP), Plant Quarantine Station, 24- Parganas, Bongaon (W. B.)	Concerned port of entry	Import of plants and plant materials for consumption only
10.	Plant Protection Officer (PP/E), Plant Quarantine Station, 25-A, Hariyala Plot, Down Area, Bhavnagar 364 001	Entry point Bhavnagar and Kandla	Import of plants and plant materials for consumption only
11.	Plant Protection Officer (E/PP), Plant Quarantine Station, Wellington Island, Kochy	Concerned port of entry	Import of plants and plant materials for consumption only
12.	Plant Protection Officer (E/PP), Plant Quarantine Station, Panitanki, Naxalbari, Rath Khola, P.O. Naxalbari, Dist. (W.B)	Concerned port of entry	Import of plants and plant materials for consumption only
13.	Plant Protection Officer (PP/E), Plant Quarantine Station, Gede Road, Nadia 741 503 (WB)	Concerned port of entry	Import of plants and plant materials for consumption only
14.	Plant Protection Officer (E/PP), Plant Quarantine Station, T.C. No. 206419, Krishnamurari Road, Kaithamukku P.O., Thiruvananthapuram 695 024	Concerned port of entry	Import of plants and plant materials for consumption only
15.	Plant Protection Officer (E/PP), Plant Quarantine Station, Harbour, Visakhapatnam	Concerned port of entry	Import of plants and plant materials for consumption only

Continued...

...*Continued*

No.	Issuing Authority	Jurisdiction	Authorised to Issue Permits for
16.	Senior Plant Pathologist, Plant Quarantine Station, Cargo Air Terminal Complex, Begumpet, Hyderabad	Concerned port of entry	Import of plants and plant materials for consumption only
17.	Deputy Director (E/PP), Central Integrated Pest Management Centre, Koramangala Road, White field, Bangalore	Concerned port of entry	Import of plants and plant materials for consumption only
18.	Plant Protection Officer (E/PP), Central Integrated Pest Management Centre, 16 Professor Colony, Bhanwar Kua Main Road, Indore 452 001	Concerned port of entry	Import of plants and plant materials for consumption only
19.	Plant Protection Officer (E/PP), Central Integrated Pest Management Centre, B-16, Mahatmanagar Extension, Lucknow 226 020	Concerned port of entry	Import of plants and plant materials for consumption only
20.	Plant Protection Officer (E/PP), Central Integrated Pest Management Centre, New Punai Chowk Near Dr. Dutta House, Patna 800 023	Concerned port of entry	Import of plants and plant materials for consumption only

APPENDIX-IIA

List of Prohibited/Restricted Plant Species Covered Under Schedule-II of PFS Order, 1989

No.	Plants, Seeds and Other Propagating Materials	Countries From Where Import is Prohibited	Pests for Which Additional Declarations in Official Phytosanitary Certificates are Required	Special Conditions for Import
1.	All species of *Allium* (onion, garlic, leek, chive, shallot etc.)	—	Smut (*Urocystis cepulae*), stem and bulb nematodes (*Ditylenchus dipsaci*)	—
2.	Cacao and all species of family Sterculiaceae and Bombaceae	West Indies, Africa and Sri Lanka	Pod rot (*Monila roreri*), mealy pod (*Trachysphaeria fructigena*), witches broom (*Crinipellis perniciosus*) and swollen shoot virus	Growing imported seeds and plants under post entry quarantine
3.	All species of *Citrus* (lemon, lime, orange, grapefruit etc.)	—	"Mal Secco" (*Deuterophorna*) *tracheiphila*)	Growing imported seeds and plants under post-entry quarantine
4.	Coconut seeds and plants (all species of *Cocos*)	—	Red Ring nematode (*Rhadinaphelenchus cocophilus*), lethal yellowing, *Cadang cadang,* bronze leaf wilt, Guam coconut disease, leaf scorch, coconut leaf miner (*Promecotheca cumingi*)	i. Consignments can only be imported by Director, Central Plantation Crops Research Institute, Kasaragod, (Kerala) ii. Imported material shall be grown in individual containers under post-entry quarantine

Continued...

...*Continued*

No.	Plants, Seeds and Other Propagating Materials	Countries From Where Import is Prohibited	Pests for Which Additional Declarations in Official Phytosanitary Certificates are Required	Special Conditions for Import
5.	Coffee (a) Plants, seeds (all species of of *Coffea*)	—	American leaf spot (*Omphalia flavida*), virus diseases and coffee berry borer (*Hypothenemus hampei*)	Consignments of seeds, plants can only be imported by Director,Coffee Research Station, Chikmagalore, Karnataka
	(b) Coffee beans	Srilanka. Africa, South America	Coffee berry borer (*Hypothenemus hampei*)	—
6.	Cotton seeds (all species of *Gossypium*)	—	Black arm (*Xanthomonas malvacearum*) and stunt (*Glomerella gossypii*)	Consignment of seeds can only be imported by Director, Central Institute for Cotton Research, Nagpur, Maharashtra
7.	Forest seeds (all species of *Pinus*, *Ulmus* and *Castanea*)	—	*Cronartium ribicola, Endothiaparasitica, Ceratocustis ulmi, Dothiostroma pini, Lophodermium pinastris*	Consignments can only be imported by Director, Biological Research, Forest Research Institute, New Forest Post, Dehradun or any organisation under Central or State Government.
8.	Groundnut seeds (Peanut) all species of *Arachis*	—	i. Production of seeds in areas free of *Puccinia arachidis* and *Sphaceloma arachidis*	i. Consignments can only be imported as decorticated seeds

Continued...

...*Continued*

No.	Plants, Seeds and Other Propagating Materials	Countries From Where Import is Prohibited	Pests for Which Additional Declarations in Official Phytosanitary Certificates are Required	Special Conditions for Import
			ii. Inspection of parent crops in active growing season and certification for freedom from peanut mottle, peanut stunt and marginal chlorosis viruses	ii. Consignments can only be imported for research work iii. Consignments originating from North and South America shall be grown in intermediate post-entry quarantine facility in non-groundnut growing countries and healthy seedlings or cuttings shall only be imported
9.	Lucerne (all species of *Medicago*)	—	Bacterial wilt (*Corynebacterium insidiosum*)	—
10.	Potato (all species of *Solanum*)	—	Wart (*Synchytrium endobioticum*), cyst nematodes (*Globodera pallida* and *Globodera rostochiensis*), *Leptinotarsa decimlineata* and freedom from virus diseases	i. Consignments can only be imported by Director, Central Central Potato Research, Institute, Shimla, Himachal Pradesh

Continued...

...*Continued*

No.	Plants, Seeds and Other Propagating Materials	Countries From Where Import is Prohibited	Pests for Which Additional Declarations in Official Phytosanitary Certificates are Required	Special Conditions for Import
				ii. Consignments shall be grown under post-entry quarantine
11.	Rubber (all species of *Hevea*)	America or West Indies	South American leaf blight (*Mycrocyclus ulei*), *Sphaerostilbe repens*	—
12.	Sugarcane (all species of *Saccharum*)	Fiji, New Guinea, Australia, Philippines	Leaf scald (*Xanthomonas albilineans*), Gummosis (*Xanthomonas vasculorum*), Sereh, white leaf, downy mildew and chlorotic streak	i. Consignments shall be imported only by Director, Sugarcane Breeding Institute, Coimbatore, Tamil Nadu ii. Consignments shall be grown under post-entry quarantine
13.	Sunflower (all species of *Helianthus*)	Argentina, Peru	Downy mildew (*Plasmopara halstedii*), *Orobanche cumana*	i. Consignments imported for sowing and planting shall be be grown under post- entry quarantine ii. Consignments imported for consumption shall be utilised

Continued...

...*Continued*

No.	Plants, Seeds and Other Propagating Materials	Countries From Where Import is Prohibited	Pests for Which Additional Declarations in Official Phytosanitary Certificates are Required	Special Conditions for Import
				under supervision of Plant Protection Adviser or any officer authorised by him
14.	Tobacco (all species of Nicotiana) i. Tobacco seed	—	Blue mould (*Peronospora tabacina*)	i. Tobacco seed for sowing and planting shall only be imported by Director, Central Tobacco Research Institute, Andhra Pradesh, for research ii. Consignments shall be grown under post-entry quarantine
	ii. Unmanufactured tobacco	—	Tobacco moth (*Ephestia elultella*)	—
15.	Berseem (all species of *Trifolium*)	—	Bulb and stem nematode (*Ditylenchus dipsaci*)	—
16.	Wheat (all species of (*Triticum*)	—	Ergot (*Claviceps purpurea*), dwarf bunt (*Tilletia contraversa*), spikelet rot	i. Consignments r for sowing and planting shall only be

Continued...

...Continued

No.	Plants, Seeds and Other Propagating Materials	Countries From Where Import is Prohibited	Pests for Which Additional Declarations in Official Phytosanitary Certificates are Required	Special Conditions for Import
	i. for sowing and planting granary weevil		(*Pseudomonas atrofaciens*) and Plant Genetic (*Sitophilus granarius*)	imported by Director, National Bureau of Resources, New Delhi ii. Consignments shall be grown in post-entry quarantine
	ii. for consumption	—	Granary weevil (*Sitophilus granarius*)	i. Ergot not to exceed 0.05% limit
	Note: Import of wheat preferably be undertaken during summer			ii.Dwarf bunt not to exceed 0.005%
17.	Paddy seeds/ rice	—	Granary weevil (*Sitophilus granarius*)	—
18.	Cuttings/ saplings/bud-wood of flowers and ornamental plants	—	Exotic pests and diseases to be specified in the import permit	Consignments shall grown in approved post-entry quarantine
19.	Planting material/ seeds of fruits	—	Exotic pests and diseases to be specified in the in the import permit	i. Import to be permitted selectively by the Central Govt. on the

Continued...

...*Continued*

No.	Plants, Seeds and Other Propagating Materials	Countries From Where Import is Prohibited	Pests for Which Additional Declarations in Official Phytosanitary Certificates are Required	Special Conditions for Import
				recommendations of Director of Horticulture/ Agriculture of a e State Government ii. Consignments to be grown under post-entry quarantine conditions, to be specified by the Plant Protection Adviser

APPENDIX IIB

List of Restricted Plant Species Covered Under New Seed Policy for Which PPA Specifies Additional Declarations and Special Conditions

No.	Name of Plant	Additional Declarations for Quarantine Pests Required to be Incorporated in PSC	Special Conditions for Import
1.	Banana (*Musa* sp.)	1. *Burkholderia (Pseudomonas) solanacearum* Race 2 (Moko wilt) 2. *Mycosphaerella fijiensis* (Black leaf streak, and 3. *Radopholus similis* (Burrowing nematode)	1. Freedom from soil 2. Growing under post-entry quarantine
2.	Barley (*Hordeum vulgare*)	1. *Pseudomonas syringae* p.v. *atrofaciens* (Glume rot) 2. *Claviceps purpurea* (Ergot) 3. Barley stripe virus	1. Freedom from weed seeds 2. Imported seed subject to grow-out test
3.	Beans (*Phaseolus* sp.)	1. *Curtobacterium* (*Corynebacterium*) *flaccumfaciens* p.v. *flaccumfaciens* (Bacterial wilt) 2. *Pseudomonas savastastanoi* p.v. *phaseoli-cola* (Halo blight) 3. Bean viruses	1. Freedom from weed seeds and-soil contamination 2. Imported seed subject to grow-out test
4.	Beet Root (*Beta vulgaris*)	1. *Curtobacterium* (*Corynebacterium flaccumfaciens* p.v. *betae* (Silvering disease) 2. *Pseudomonas syringae* p.v. *acuta* (Bacterial blight) 3. *Peronospora farinosa* (Downy mildew) 4. *Phoma betai* (Leaf spot) 5. *Heterodera schachtii* (Cyst nematode	1. Freedom from weed seeds and soil contamination 2. Imported seed subject to grow-out test
5.	Berseem (*Trifolium alexandrinum*)	1. *Ditylenchus dipsaci* (Stem and bulb nematode) 2. White clover mosaic virus	1. Freedom from weed seeds and soil contamination

Continued...

...Continued

No.	Name of Plant	Additional Declarations for Quarantine Pests Required to be Incorporated in PSC	Special Conditions for Import
		3. Yellow mosaic virus 4. Red clover mosaic virus	
6.	Broad bean (*Vicia faba.*)	1. *Ascochyta fabae* (Leaf and pod spot. 2. *Ditylenchus dipsaci* (Stem and bulb nematode) 3. Broad bean stain virus 4. *Heterodera glycines* (Cyst nematode	1. Freedom from weed seeds and soil contamination
7.	Cabbage (*Brassica* sp.)	1. *Pseudomonas syringae* p.v. *maculicola* (Black leaf spot) 2. *Xanthomonas campestris* p.v. *campestris* (black rot) 3. *Phoma lingam* (Black leg)	1. Freedom from weed seeds 2. Inspection of mother crop & certification for freedom from pests indicated in column 2
8.	Carnation cuttings *(Dianthus caryophyllus*)	1. *Burkholderia* (*Pseudomonas*) *caryophylli* (Bacterial spot) 2. *Fusarium oxysporum* f.sp. *dianthi* (Fusarium wilt) 3. *Uromyces dianthi* (Rust) 4. Carnation viruses	1. Freedom from soil contamination 2. Growing imported cuttings under post-entry quarantine
9.	Carrot (*Daucus carota*)	1. *Pectobacterium* (*Erwinia*) *carotovora* (Crater rot) 2. *Xanthomonas hortorum* p.v. *carotae* (Bacterial blight) 3. Carrot viruses (red leaf and mottle dwarf	
10.	Cauliflower (*Brassica* sp.)	Additional declarations, same as for cabbage	Special conditions, same as for cabbage
11.	Chicory (*Cichorium intybus*)	1. *Pseudomonas cichorii* (Bacterial blight) 2. Chicory yellow mottle virus	
12.	Chilli (*Capsicum* sp.)	1. *Xanthomonas vesicatoria* (Bacterial scab)	

Continued...

...Continued

No.	Name of Plant	Additional Declarations for Quarantine Pests Required to be Incorporated in PSC	Special Conditions for Import
13.	Chrysanthemum (*Dendronthema* sp.) cuttings	1. *Rhodococcus (Corynebacterium fascians* (Fasciation) 2. *Apelenchoides fragariae* and *A. ritzemabosi* (Foliar nematodes) 3. *Ditylenchus dipsaci* (Stern and bulb nematode) 4. Chrysanthemum viruses, namely, tomato aspermy virus, chrysanthemum stunt virus, etc.	1. Freedom from soil contamination 2. Growing imported cuttings under plant quarantine
14.	Cow pea (*Vigna* sp.)	1. *Pseudomonas syringae* p.v *syringae* (Brown spot) 2. Cow pea viruses, namely, cowpea mosaic, ring spot and true mosaic viruses	
15.	Cucumber (*Cucumber sativus*)	1. *Pseudomonas syringae* p.v. *lachrymans* (Angular leaf spot)	
16.	Gladiolus bulbs (*Gladiolus* sp.)	1. *Burkhoderia marginalis* (Scab and neck rot) 2. *Botrytis gladioli* (Corm rot) 3. *Fusarium oxysporium* f. sp. *gladioli* (Storage rot) 4. *Septoria gladioli* (Hard rot) 5. *Stromatinia gladioli* (Dry rot) 6. *Urocystis gladioli* (Smut)	1. Freedom from soil contamination 2. Imported bulbs subject to grow-out test or post-entry quarantine
17.	Gram or chick pea (*Cicer aeratinum*)	1. *Ascochyta rabiei* (Ascochyta blight)	
18.	Grapevine (*Vitis* sp.) cuttings	1. *Agrobacterium tumefaciens* (Crown gall) 2. *A. rhizogenes* (Hairy root) 3. *Pantoea agglomerans* (Gummosis)	1. Freedom from soil contamination 2. Growing imported cuttings under post-entry quarantine 3. Inspection of stock and

Continued...

...*Continued*

No.	Name of Plant	Additional Declarations for Quarantine Pests Required to be Incorporated in PSC	Special Conditions for Import
		4. *Cryptosporella viticola* (Dead arm) 5. *Phakospora vitis* (Rust) 6. *Xytella fastidiosa* (Pierces' disease) 7. *Xylophilus ameplinus* (Bacterial necrosis and canker) 8. Fan leaf virus and related strains 9. Flavescence doree (PLO)	certification for freedom from viruses affecting grapevine Pierces' disease and flavescence doree, etc.
19.	Lettuce (*Lactuca sativa*)	1. *Septoria lactucae* (Leaf spot) 2. Lettuce mosaic virus	
20.	Maize or corn (*Zea mays*)	1. *Clavibacter (Corynebacterium) michiganense* sub. sp. *nebraskense* (Nebraska wilt) 2. *Pantoea stewartii* sub. sp. *stewartii* (Stewart's wilt) 3. *Drechsclera maydis* Race T (Southern corn blight) 4. *Claviceps gigantea* (Ergot) 5. *Physopella zeae* (Tropical rust) 6. *Protephanus truncatus* (Larger grain borer)	
21.	Narcissus bulbs (*Narcissus* sp.)	1. *Botrynia polyblastis* (Narcissus fire) 2. *Fusarium oxysporium* f. sp. *narcissi* (Basal rot) 3. *Stagnospora curtissi* (Leaf scorch) 4. Stripe virus 5. *Ditylenchus dipsaci* (Stem and bulb nematodes) 6. Bulb flies	1. Freedom from soil contamination 2. Imported bulbs subject to grow-out test or post-entry quarantine

Continued...

...Continued

No.	Name of Plant	Additional Declarations for Quarantine Pests Required to be Incorporated in PSC	Special Conditions for Import
22.	Oats (*Avena sativa*)	1. *Cochliobolus victoriae* (Victoria blight) 2. *Pseudomonas coronafaciens* (Halo blight) 3. *Pseudomonas striafaciens* (Bacterial stripe) 4. *Ditylenchus dipsaci* (Stem and bulb nematode) 5. Barley stripe mosaic virus	1. Freedom from soil and weed seeds
23.	Oil palm (*Elaeis guineensis*)	1. *Cercospora elaeidis* (Freckle) 2. *Fusarium oxysporium* f. sp. *elaeidis* (Vascular wilt) 3. *Phytomonas staneli* (Hart rot) 4. *Rhadinnaphelenchus cocophilus* (Red ring nematode) 5. *Pachymerus* sp. (Kernel borer) 6. Leaf mottle	2. The nurseries raised out of imported seed/seed sprouts subject to post-entry quarantine
24.	Peas (*Pisum sativum*)	1. *Pseudomonas savastanol* p.v. *phaseolicola* (Halo blight) 2. *Pseudomonas syringae* p.v. *pisi* (Bacterial blight) 3. *Heterodera goettingiana* (Cyst nematode) 4. pea early browning virus	1. Freedom from soil contamination 2. Imported seed subject to grow-out test
25.	Poplar (*Populus* sp.) cuttings	1. *Mycosphaerella populorum* (Canker) 2. Poplar mosaic virus 3. Poplar streak virus 4. Poplar wilt virus	1. Imported cuttings subject to post-entry quarantine
26.	Rose (*Rosa* sp.) cuttings	1. *Agrobacterium tumaefaciens* (Crown gall) 2. *A. rhizogenes* (Hairy root) 3. *Coniothyrium wernsdoriffiae* (Brand canker)	1. Freedom from soil contamination 2. Imported cuttings grown under post-entry quarantine

Continued...

...*Continued*

No.	Name of Plant	Additional Declarations for Quarantine Pests Required to be Incorporated in PSC	Special Conditions for Import
		4. *Leptospheria coniothyrium* (Common stem canker) 5. *Cryptosphaerella umbrina* (Brown canker) 6. Mosaic virus 7. Streak virus 8. Wilt virus	
27.	Soybean (*Glycine soja*)	1. *Curtobacterium flaccumfaciens* p.v. *flaccum faciens* (Bacterial wilt) 2. *Pseudomonas savastanoi* p.v. *glycinea* (Greasy spot) 3. *Diaporthe phaseolorum* var. *batatis* (Stem canker) 4. *Pernospora manshurica* (Downy mildew) 5. Soybean mosaic virus 6. Tobacco ring spot virus	1. Freedom from soil contamination and weed seeds 2. Imported seed subject to grow-out test
28.	Strawberry (*Fragaria* sp.) cuttings	1. *Xanthomonas fragariae* (Angular leaf spot) 2. *Phomopsis obscurans* (Leaf blight) 3. *Phytophthora fragariae* (Red stele disease) 4. *Phytophthora cactorum* (Crown rot) 5. Latent ring spot virus and other viruses affecting *Fragaria*	1. Freedom from soil contamination 2. Imported cuttings subject to growout test
29.	Tomato (*Lycopersicon esculentum*)	1. *Clavibacter michiganense* sub. sp. *michiganense* (Bacterial canker) 2. *Pseudomonas syringae* p.v. tomato (Bacterial leaf spot) 3. *Xanthomonas vesicatoria* (Bacterial scab)	

APPENDIX IIIA

Import Inspection Fees

(Clause 3(12) 12, Schedule III of PFS Order, 1989, as amended by S. O. 467 (E) dated 16th May 2000 and S.O. 574 (E) dated 14th June 2000)

No	Particulars of Import	Volume/Wt.	Fee
1.	Plants/planting materials, including cutting, saplings, bud wood etc., requiring Post-entry Quarantine	1. Up to 10 numbers 2. Above 10 numbers	Rs. 250 Rs. 250 plus @ Rs. 1.60 per plant
2.	Plants/planting material including bulbs, tubers, corms, rhizomes etc., not requiring Post-entry quarantine.	1. Up to 100 numbers 2. Above 100 numbers	Rs. 100 Rs. 100 plus Rs. 0.40 per plant
3.	Seeds for sowing	1. Up to 1 kg 2. Above 1 kg up to 100 kg 3. Above 100 kg up to 500 kg 4. Above 500 kg	Rs. 100 Rs. 100 plus Rs. 50 per addl. kg Rs. 5050 plus Rs. 15 per addl. kg Rs. 11050 plus Rs. 5 per addl. kg
4.	Plant material, seeds and fruits for consumption	1. Up to 2 kg 2. Above 2 kg up to 100 kg 3. Above 100 kg up to 1000 kg 4. Above 1000 kg tonne for pulses	Rs. 50 Rs. 50 plus Rs. 5 per addl. kg or part thereof Rs. 550 plus Rs. 2 per addl. kg Rs. 2500 plus Rs. 50 per addl. and Rs. 75 per addl. tonne in respect of others

Note: Fraction of kg/tonne may be rounded off to the nearest unit.

APPENDIX IIIB

Fumigation/Disinfection/Disinfestation/Supervision Charges

No	Particulars of Import	Volume/Wt.	Fees
1.	Plants/Planting Products/ Planting materials/Dry fruits/ Fresh fruits/ Vegetables/ Seeds/ Soil/earth/clay	**A. On Volume Basis**	
		(i). Up to 5 cu m	Rs. 600
		(ii). Above 5 cu m	Rs. 600 plus Rs. 300 per addl. cu. m. or part thereof
		B. On Container Basis	
		20 ft container (33 cu m)	Rs. 2400
		40 ft container (66 cu m)	Rs. 4500
		C Supervision Charges (The importer shall arrange for fumigation, disinfestation of consignment at his cost, under the supervision of the Plant Protection Advisor or an officer authorised by him on this behalf	Rs. 500 per day per consignment

Flower seed production trials

Mesembryanthemum criniflorum

Papaver rhoeas

Hollyhock 5 hybrid Pusa Pink Beauty

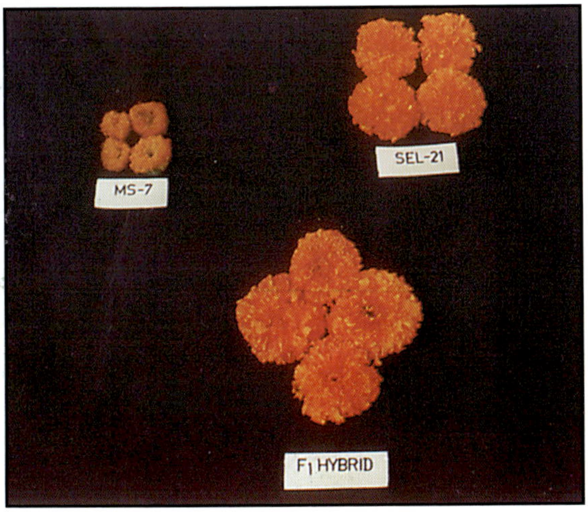

African Marigold

African Marigold b

African Marigold c

African Marigold d

Preparation for placing rock wool blocks in tunnel type computer controlled polyhouse

Inside view of a tunnel type polyhouse with Rose plants planted in rock wool blocks

Crop grown in greenhouse

Side view of greenhouse

Greenhouse exterior

Greenhouse interior with growing crops

Plants in greenhouse protection

Thriving Gerbera production under greenhouse

Inner view of greenhouse

Plantation in greenhouse

Greenhouses in a row

Various plants inside greenhouse

Polyethylene film greenhouse

Polycarbonate sheet greenhouse

INDEX

L

M